Sympathetic Attractions

James Barry: *Commerce, or the Triumph of the Thames.* This painting is the fourth in a six-part series still on display at the Society for the Encouragement of Arts, Manufactures, and Commerce. Using the language of classical allegory, Barry celebrated England's commercial progress in this picture and his accompanying text. Father Thames holds a mariner's compass in his left hand. Barry added the naval pillar in the background in 1801 to commemorate England's maritime victories. Royal Society of Arts, London.

Sympathetic Attractions

MAGNETIC PRACTICES, BELIEFS, AND SYMBOLISM IN EIGHTEENTH-CENTURY ENGLAND

PATRICIA FARA

PRINCETON UNIVERSITY PRESS

PRINCETON, NEW JERSEY

Published by Princeton University Press, 41 William Street, Princeton, New Jersey 08540
In the United Kingdom: Princeton University Press, Chichester, West Sussex

---

Library of Congress Cataloging-in-Publication Data

Fara, Patricia.
Sympathetic attractions : magnetic practices, beliefs, and symbolism in
Eighteenth-Century England / Patricia Fara.
p. cm.
Includes bibliographical references and index.
ISBN 0-691-01099-4 (cloth : alk. paper)
1. Magnets—England—History—18th century.
2. Science—England—History—18th century.
I. Title.
QC750.4.F37 1996
306.4′5—dc20 96-10785 CIP

This book has been composed in Times Roman

Princeton University Press books are printed on acid-free paper and meet the guidelines for
permanence and durability of the Committee on Production Guidelines for Book
Longevity of the Council on Library Resources

Printed in the United States of America
by Princeton Academic Press

1 2 3 4 5 6 7 8 9 10

TO THE MEMORY OF CHARLENE GARRY

CONTENTS

LIST OF FIGURES AND TABLES ix

ACKNOWLEDGMENTS xi

ABBREVIATIONS xiii

INTRODUCTION 3

CHAPTER ONE
Mapping Enlightenment England: Practitioners and Philosophers 11

CHAPTER TWO
"A Treasure of Hidden Vertues": Marketing Natural Philosophy 31

CHAPTER THREE
The Direction of Invention: Setting a New Course for Compasses 66

CHAPTER FOUR
An Attractive Empire: Mapping Terrestrial Magnetism 91

CHAPTER FIVE
Measuring Power: Patterns in Experimental Natural Philosophy 118

CHAPTER SIX
God's Mysterious Creation: The Divine Attraction of Natural Knowledge 146

CHAPTER SEVEN
A Powerful Language: Images of Nature and the Nature of Science 171

CONCLUSION 208

APPENDIX
Magnetic Longitude Schemes 215

NOTES 219

BIBLIOGRAPHY 269

INDEX 319

FIGURES AND TABLES

FIGURES

Frontispiece James Barry: *Commerce, or the Triumph of the Thames*

1.1 William Hogarth: "Frontis-Piss" 26
2.1 Thomas Tuttell's Magnetic Playing Card 32
2.2 Gowin Knight 38
2.3 Artificial Magnets 41
2.4 Gowin Knight's Machine for Making Artificial Magnets 43
2.5 The Countess of Westmorland's Loadstone 48
2.6 Loadstones 49
2.7 John Bennett's Trade Card 51
2.8 George Adams' Magnetic Cabinet 56
2.9 James Graham and Gustavus Katterfelto 57
2.10 Temple Henry Croker's Magnetic Perpetual-Motion Machine 59
2.11 Magnetic Rational Recreations 63
3.1 Gowin Knight's Steering Compass 68
3.2 A Traditional Steering Compass 76
3.3 Gowin Knight and John Smeaton's Azimuth Compass 80
3.4 Gowin Knight's Certificate 82
4.1 Edmond Halley's 1701 Magnetic Chart of the Atlantic Ocean 94
4.2 The Boundaries of Magnetic Surveys 97
4.3 Joseph Banks as a Botanic Imperialist 104
4.4 Captain Cornwall's Measurements of Magnetic Variation 107
4.5 Captain Middleton's Measurements of Magnetic Variation 107
4.6 Edmond Halley's World Chart of Magnetic Variation 111
4.7 Magnetic Illustrations from Thomas Young's *Lectures* 114
4.8 John Lorimer's Thought Experiment 115
4.9 Adam Walker's Visualization of the Earth's Internal Magnetic Structure 116
5.1 John Canton's Method of Making Artificial Magnets 119
5.2 William Gilbert's Illustration of a Dip Circle 134
5.3 Dip Circle of Edward Nairne and Thomas Blunt 134
5.4 Ralph Walker's Variation Compass for Navigators 136
5.5 Gowin Knight's Variation Compass for Natural Philosophers 136
5.6 Variation Compass Belonging to Henry Cavendish 137
6.1 Athanasius Kircher's Magnetic Kingdom of Nature 148
6.2 Portrait of Edmond Halley 154
6.3 Edmond Halley's Model of Earth's Magnetic Structure 154
6.4 "A Description of the Coledge of Physicians" 160
6.5 Laputa, the Flying Magnetic Island 162
7.1 Edmond Halley's Circulating Effluvia 178
7.2 "I am my Beloveds, and his desire is towards me" 186
7.3 Franz Mesmer's Baquet 196
7.4 "ANIMAL MAGNETISM—The Operator putting his Patient into a Crisis" 198

7.5 "Animal Magnetism . . . or Count Fig in a Trance" 201
7.6 "Billy's Gouty Visit, or a Peep at Hammersmith" 203
7.7 Caricature of Magnetic Treatment 205

Tables

3.1 Eighteenth-Century Compass Prices 83
5.1 Measurements of Magnetic Force up to 1750 126

ACKNOWLEDGMENTS

ABOVE ALL, I should like to thank Jim Secord for his constant enthusiasm and for being so generous with his time. Without his continual help and encouragement, I would never have started this book, let alone finished it. I am also especially grateful to Simon Schaffer for reading and discussing several draft chapters, and to Jan Golinski, Chris Lawrence, and Larry Stewart for detailed comments on an earlier version of the entire text.

Several people made detailed and extremely valuable comments on particular aspects of this book, notably Jim Bennett, Anthony Fanning, John Gascoigne, Anne Secord, Alice Walters, and Alison Winter. In addition, I have benefited greatly from conversations and correspondence with Donna Andrew, Michael Bravo, Tony Campbell, Geoffrey Cantor, Alan Clarke, Patrick Curry, Judith Field, Alan Gauld, David Gooding, Beryl Hartley, Frank James, Stephen Johnston, Stephen Lloyd, Alan Morton, Roy Porter, Stephen Pumfrey, Keith Schuchard, and Richard Yeo.

My thanks go also to Roger Pring for his photographic contributions, and to close friends who have tolerated me while I was writing this book.

ABBREVIATIONS

Arderon correspondence	The correspondence between Henry Baker and William Arderon from 1744 to 1767. Four volumes. National Art Library (Victoria and Albert Museum), Forster 47.c.11–14
Baker correspondence	The correspondence of Henry Baker. Eight volumes. John Rylands Library, Manchester University, MS/9
Banks	Sophia Banks' collection of broadsides. British Library, pressmark L.R.301.h.3
Canton papers	The correspondence and papers of John Canton. Three volumes. The Royal Society, London
Catcott correspondence	The correspondence of Alexander S. Catcott and Alexander Catcott. Bristol Central Library, reference 26063
Lysons I	Five-volume set of Lysons collectanea at the British Library, pressmark c.103.k.11
Lysons II	Two-volume set of Lysons collectanea at the British Library, pressmark 1881.b.6
PRO	Public Record Office, London
RGO	Royal Greenwich Observatory (Board of Longitude papers at Cambridge University Library)
RSJB	Royal Society Journal Book (copy)
Wilson correspondence	The correspondence and papers of Benjamin Wilson, 1743–1788. British Library Additional Manuscripts 30094

Sympathetic Attractions

INTRODUCTION

In 1783, the Society for the Encouragement of Arts, Manufactures, and Commerce welcomed Samuel Johnson and other distinguished visitors to admire the six large murals which it had commissioned from James Barry for its new meeting rooms near the Strand. Barry, who intended these paintings to restore the flagging reputation of British art, had designed a series portraying the progressive stages of human culture. He called his fourth illustration, reproduced as the Frontispiece, *Commerce, or the Triumph of the Thames*, choosing the imagery of classical allegory to celebrate London's central role in Britain's expanding trading empire. Barry explained that famous British navigators, national heroes like Francis Drake and James Cook, are dressed as Tritons. They are transporting Father Thames, who holds in his left hand a "mariner's compass; from the use of which, modern navigation has arrived at a certainty, importance, and magnitude, superior to any thing known in the ancient world; it conquers places the most remote from each other; and Europe, Asia, Africa, and America, are thus brought together, pouring their several productions into the lap of the Thames." Mercury, the god of commerce, hovers overhead as Nereids contribute manufactured goods from England's developing industrial cities.[1]

This hymn to trade expresses visually common Enlightenment declarations of confidence in economic progress through invention. Eighteenth-century connoisseurs valued historical painting as the highest form of art, the genre most appropriate for judging ethical issues and conveying messages of enlightened advancement through learning from previous errors. Like many of his contemporaries, Barry accorded a key function in commercial development to the magnetic compass, frequently hailed as a Renaissance innovation equal in importance to the printing press and gunpowder. Barry consciously couched his picture in mythological terminology, placing it within a series depicting how moral improvement runs concurrently with technical advances. During the eighteenth century, such rhetorics of tightly linked religious and material progress increasingly predominated over pessimistic visions of decline; their pervasive variety exemplifies the interconnectedness of cultural spheres which are now regarded as distinct. Opponents of international trade and commercial expansion did not protest solely on economic or political grounds but were preoccupied with the country's moral and intellectual deterioration under a corrupting régime of luxury.[2]

Such long-forgotten considerations molded eighteenth-century perspectives. Similarly, for Johnson and his fellow public benefactors, flatteringly portrayed by Barry in the adjacent picture, the magnetic compass lying at the heart of this image signalled connotations concealed from modern viewers accustomed to compartmentalized knowledges and practices and no longer sensitive to traditional symbolism and classical allusions. Compasses guided not only the hardy mariners exploring the terrestrial globe but also humble pilgrims voyaging through life towards God. Compass needles represented the constancy of human souls irresistibly drawn to the divine center of attraction, while magnetic powers sympathetically bonded people to each other and to the rest of the universe. When Johnson wrote a magnetic satire, he relied on his readers' familiarity with the long-standing sexual associations underpinning his polyvalent references to magnets.[3] Furthermore, just as writers and artists gained inspiration from mythology for their moral lessons, so too investigators of magnetic behavior turned to the earth's past for understanding the current patterns of terrestrial magnetism which governed compass performance.

Magnets might initially seem rather unpromising tools for dissecting the intricacies of cultural debates. Their power lies in the multiplicity of meanings carried by magnetic vocabulary, meanings which were context-dependent and which continually shifted. Writing from the vantage point of the late twentieth century, it is, of course, impossible to escape from our own concerns. Thus our recent preoccupations with issues related to consumption and gender are rooted in current disenchantments. Historians and literary critics perceive eighteenth-century England as the primary site for fundamental transformations which vitally affected modern life. Similarly, the postwar expansion of interest in the history of science stems from an increasing awareness of the need to examine the pervasive influence of science and technology. As a consequence, since magnets are now largely viewed as objects of scientific interest, one immediate appeal of examining their locations in eighteenth-century discourses is to reveal some of the countless negotiations which contributed to the construction of magnetism[4] as one of the new public sciences governed by the Royal Society.

One aim of this book is to provide insight into the processes through which science has come to play such a powerful role in society. Historians of science have often glossed over the eighteenth century as an inert buffer, a chronological necessity separating the exciting periods of the Scientific Revolution and the Victorian age of invention. This retrospective judgment was initiated by nineteenth-century polemicists writing history to serve their own ends. Charles Babbage, for example, eagerly promoting the interests of the Cambridge analysts, punned on the Newtonian calculus notation when dismissing the University's adherence to the "Dot-Age" of mathematics. Charles Lyell advertised his own approach to geology by sneering at his immediate predecessors. Such interpretations have had a long-lasting influence, particularly those expressed by William Whewell. Paralleling his neglect of other fields, he failed to name any Englishman researching into magnetic phenomena between Edmond Halley and Peter Barlow, contemporaries respectively of Isaac Newton and Michael Faraday.[5] This omission has been perpetuated by subsequent historians of magnetism.[6] But new anxieties inform new narratives. Herbert Butterfield famously decreed that the Scientific Revolution reduced the Reformation and the Renaissance to mere episodes, but historians have subsequently reappraised this influential evaluation. As traditional historiographies crumble under novel methodologies, historians of science are viewing the eighteenth century with refreshed eyes. Perceiving it as a period whose importance for the establishment of disciplinary science has been insufficiently recognized, they are urging intensive primary research and more sympathetic analyses.[7]

Modern cultural historians are concerned to avoid imposing today's categories on past activities. From this perspective of a new historicism, critics of scientific histories point to the dangers of evaluative anachronism—selecting, narrating, and assessing past events with criteria informed by anachronistic views of their value and significance. So while the authors of the best recent accounts for the eighteenth century of single scientific topics—electricity, light, and geology, for instance—carefully articulate participants' contemporary interests, such studies are vulnerable to accusations of evaluative anachronism because of their focus on the writers, ideas, and practices that retrospectively seem to have contributed to present scientific institutions and beliefs.[8] Magnetism provides an excellent example illustrating how what people now recognize as scientific disciplines were often in the past diffuse collections of practices, whose future coherence was not readily apparent. But our rejection of grand narrative histories—those Whiggish celebrations of progressive science endorsed by the successors of Barry and his rhetorical contemporaries—imposes limitations. Microscopic dissection of specific episodes fragments the historical picture by generating a multiplicity of interpretations at the expense of overall syntheses. Bracketing scientific practices with other cultural activities in their local context has demoted the privileged nature of scientific knowledge. Nevertheless, one drawback of demystifying science is that it renders mysterious the processes by which it has become so powerfully prestigious.[9]

We need, therefore, to experiment with new approaches to constructing historical narratives. We need pictures painted with a brush broad enough to portray the large-scale changes contributing to the emergence of our present scientific consumer society, but which also capture images of life as perceived by people at the time. We need to incorporate fine-brushed renderings of individual scenes into panoramic cultural landscapes, so that we can investigate the sculptural processes of transformation over time. We also need to transcend our bicultural demarcation of the world and explore how what are commonly perceived as distinct discourses—the Arts and the Sciences—are mutually fashioned by each other.[10]

This study of magnetic concerns in the eighteenth century aims to contribute towards such a revised historical vision. By examining the shifting identities of English practitioners and their beliefs, this book retrieves the differing implications of magnets and compasses. At the same time, it reveals the contested rise to public power of natural philosophers, the men who became the accredited scientific custodians of magnetic expertise. And by contextualizing these accounts within the massive transformations of eighteenth-century English life, the book illustrates how modern society is intermeshed with science's history.

Three centuries ago, there were no red-coated steel horseshoes, no lengthy chapters on magnetic phenomena in physics textbooks, no electromagnetic equipment, and no vibrating fields of force providing a basis for sexual metaphors. English natural philosophers, struggling to validate their activities in a frequently hostile milieu, had abandoned their earlier focus on magnetic phenomena. Practical magnetic expertise lay mainly in the maritime community, in the hands of men admired for their skill as they navigated the oceans to increase England's trading wealth and international possessions. Their compasses were made of poor materials. The crudely fashioned iron needles were magnetized by lumps of natural loadstone, inferior in quality to the dark blue ores from Ethiopia for which wealthy virtuosi paid huge sums of money to enlarge their cabinet collections. Most people who wondered about such things believed that the universe was bonded magnetically and that the tides could be explained by the magnetic attraction between the earth and the moon. People remarked that, like God's power over humans, magnetic forces stretched mysteriously through space, undetectable except for their uniquely specific action on iron. Writers linked the sympathetic nature of magnetic activity with other relationships of attraction, such as the simultaneous maturation of wines and growing grapes or the ability of snakes to freeze their prey. Therapists tapped magnetic influences to guard against miscarriage or draw out the pains of gout, childbirth, and toothache.

In contrast, by the end of the eighteenth century, the Royal Society was an influential, prestigious organization intimately involved in a variety of governmental and commercial operations. Most books on natural philosophy included a separate—albeit brief—section describing magnetic experiments and the earth's magnetic patterns. London's craftsmen were selling a range of delicate, precise measuring instruments incorporating permanently magnetized steel needles; observers were sending in accurate readings to the Royal Society from purpose-built laboratories in remote corners of the world. Leisured consumers of polite entertainment could learn about the behavior of compasses and magnets at popular lectures, or purchase elaborate parlor tricks. Authors were embellishing their literary texts with magnetic imagery newly enriched by the practice of animal magnetism.

The very variety of these changes demonstrates the potential offered by magnetic practices, ideas, and meanings for exploring the cultural shifts of the eighteenth century. Examining a single concept in such detail for a particular period and country enables grander historical themes, such as the emergence of the public sphere, the creation of a consumer society, and the construction of gender differences, to be fleshed out and refined. So one function of this book is to complement, unify, and extend the recent expositions by Larry Stewart and Jan Golinski of the rise of a powerful public science. The myriad negotiations underlying this transition are exemplified by the contested transfer of publicly accredited magnetic expertise from navigators to natural philosophers. But since this study emphasizes the significance of magnetic compasses for international navigation and trade, it also amplifies other important analyses by eighteenth-century historians, such as John Brewer's account of the rise of the fiscal-military state and Linda Colley's insistence on the vital role of commercial interests in the forging of the British nation. It thus underlines the integration within eighteenth-century concerns of what are retrospectively perceived as scientific interests.[11]

Furthermore, magnets furnish valuable historical leverage because they were symbolically laden. In his analysis of inhibition, a word drawn from popular usage but which acquired specialized definitions in psychology and physiology, Roger Smith has illustrated how science is intermeshed with other activities by a common web of language.[12] Similarly, investigating magnets and compasses not only reveals the metaphorical force of literary imagery embodying beliefs about the natural world but also casts light on how words can carry social values into the heart of science. Constructing magnetism as a public science entailed imposing gender, religious, and linguistic restrictions on its practitioners. Reciprocally, in mutual processes of interactive shaping, establishing magnetic science affected how such distinctions were themselves defined. Two features of magnetic connotations qualify them particularly well for this type of analysis: their variety and their selectiveness. The rich diversity of magnetic implications, so invaluable for eighteenth-century commentators, also provides us with multiple entrées into their debates. Yet at the same time, magnetic references are not overabundant. Words such as "electricity" and "light," which at first sight seem ideal for such a project, are less well suited because their very frequency blurs their specificity as signifiers. Writers carefully honed their magnetic imagery, enabling us to gain greater insight both into their own preoccupations and also into their collective influence on our own lives.

Because this book examines historical topics, including commercialization, imperialism, and the function of language, the material is arranged thematically rather than chronologically. In broad terms, after the first introductory chapter, the focus of analysis moves from practical domains of invention, experiment, and navigation to more theoretical ones of interpretation and polemical argument. Each chapter is at one level self-contained, viewing the changes in magnetic practices and meanings from a different analytical perspective. But just as contemporary participants perceived their lives as an integral whole, so too the chapters in this book are bound together as a kaleidoscopic image of a transforming eighteenth-century England.

The opening chapter surveys the magnetic practitioners and philosophers whose ideas and activities form the basis for the subsequent analyses. Different people valued different types of magnetic knowledges; for diverse reasons, they found magnetic phenomena interesting because they could enroll them for furthering particular objectives. People held conflicting attitudes not only because they were engaged in a variety of activities but also because their ideas were framed by contrasting religious, political, and philosophical orientations. Although this book intentionally transcends modern disciplinary boundaries, any study must necessarily be limited by its own boundaries of place and time. So this first chapter maps these limits, emphasizing the distinctiveness of Enlightenment England and exploring how that elastic label, the eighteenth century, can be tailored to fit magnetic affairs.

The second and third chapters focus on the fruitful alliance of commerce and natural philosophy, illuminating the roles of entrepreneurial natural philosophers in constructing a bureaucratic scientific consumer society. As they produced and marketed magnetic texts and equipment, these innovative lecturers, inventors, and instrument makers advertised magnetic research as a commercially beneficial endeavor. By stressing the value of their redesigned compasses in improving navigation, they reinforced public perception of the advantages which the pursuit of natural philosophy could yield for the expanding British nation, so dependent on international trade and territorial conquest. A biographical analysis of Gowin Knight, the Royal Society's mid-century leading magnetic expert, exemplifies how inventive natural philosophers contributed towards the foundation of professionalized disciplinary science operating in a public sphere. Like their colleagues working in other fields, they helped to transform a stable hierarchical society dominated by patronage into one governed by the state, with established career structures for men of enterprise.

International aspects of this theme are developed in the following chapter, which examines the entrenchment of natural philosophy within imperializing interests of territorial expansion and mercantile exploration. In the 1690s, Halley set the agenda for a century of data collection and interpretation by advertising the value to navigation of his own theory-laden map of terrestrial magnetism. As natural philosophers sought to demonstrate their mastery of the terraqueous globe through mensuration, they recruited maritime practitioners as research assistants and appropriated naval technologies of representation. Taking advantage of the convergence of commercial, governmental, and natural philosophical interests, the Royal Society maneuvered itself into a strategic position for allocating public funds.

The fifth chapter describes contests for cultural authority between communities of practitioners endorsing differing approaches to the natural world. These conflicts exemplify larger cultural debates about the value of calculation and measurement. In alliance with the London instrument trade, natural philosophers sought to consolidate their growing status as experts by marketing the conclusions of their magnetic research as organized sets of knowledge. Promoting magnetic studies for their practical value, they imposed precision instruments and calculating techniques on navigators and proclaimed their own authority in maritime affairs.

These analyses combine to reinforce one of the book's major arguments, that the creation of a powerful public science was rooted in Britain's mercantile imperial expansion; it was entrenched in the development of a consumer society regulated by a centralized government. Another aim of this book is to examine how specialists forging their own areas of authority mutually fashioned their identities by distinguishing themselves from other sectors of society. Constructing new disciplines—such as the science of magnetism—entailed molding gender and class distinctions as well as differentiating the style and content of academic discourses.

The last two chapters enrich both these themes by examining how polemicists took advantage of the ambiguities of magnetic vocabulary. The sixth chapter concentrates on people's changing attitudes towards the past. Explorations of the earth's magnetic history were intertwined with research into earlier languages and societies. Just as the theories constructed by natural philosophers were shaped by religious commitments, so too narratives of cultural transformation were framed by allegiances to various accounts of the natural world. Protagonists in controversies about the relationship between divine intervention and human inventiveness drew on rival versions of the origins of the compass. Advocates of commercial expansion through technical advance cited magnetic discoveries as paradigms of progress, whereas critics of natural philosophy emphasized the unsolved mysteriousness of magnetic powers.

The final chapter focuses on language. It simultaneously analyzes debates about language itself and demonstrates how natural philosophers used linguistic criteria to demarcate acceptable practices and practitioners. The chapter closes with a case study illustrating the ideological power of language: the rhetorical denunciations of a novel medical therapy, the animal magnetism based on Franz Mesmer's Parisian success, consolidated the boundaries of science and modified English magnetic symbolism. In texts of natural philosophy, religion, and literature, writers deliberately exploited multivalent magnetic vocabulary to reinforce their arguments tropologically. People's opinions about language were grounded in their approach towards learning about God's creation. Whereas many natural philosophers declared that metaphorical terms should be eliminated from their accounts, magnetic imagery resonated amongst descriptions of the physical, spiritual, and human worlds. Examining magnetic figurative language illuminates how the shaping of scientific disciplines entailed reforming society. As part of their bid for legitimation as experts, natural philosophers increasingly policed the form and content of the new magnetic science, thus determining who should be accredited participants.

Writers have continued to find magnetic metaphors attractive vehicles for persuasive exposition. E. P. Thompson visualized eighteenth-century society as two opposing magnetic poles, the aristocracy and the gentry versus the crowd. Thirteen years later, defending this dualism against his critics, he sustained this bipolar model over several pages of metaphorically inspired riposte, insisting that it was an ideological force continually voiced by the historical actors themselves.[13] Like the tropological language of eighteenth-century writers, Thompson's image expresses a cultural judgment in contemporary scientific vocabulary—in his case, a force field. Although many historians have repainted Thompson's picture of class struggle, it has indelibly affected their own historical narratives—including this one—both directly, and indirectly through other secondary texts. Similarly, eighteenth-century authors deploying magnetic imagery relied on contemporary theories but also influenced a wide range of cultural controversies whose outcomes are still resonating today.

Diverse characteristics of modern life, including our scientific institutions, our literary texts, and our economic systems, are woven together through their shared provenance. Our historical categories are still too constrained by academic disciplinary definitions. This book could perhaps be described as a history of science, because one of its objectives is to narrate the construction of magnetism as a scientific discipline. But this description would be too narrow, because it would have been articulated from a twentieth-century perspective. With their multiple associations, magnetic phenomena provide a marvellous opportunity to enter into the eighteenth-century world before the fabrication of limiting labels like science, theology, economics, and literature. Despite warnings of Whiggishness, the goal of understanding how modern society has been formed is a legitimate one for historians: but it can only be approached by working forwards from the past using reconstructions which have been stripped clean of retrospective awareness of the intervening centuries. Because this book recreates predisciplinary magnetic experiences, it is a truly interdisciplinary history.

CHAPTER ONE

Mapping Enlightenment England: Practitioners and Philosophers

EPHRAIM CHAMBERS has won renown as the English initiator of Enlightenment projects to systematize rational learning in comprehensive encyclopedias. He followed earlier classifiers by visualizing the organizational scheme of his *Cyclopædia* as a map of knowledge. Employing geographical metaphors, he claimed to guide his readers through the wilderness of knowledge, indicating the flourishing growth of well-tilled areas of expertise and marking where the limits of the *terra cognita* could be extended. Like the French encyclopedists who imitated his enterprise, Chambers expressly articulated the arbitrariness with which he had divided his two largest territories, the arts and the sciences, into their subordinate provinces. It was time, he argued, to demolish the partitions established by earlier discoverers and explore those rich tracts which had for too long lain neglected beyond the pale.

Examining eighteenth-century maps and trees of knowledge underlines the mutability of classification systems. Enlightenment plans differed not only from each other but also from modern layouts of disciplinary concerns. Following Chambers' map, an intellectual traveller would arrive at the domain of natural philosophy by following the signpost marked "Rational," which also led to religion, mathematics, and metaphysics. But—unexpectedly for readers today—the lands of optics, pneumatics, and astronomy did not lie along this road: they could only be reached by backtracking to a major fork and then heading off in the same direction as voyagers to falconry, alchemy, and sculpture. One tree of this period showed the fruit of "Physicks" dangling from a small branch next to "Thaumaturgicks," the science of conjuring and wonders. Another chart compiler equated natural philosophy with physics and geology, the study of "this Earth with its Furniture." But the French *philosophes* trimmed the tree to fit their rhetoric of the primacy of reason, shown in the *Encyclopédie* as the sturdy central trunk carrying two leafy limbs of mathematics and physics.[1]

None of these mapmakers surveyed any territory labelled either electricity or magnetism. Most historians, reared in a world of electromagnetic fields, unthinkingly bracket together the study of electrical and magnetic forces. But throughout the eighteenth century, people viewed them as two distinct powers of nature. This ontological differentiation belonged to a constellation of features which distinguished electrical and magnetic practices. Another pervasive misperception is to view the eighteenth century as the progressive development of Newtonian ideas entailing the mathematization of philosophical pursuits. Generations of polemicists have rendered Newton's image into a powerful icon whose influence is hard to escape. Perpetuating their obliteration by Newtonian rhetoricians, modern writers too often ignore the voices of those eighteenth-century critics who opposed Newton's theories. Many of these regretted that natural philosophy was unwarrantedly invading moral domains or judged that mathematics was an inappropriate tool for investigating the natural world. As late as 1833, Whewell felt it necessary to invent the word "scientist" in order to legitimate his colleagues' activities. He wanted to accord status to men like Faraday, consolidating the bonds between electrical and magnetic phenomena. Studying the construction of magnetism as a scientific discipline—a demarcated cartographical entity—contributes to understanding how the foundation of professionalized science formed part of the cultural transformations reflected by the shifting territories of knowledge. Retrieving marginalized historical actors, and viewing the creation of disciplinary magnetism as a contested process, rectifies the glib reiteration of Newtonian rhetorics. Separating out the anachronistic fusion of electrical and magnetic interests exemplifies the historical rewards of using contemporary rather than modern criteria for envisaging terrains of the past.[2]

Magnetic concerns were scattered across any geographical or cultural map of the eighteenth-century world. People living in different places and engaged in different activities held various types of magnetic knowledge. English natural philosophers were not engaged in the same types of magnetic projects as their Continental counterparts, while within England natural philosophers disagreed over the best approach to learning about God's magnetic creation. It was only towards the end of the century that magnetism became a coherent subject of scientific study. Until then, magnetic information was valued for its usefulness—by practical men such as navigators, surveyors, and miners, as well as by natural philosophers justifying their investigations of the natural world.

As the century opened, maritime practitioners displayed the greatest interest in magnetic behavior. Navigation manuals discussed the problems posed to international travel by the vagaries of terrestrial magnetism and described the types of compass appropriate for voyages overseas. Elsewhere, writers categorized magnetic information and devices in ways alien to modern opinions. So when he catalogued the Royal Society's Museum collection in 1681, Nehemiah Grew classified the magnetic equipment alongside scale models and canoes.[3] For Chambers, magnets lay within mineralogy, the study of the earth's history forming part of descriptive natural history rather than rational natural philosophy. Books on natural philosophy gave magnetic attraction only a passing mention amongst forces such as cohesion, gravity, and electricity. If mentioned at all, magnetic phenomena belonged in mechanics. Benjamin Worster, for instance, included magnetic demonstrations in his lectures at Thomas Watts' Academy in London during the 1720s. Author of a text which remained influential for many years, he squeezed in "Experiments with the Load-Stone" amongst discussions of tides, phases of the moon, and bodies descending inclined planes.[4] At a popular level, magnetic and sympathetic attraction were virtually synonymous, entering into accounts of weapon salves, religious enthusiasm, and medical therapy.

As people utilized magnetic powers in different ways, élite natural philosophers defined a boundary for the new scientific discipline of magnetism. Inside it they placed their own types of experiments and ideas, including the information about terrestrial magnetism they had taken over from navigators and subsequently expanded. Beyond this limit they set older associations with sympathies, medicine, and the occult, together with the practice of animal magnetism which flourished in London during the late 1780s. Natural philosophers explored the relationship of this new territory of magnetism to other categories of knowledge about the natural world, removing it from the domains of mechanics and navigation to place it uncertainly next to electricity.

This opening chapter explores this book's own boundaries by mapping attitudes towards investigating and utilizing magnetic effects. It is divided into two sections. The first emphasizes the distinctiveness of magnetic practices in England, which differed markedly from those of its Continental neighbors. Magnetic research exemplified Voltaire's famous quip contrasting the full world of vortical Paris with the empty space of Newtonian London.[5] Unlike today, activities related to magnetic and electrical phenomena were sharply differentiated from each other. The second section focuses on the beliefs and practices of different groups of people writing about magnetic phenomena: it compares their opinions about magnetic behavior and explores their interaction with other communities. Mathematical practitioners, the traditional experts on navigational techniques, made vital contributions to magnetic investigations. Very few natural philosophers wrote extensively on magnetic topics, but as they polemically deployed their ideas, experiments, and inventions to consolidate their own positions, and to attack those held by others, they affected the outlooks of their contemporaries and their successors in many fields.

Mapping the Boundaries of Magnetic England in the Eighteenth Century

Participating in the recent revival of interest in the eighteenth century, now perceived as a key period in Britain's history, eminent scholars have chosen to demarcate the century's start and finish in various ways. Some have selected dates of important political events, such as the Glorious Revolution of 1689 or the Reform Bill of 1832, to bound eighteenth centuries of varying lengths.[6] But events unroll continuously, so that confining them in the strait-jacket of dates highlights the historian's authorial role. In particular, historians of science, concerned to move away from heroic accounts of scientific discovery, are wary of investing too much importance in specific episodes. There was no single magnetic practice, and no such precise chronological limits can be set for this study. However, significant transformations in attitudes and practices make it analytically useful to describe a magnetic eighteenth century stretching from roughly 1690 to 1800.

Much of the seventeenth century had been dominated by William Gilbert's *De Magnete*, conveniently published in exactly 1600. Gilbert's audiences immediately interpreted his book as a magnetic philosophy, one in which he successfully demarcated magnetic power from assorted Renaissance sympathies, elevating it into the central principle of his vitalist terrestrial cosmology. Various maritime, philosophical, and religious communities sustained intense interest in his work. Natural magicians appropriated Gilbert's authority as endorsement for their practices, Jesuits enrolled his philosophy to underpin their cosmological arguments, and natural philosophers sought greater understanding of the patterns of terrestrial magnetism, so important for commercial navigation. In Restoration England, mechanical philosophers turned to magnetic phenomena for furnishing a model of cosmic forces, a solution to the longitude problem, and evidence of a corpuscular mechanistic nature. But after the mid 1680s, their interest collapsed: their research into mineral magnets became subsumed into a more general study of effluvia, and they found measurements with magnetic instruments insufficiently accurate for substantiating theoretical predictions of the changes in terrestrial magnetism.[7]

This lull in English magnetic research forms a natural break between two centuries which can be broadly characterized as Gilbertian and Halleyan. During the eighteenth century, people were generally less interested in the properties of loadstone or magnetic bars than in terrestrial magnetism and compasses, potential resources for improving navigation. Halley first presented his ideas about terrestrial magnetism to the Royal Society in 1683, but his paper of 1692 was far more influential: Figure 6.2 portrays him holding an explanatory diagram. He proposed a nuclear model of the earth's internal structure, one which he intended to account for the behavior of compasses as ships travelled round the world. After two Atlantic voyages of exploration and magnetic measurement, he marketed a chart based on his ideas, claiming it would be invaluable for international navigators. Throughout the eighteenth century, writers on terrestrial magnetism invariably referred to Halley. Although natural philosophers criticized his theories, they devised no viable competitor, and his suggestions were being incorporated into revised models well into the nineteenth century. But although Halley's name continued to dominate theoretical discussions, the organizational frameworks within which data were collected and represented changed substantially. In 1801—when Matthew Flinders set out for Australia—governmental, scientific, and commercial institutions were cooperating to an unprecedented degree. Flinders' compass experiments heralded a new era in international magnetic measurement.

The boundaries for the investigations into mineral magnets are less clearly defined. Natural philosophers undertook systematic programs of methodical experimental research, often in the hope of gaining financial rewards. After the abandonment of Gilbertian magnetic philosophy, some natural philosophers in Oxford and London continued to explore the mechanical properties of iron and loadstone. They tried to derive a simple law of magnetic force paralleling Newtonian gravity, and sought ways of making iron and steel more strongly magnetic. The pattern changed radically in the middle of the century, when Knight successfully marketed powerful artificial magnets made from steel bars. This innovation stimulated further research, the manufacture of more precise measuring instruments, and the incorporation of magnetic topics into educational courses. By 1787, magnetism had acquired sufficient identity as a topic of study for Tiberius Cavallo, Knight's successor as the Royal Society's magnetic expert, to judge it worthwhile to devote an entire book to the subject. Again, the year 1801 acts as an appropriate symbolic flag marking the advent of a new magnetic era: John Robison published novel techniques in the *Encyclopædia Britannica* for analyzing magnetic phenomena.[8]

Well before the Napoleonic initiation of overtly nationalistic sciences, natural philosophy was firmly tied to chauvinist ideologies and rooted in political rivalries. John Desaguliers, for instance, was an entrepreneurial projector and experimental demonstrator at the Royal Society who commemorated the accession of George II with a long poem celebrating the Newtonian system. In his book of popular lectures, he used the enigma of magnetic phenomena for reinforcing his patriotic denunciation of French Cartesianism: "It is to Sir Isaac Newton's Application of Geometry to Philosophy, that we owe the routing of this Army of Goths and Vandals in the Philosophical World." Although Enlightenment philosophers were keen to proclaim the existence of an invisible Republic of Letters, protagonists with other interests propagated nationalist messages of superiority rather than supranational unity. As John Bull and Britannia became familiar symbols creating and reinforcing a united British identity, natural philosophers stressed their contribution to Britain's commercial supremacy.[9]

Continental authors frequently vaunted English practices as the embodiment of Enlightenment ideals, admiring the country's liberal democracy and progressive natural philosophy. But the English themselves did not wholeheartedly embrace what are now portrayed as major Enlightenment values: many of them clung to a traditional hierarchical society bonded with the Church of England. Modern writers are increasingly recognizing this distinctiveness of Enlightenment England and exploring the gaps between ideologies and actualities.[10] In addition, for the last couple of decades, historians of science have emphasized the importance of local concerns in the closure of scientific controversies. Many have undertaken detailed analyses of specific episodes in very limited spaces, such as a single laboratory. But even on a larger scale, location remains an important analytical tool. Recent scholars synthesizing closely focused studies to provide a broader picture have restricted their accounts to England, acknowledging that the country's geographical boundaries corresponded with cultural ones.[11]

The magnetic activities of English natural philosophers differed from those of their Continental counterparts in several important ways. With no state funding or control, magnetic researchers in England were interested in experimental demonstration and in marketing magnetic texts and apparatus. Magnetic theories had been central to René Descartes's bid to displace occult explanations by mechanical ones, and European natural philosophers continued to construct neo-Cartesian magnetic theories. Newton, on the other hand, had made only passing and contradictory references to the behavior of magnets. One book reviewer, scathingly condemning a French neo-Cartesian author, articulated a common English attitude: "This theory consists in an imaginary motion of magnetic vortices, which he impels, directs, and diverts, just as it will suit the several phenomena of his experiments. An hypothesis like this, altho' ingeniously forged, and pliable as the iron rods in his own experiments, doth not fall within the compass of my design, which is to exhibit nature, and not to transcribe romances."[12] English magnetic researchers eschewed theoretical speculation and publicly promoted the value of experiment. They prized the utilitarian benefits of loadstone more highly than postulating complex explanatory mechanisms. In France, magnetic phenomena were seen to be so important that for Jean-Jacques Rousseau they provided the pre-adolescent Emile's first introduction to studying nature. But in England, they scarcely received a mention in Tom Telescope's lessons, the century's publishing success in the new genre of children's didactic literature. Tom's end-of-century successor was called William Magnet, but he emphasized the importance of loadstones for navigation, not for natural philosophy.[13]

During the first half of the century, there were three important centers of magnetic research: Paris, London, and Leiden. The reports produced by the Paris Academy were very different from those of the English and Dutch experimenters, who were in close contact with each other.[14] Although French investigators did not accept Cartesian ideas uncritically, they directed many of their experiments towards developing theories of circulating subtle matter to explain magnetic behavior.[15] In contrast, in England during this period, natural philosophers concentrated on experiment. They examined the mechanical properties of magnets, searched for a regular law of magnetic attraction, and investigated terrestrial magnetism.

Knight's mid-century introduction of cheap, powerful artificial magnets stimulated researchers in Europe as well as in England, but their fields of inquiry remained distinct. Continental natural philosophers, prompted by the topics set in the Academy prize competitions, developed ever more complex magnetic fluid theories and used increasingly mathematical techniques of analysis. They explored methods of improving the quality of artificial magnets for constructing instruments designed to support their theoretical conjectures. Most famously, Franz Aepinus published his mathematical account of electrical and magnetic fluids, and Charles Coulomb verified experimentally the inverse-square law he had derived theoretically.[16] In England, on the other hand, artificial magnets were promoted for their practical value, most significantly for the benefits they promised to navigation. English natural philosophers continued to regard magnetic theories as fruitless hypotheses; they repudiated mathematical methods and manifested little interest in Continental results.[17] They concentrated on performing experiments which could contribute to improving navigational safety, such as analyzing the shifting patterns of terrestrial magnetism. Collaborating with instrument sellers, they examined the nature of magnetic materials to help design more accurate equipment for the international instrument trade. They wrote texts for polite consumption which stressed the usefulness of their practical investigations and the importance of compasses for the country's commercial expansion.

The patterns of magnetic research in England and the Continent were greatly affected by the contrasting degrees of state support given to intellectual activities. As one of Johnson's biographers disapprovingly expressed it, in France "[g]enius was cultivated, refined, and encouraged," whereas the English public was "engrossed by . . . trade and commerce, and the arts of accumulating wealth."[18] French natural philosophers working in state institutions received financial backing for their theoretically oriented investigations of magnets. Awards offered by European academies stimulated research: in 1746, Leonhard Euler and Daniel Bernoulli were amongst competitors submitting Cartesian-based essays on magnetic phenomena.[19] In addition, European universities and medical societies sponsored research into the therapeutic benefits of artificial magnets. The French government established naval academies and institutes for maritime research long before public money became available in Britain for equivalent organizations. It set up a centrally coordinated program to collect and map the magnetic observations of maritime travellers.[20]

French natural philosophers were not engaged in the same symbiotic, commercialized relationship with instrument makers as their English counterparts. London craftsmen were esteemed for their skills, the most eminent being Fellows of the Royal Society. They enjoyed a worldwide reputation for their instruments and operated with virtually no legal constraints on their activities. French instrument makers, of lower social status but more closely regulated, visited England to acquire practical expertise. Some of them imported innovations such as new processes of steel manufacture or instrument designs for development under license.[21] English entrepreneurial philosophers interested in magnets and compasses were obliged to find their own sources of finance in order to survive in a competitive environment. Working in collaboration with instrument makers, they produced compasses and other magnetic equipment which could be marketed. Compilers of magnetic charts relied on volunteers to contribute measurements, or persuaded organizations like the Royal Navy and the East India Company to back them. English medical innovators advertised therapeutic magnetic devices similar to those under academic scrutiny on the Continent, but were marginalized as mercenary quacks by rival practitioners. Natural philosophers tailored their texts and their experiments to fit polite purchasers, stressing the educational value or utilitarian benefits of their studies. By presenting themselves as the authoritative providers and dispensers of useful knowledge, they contributed to the construction of magnetism as a public science.

English natural philosophers and their audiences were far less interested in magnetic phenomena than in electrical ones. Joseph Priestley had contemplated writing a history of magnetism, but in what he perceived as the more profitable enterprise of selling a book on electricity, relegated the subject to a two-page rendition of Aepinus' ideas gleaned at second hand from Richard Price.[22] Like many other natural philosophers, Priestley was keen to benefit from the spectacular nature of electrical wonders, which lent themselves so admirably to public displays as well as to philosophical speculation. At the beginning of the century, Francis Hauksbee, Newton's experimental collaborator at the Royal Society, had demonstrated the glowing glass globe of his new electrical machine, a piece of equipment which other natural philosophers appropriated as central stage prop for their public displays. Particularly after the mid-century invention of the Leiden jar, popular interest in electricity rocketed as philosophical performers dramatically displayed their domination of this recently discovered power of nature.[23] Their shows included suspended boys, kissing Venuses, and sparkling water fountains: oscillating compass needles and patterns of iron filings offered meager pulling power for the fascinated attention of experimenters and audiences alike. Unlike the electrical machine and the Leiden jar, novel magnetic instruments were not producing new phenomena but enabling familiar ones to be replicated more easily and more reliably.

Electrical performances also dominated the private stage of the Royal Society. The Fellows actively explored this new phenomenon, one scarcely mentioned by earlier writers on magnetic topics like Gilbert and Descartes. The *Philosophical Transactions* published up to six articles a year on electricity during the second half of the century, about two of them by new authors.[24] The pattern of papers on magnetic topics was very different. There was rarely more than one paper a year, and research into the properties of magnets dwindled away in the second half of the century after the introduction of artificial magnets. The incidence of papers on terrestrial magnetism dropped during periods of war, since they frequently comprised reports of navigators' measurements made on international voyages; however, some natural philosophers at home did engage in projects to take readings systematically with redesigned instruments.

A few natural philosophers did attempt to establish links between electrical and magnetic effects. They knew that thunderstorms could ruin a ship's compasses or magnetize sets of cutlery. Benjamin Franklin, later to be idolized in France as the revolutionary electrician who "snatched lightning from the sky and the scepter from the tyrant," mimicked experimentally the effects of lightning on compass needles. Some Leiden jars had adjustable dials with four settings: detonating cannon, altering a compass needle, killing small animals, and melting wire.[25] When Knight introduced his magnetic bars, his immediate circle of colleagues at the Royal Society initially incorporated them into their electrical research. Henry Baker, internationally renowned for his work on microscopes and crystallization, electrified his own son to produce colorful streams of light from a magnet he was holding in one hand. But although such performances fleetingly captivated the Fellows' interest, they were soon dropped from electrical repertoires.[26]

Writers carefully drew up lists comparing the characteristics of electrical and magnetic effects and almost universally concluded that their differences outweighed their similarities.[27] Since experimenters frequently incorporated their own bodies within their trials, they perceived one important distinguishing criterion to be the contrasting physical experiences produced by electrical and magnetic equipment. Electrical therapy, initiated in Italy, enjoyed far higher prestige in England than magnetic medicine. The Royal Society endorsed research, publishing several enthusiastic papers about novel treatments, while London's Electrical Dispensary treated three thousand patients in ten years. But there was virtually no press coverage of the medical cures with artificial magnets recommended on the Continent, even though they promised a far less painful and dangerous route to health. Respected men as diverse as Joseph Priestley, the Birmingham chemist and radical Unitarian minister, and John Wesley, the evangelical preacher, proclaimed the virtues of electric therapy but virulently condemned the claims of magnetic therapists and animal magnetizers.[28]

As natural philosophers used their spectacular investigations of electricity for enticing polite audiences, their experimental knowledge rapidly increased, and they proposed numerous theoretical explanations. Electricians focused on displaying and testing theoretical ideas in experimental rooms and lecture halls, vociferously seeking to legitimate an authoritative role for themselves in the face of bitter condemnations of the alliance between electricity, commerce, and theatrical display. Unlike magnetic research, with its preexisting utilitarian value, they were forced to promote new inventions such as lightning conductors. Johnson, who was personally involved in magnetic schemes to improve navigation, lampooned the Royal Society for seeking public acclaim rather than public benefit: "The curiosity of the present race of philosophers having been long exercised upon electricity, has been lately transferred to magnetism . . . if not with much advantage, yet with great applause."[29]

In contrast, writers on magnetic topics quietly and gradually gathered global navigational expertise into their land-based domain of natural philosophy. Drawing on older experimental results, magnetic investigators concentrated on improving instruments and, by enrolling naval men as imperial collaborators, collecting measurements of terrestrial magnetism from around the globe. Whereas patterns of iron filings could hardly compete with dramatic electrical experiments, magnetic studies had long been perceived as important because of the value of compasses for navigating the oceans and negotiating deserts, jungles, and mines. Unlike electricians, magnetic researchers concentrated not on innovation but on carefully repeating relatively unexciting experiments with increasing precision. While electrical theories proliferated, most authors stressed the futility of generating unsatisfactory explanations of the mysterious magnetic phenomena defying natural philosophers' attempts to decipher nature's hidden mysteries.

Natural philosophers constructed and used electrical models differently from magnetic ones. The Leiden jar was a philosophical invention, a demonstration device and central stage prop for theatrical displays. Since its operation defied contemporary explanations, it also functioned as a hardware resource stimulating new theories. Electricians used their equipment as theory-laden conceptual analogies for thunderous clouds and lightning storms.[30] But experimenters did not need to create magnetic effects. They were tapping in their demonstration rooms the same natural power on which navigators had long relied for traversing the oceans. By using terrellae made of loadstone, philosophical voyagers could, in the comfort of their own homes, use these almost naturally occurring iconic models to explore empirically the magnetic characteristics of the wider world. Cavallo pointed out how "philosophers have appropriated to the magnet the poles, the equator, and the meridian, in imitation of the terraqueous globe."[31] By recruiting maritime travellers to send them observations, natural philosophers could condense the earth's magnetic patterns into a chart on a single sheet of paper and summarize its history in tables. Although cartographers of knowledge did not recognize a territory of magnetism, terrestrial surveyors were mapping the earth's magnetic characteristics.

Practitioners and Philosophers

The *Encyclopædia Britannica* did not classify magnetism as a university subject until the middle of the nineteenth century.[32] Like the construction of other areas of specialized expertise, the forging of this new science formed part of multiple transformations taking place during the eighteenth century. Individual natural philosophers took many steps which collectively contributed to the public endorsement of their status as creators and guardians of magnetic knowledge. They defined a new science of magnetism which excluded older connotations redolent of magic and the occult but included reformulated versions of practical navigational expertise. As with other sciences, they deemed that magnetic texts and experiments should conform to certain styles of presentation and method. This content of magnetic science both shaped and was shaped by the ways in which other new specialists were simultaneously delineating their own fields. Similarly, the emergence of accredited practitioners of magnetic science affected and was affected by larger processes involved in constructing a new public sphere.

As natural philosophers sought prestige and financial backing, they advertised the usefulness of their experiments by emphasizing their practical value. Despite its regal endorsement, critics often mocked the early Royal Society as a gentlemanly club of scatty virtuosi dreaming up unrealizable projects. Particularly in the first half of the eighteenth century, satirists frequently parodied their activities. The Laputan episode in Jonathan Swift's *Gulliver's Travels* provides one of the most famous examples, but many of his contemporaries articulated similar reservations—men like Alexander Pope, Henry Fielding, and William Hogarth.[33] The Society was forced to compete with other clubs to establish itself as a prestigious and useful organization. From its very inception in 1660, the Fellows had been concerned to demonstrate the important improvements they could make in navigation, vital lifeline of an insular maritime nation. Natural philosophers continued to stress the value of their magnetic studies for England's expanding commercial empire. Thus George Adams, a prominent instrument maker and popularizer in the last quarter of the eighteenth century, explained that compasses were "of the utmost service to mankind in general, but more particularly to an Englishman; the riches and power of whose country depend on navigation." Magnetic goods had to compete in a varied consumer market, and lecturers and writers on natural philosophy cooperated closely with individual instrument sellers. They recommended each other's products in mutually profitable advertising ploys which are sometimes hard to distinguish from plagiarism. By the end of the century, information about magnetic experiments and instruments was included in numerous books designed to capture the interest and the money of polite audiences.[34]

English natural philosophers insisted that the only chance of solving the mystery of magnetic activity lay in experiment rather than in theory. Desaguliers instructed his readers that "unless we can demonstrate what we explain, it is better to own our Ignorance. . . . If ever we come to know the Causes of the various Operations of Magnetism, it will sooner be owing to a Comparison of the Experiments and Observations of Norman, Pound, Lord Paisley [and others] who acknowledge themselves ignorant of the Causes of those surprizing Effects . . . than to twenty Hypotheses."[35]

Desaguliers's emphasis on the power of demonstration exemplifies the crucial influence of mathematical practitioners on English natural philosophy during the seventeenth century. Mixed mathematicians—men like surveyors, navigators, and astronomers—had traditionally approached nature differently from natural philosophers. They were interested in quantitative rather than qualitative results, in measurement and description rather than causal explanation. Such divergences were more fundamental than the social distinctions separating gentlemanly natural philosophers from artisans and craftsmen. The study of terrestrial magnetism provides an excellent example of how the practical tradition of the mathematical sciences converged with the theoretical approach of natural philosophy. In his *De Magnete*, often cited as a key text, Gilbert appropriated instruments developed by mathematical practitioners. Through stressing that the problem of understanding how the earth's magnetic patterns varied could be tackled by measurement, he provided a model for new experimental practices.[36] At the early Royal Society, Robert Hooke—

that prestigious but impoverished inventor who himself straddled social boundaries—consciously developed mathematical instruments for displaying natural phenomena. His insistence that demonstrations carried explanatory as well as illustrative power influenced subsequent magnetic researchers far more than either of Newton's major texts, the Latin *Principia Mathematica* and the English *Opticks*. Hooke sought knowledge through making traditional instruments more accurate, including magnetic compasses. He investigated magnetic phenomena quantitatively and cooperated with Halley to improve compass design.[37]

Instrument makers continued to play a central role in magnetic research throughout the eighteenth century. They modified older instruments, essentially unchanged since Gilbert's time, to make them more suitable for natural philosophers concerned with precision measurement as a route to formulating laws of nature. Eminent contributors to London's international leadership of the instrument trade, these craftsmen were valued for their innovations. Although experimental philosophers were wary of engaging in manual labor, several skilled instrument makers were Fellows of the Royal Society. For example, the clockmaker George Graham not only redesigned compasses for measuring terrestrial magnetism but also undertook long series of experiments. Prestigious natural philosophers often named their close collaborators, thus giving free publicity to leading instrument sellers expanding their craft workshops into major retail outlets.[38] Inevitably, many of their reports concealed the existence of assistants, whose vital contributions passed unrecorded. Henry Horne, an expert on iron ores, castigated John Canton, a schoolteacher who rose to prominence at the Royal Society, for condescendingly describing him as "the smith that I *employed*." Readers of Cavallo's authoritative text would not have suspected that he had turned to a locksmith for advice on his magnetic experiments.[39]

Natural philosophers came to supplant the traditional keepers of practical knowledge concerning the earth and the skies. Since the vagaries of terrestrial phenomena affected agriculture and travel as well as people's emotional and physical well-being, anyone who could offer reliable information stood to gain both financially and socially. For their data on minerals, weather patterns, plants, ocean currents, star configurations, and magnetic fluctuations, natural philosophers initially depended on the observations of experienced practitioners using traditional artisans' tools designed to solve practical problems. Gradually they maneuvered themselves into a position of being publicly perceived as the authorities in new sciences such as geology, meteorology, botany, pharmacology, and magnetism.

The Fellows of the early Royal Society engaged in an active propaganda campaign to promote their activities. One of the most famous texts was Thomas Sprat's *History of the Royal Society*, whose frontispiece encapsulates the founders' Baconian rhetoric of progress through experiment. As Steven Shapin and Simon Schaffer have argued, this symbolically laden image features Robert Boyle's air pump, emblematic of the Society's powerful material, literary, and social technologies for establishing facts of nature. But the pump lies in the background, while numerous mathematical instruments are prominently displayed in several places. Natural philosophers appropriated these practical instruments for measuring terrestrial phenomena; they used them for emphasizing the contribution they would make to the public welfare by coming to understand and control global effects. To capture public support, they continued to disseminate visual evidence of their authority over terrestrial nature as well as over traditional techniques and practitioners. For instance, the Copley Medal, the Society's annual award to distinguished Fellows, displayed Pallas standing amidst mathematical instruments, including a magnetic compass and a globe.[40] As the Royal Society successfully established a more prestigious public image, it had to distinguish itself from the skilled practitioners whose expertise it was appropriating. In about 1770, the Fellows decided to modify a new drawing for a diploma to be given to foreign members. One of the objectors argued that "such a design, in regard to taste, was never seen. For on the top of it, was the Sun, moon, planets, stars, comets, meteors, lightning, Rainbow &c. &c. Down each side were suspended a variety of mathematical apparatus's and other things, that were supposed to be discovered by particular members of the Society: . . . it looked like a show board of some mathematical instrument makers."[41]

Compasses and magnets, so vital for navigation, formed an important part of the Fellows' self-promotional exercises. They proudly showed visitors Christopher Wren's large magnetic sphere, a six-inch terrella set into a table surrounded by thirty-two glass-covered compass needles. This apparatus was designed for studying terrestrial magnetism, and it made an impression even on the skeptical London commentator Ned Ward: "[I]t made a paper of Steel Filings prick up themselves one upon the back of another, that they stood pointing like the Bristles of a *Hedge-Hog*; and gave such Life and Merriment to a Parcel of Needles, that they danc'd the *Hay* by the Motion of the Stone, as if the Devil were in them."[42]

Instruments were central to the tussles between navigators and natural philosophers for proprietorship of magnetic expertise. Enrolling navigators as assistants, magnetic researchers comfortably based in London sent out ships and equipment to record, construct, and control the whole world as their extended laboratory. By the end of the century, these metropolitan men had successfully established themselves as experts, introducing new equipment and practices as part of extensive naval reforms. The effectiveness of their doctrine of navigational improvement has all but obliterated the resistance of experienced oceanic voyagers. Maritime practitioners resented this erosion of their traditional skills and criticized natural philosophers' instruments and techniques as being inappropriate for use at sea. As one of them feelingly complained, a navigator needs better compasses and other instruments than "those false ones that are now held out only to guide him to destruction."[43]

Historians often depict the eighteenth century as a period when natural philosophers consolidated and elaborated Newton's legacy, but, like other actors' categories, both "natural philosophy" and "Newtonianism" are problematic terms. The practices and attitudes of people labelling themselves natural philosophers varied widely. For instance, some of them were engaged in an Aristotelian search for causes, while others emphasized the experimental display of effects. High Church Trinitarians insisted on the importance of scriptural texts, but natural theologians interpreted nature's book for evidence of God's beneficent design. The Newtonian image came to cast such a powerful aura that it has lost its value as a distinctive label, since writers of many persuasions paid tribute to his genius. Self-styled Newtonian natural philosophers held varying beliefs and chose to pursue different types of activity. Thus while many French natural philosophers developed the mathematical aspects of Newton's work, as presented in the *Principia Mathematica*, most of his English successors pursued the experimental agenda set out in the more accessible *Opticks*. Because Newton had bequeathed contradictory assertions in his legacy of natural philosophy, subsequent natural philosophers could adhere to quite different ontologies yet still claim to be Newtonian.[44]

Although many English writers celebrated Newton as a national hero, natural philosophers realized that his theories were not universally accepted. Often employing the terminology of warfare and religious sectarianism, propagandists strove to establish his authority. Thus Newton's "disciples" condemned the Reverend William Jones of Nayland as a "heterodox caviller" because of his attacks on "Newton's doctrine." Particularly militant protagonists used labels like Newtonian, Cartesian, and Hutchinsonian as terms of abuse for denigrating their opponents. Jones—frequently castigated as a Hutchinsonian—was privately advised by his ally the Archbishop of Canterbury that he should resign himself for the present to being "reputed *an heretic in Philosophy*." As part of his vigorous riposte, Jones ridiculed Newtonian exegetes by highlighting their contradictory opinions about the nature of gravitational attraction. Like other actors' categories, because "Hutchinsonian" and "Newtonianism" functioned both as rallying calls and as pejorative descriptions, they embraced diverse positions.[45]

Only a handful of those natural philosophers describing themselves as Newtonian wrote extensively about magnetic phenomena; none of them dedicated themselves exclusively to such studies. They proffered diverse explanations. These included not only an action like gravity operating through space but also models which are not nowadays associated with Newton's name, such as streams of particles called effluvia, or subtle fluids, supposedly weightless and invisible. Those self-styled Newtonians who did focus on magnetic nature made their results serve various ends. Knight, for example, used his magnetic research to establish his reputation as an élite natural philosopher at the Royal Society. But although continuing to benefit financially by marketing magnetic bars and compasses, he ceased active research. Several well-known natural philosophers engaged briefly in magnetic experimentation—including William Whiston, John Canton, and Henry Cavendish—but this work formed part of larger projects, such as aiming to win some of the money offered by the Longitude Board and compiling systematic meteorological information.

Newton's allies ensured that his works reached a wide audience in post-Revolutionary England, but their dissemination of this new approach to natural philosophy was embedded in political and religious disputes. They had to engage in active campaigns to establish Newton's prestige and the validity of his ideas.[46] Figure 1.1 provides a striking illustration of how issues in theology and natural philosophy were inextricably tied together. Playing to his contemporaries' double fascination with puns and bodily functions, William Hogarth designed this "Frontis-piss" for a biblical commentary on Hebraic points, although the telescope lying across Newton's book suggests initially that the topic was natural philosophy. Hogarth was referring to the antagonism towards Newton's works articulated by John Hutchinson, whose book *Moses's Principia* (1724) is lying next to Newton's *Principia Mathematica.* As his title indicates, Hutchinson attacked Newton's use of mathematics for deriving God's laws, teaching instead that natural truths, directly dictated by God, could be discovered by retrieving the original Hebrew version of the Bible. He rejected Newtonian gravity acting through space; for him, scriptural language united weight, light, and the glory of God in the physical and the spiritual worlds. Hogarth is portraying Hutchinsonian glory being drowned by the force of Newtonian gravity. A Newtonian apologist described how, in this "burlesque print," the cabbalistic disciples of "the old black Diogenes Hutchinson" were, like "a parcel of rats . . . knawing Sir Isaac's books . . . and above, Mother Mid-night drowns 'em in a deluge."[47]

The Hutchinsonian philosophers were singularly outspoken in their attacks on the prevailing Newtonian orthodoxy, but they were certainly not alone. Newton's diverse supporters were so successful at consolidating his iconic image as the undisputed hero of English natural philosophy that historians have largely ignored the voices of his critics. Hostile opponents exploited a magnetic chink in the Newtonian defenses which now provides a window for inspecting these marginalized dissidents. Since allegiances in natural philosophy were closely related to political and religious attitudes, it is highly pertinent that Tory High Churchmen expressed the strongest and most frequent criticisms. This does not, of course, mean that all High Church Tories opposed all of Newton's teachings, in the same way that there was no homogeneous Whig Latitudinarianism uniquely bonded with the establishment of his views. The existence of a pro-Newtonian group of Scottish Tory physicians at the beginning of the eighteenth century exemplifies the absence of any coherent politico-religious ideology of Newtonianism. By the end of the century, natural philosophers calling themselves Newtonian were drawing eclectically on various sources to construct their own theories.[48] Nevertheless, appreciating

FRONTIS-PISS.

Figure 1.1. William Hogarth: “Frontis-Piss.” Hogarth designed this image in 1763 as the punning frontispiece for a book on Hebraic points. The black rats represent the followers of John Hutchinson vainly trying to destroy Newton’s natural philosophy. Hutchinson believed that information about the natural world could be gained by retrieving the original Hebrew version of the scriptures dictated by God. British Museum.

the constellation of beliefs associated with Tory High Church Anglicanism provides a useful approach for exploring the shared characteristics of individual critiques located within specific contexts.

The varied practitioners and theoreticians sheltering under the Newtonian umbrella fashioned their protective shield against a perceived other of anti-Newtonians. Historians have identified several important constituencies. For example, as the century opened, Newton's henchmen had to counter objections from Tories intimately involved in the political intrigues of contemporary Cambridge, men for whom the Glorious Revolution provided a bitter reminder of the inability of human reason to fathom divine wisdom. Eminent nonjurors like the Cartesian Roger North linked the new philosophy with threateningly heterodox theological views.[49] In London, surgeons' critical attitudes towards Newton's theories were enmeshed with their Tory political affiliations through their resistance to social and epistemological reform. On the other hand, Tory High Church authors like Johnson and Jonathan Swift were concerned not so much with the content of Newton's books as with what they deemed the dangerous incursion of rational speculation into moral and religious territory.[50] George Berkeley, Bishop of Cloyne, is famous today because the academic discipline of philosophy has labelled him a leading English empiricist, but eighteenth-century readers recognized him as a spirited defender of revealed religion, the learned skeptical denouncer of Newton's calculus. Modern scholars more frequently interpret *The Analyst* for its mathematical than for its religious implications, and largely ignore *Siris*, ostensibly merely a recommendation of tar water as a universal panacea. But in these texts, Berkeley was attacking the rationalist pretensions of freethinking mathematicians and challenging the authority of Newtonian physicians by dispensing a curative morality as the High Anglican antidote to materialist irreligion.[51] In addition, *Siris* is a particularly significant book because several marginalized natural philosophers cited it as a major source for their own ideas, most importantly the Hutchinsonian Jones of Nayland. It thus acts as a material thread tying together disparate factions, reinforcing the value of conceptualizing a broadly High Church–Tory perspective for illuminating the relationships amongst natural philosophy, politics, and religion.

Berkeley's son George actively participated in the most coherent High Church group, the Hutchinsonians. Although few in number, this brotherhood contributed greatly towards fashioning the increasingly orthodox position of the Anglican Church during the reign of the pro-Tory George III. Through assiduously recommending their friends to positions of power, these theological colleagues exerted a growing influence. Hutchinson's immediate coterie had included Julius Bate and Robert Spearman, who edited the twelve volumes of his collected works, and Alexander Catcott of Bristol, a theologically inspired investigator of the earth's geological history. The movement gathered strength at the predominantly Tory Oxford University around the middle of the century, where prominent members included Walter Hodges, the provost of Oriel College, and George Horne, who became president of Magdalen and vice-chancellor of the University, and then Bishop of Norwich.[52]

Jones, subsequently the author of two Hutchinsonian texts on natural philosophy, was also in this Oxford group. He belonged to a concealed network of men who drew on Hutchinson's theological and philosophical thought as they corresponded with each other.[53] Like Bishop Horne, he deemed it advantageous to deny his Hutchinsonian allegiance for many years.[54] Jones' theological texts (particularly the *Catholic Doctrine of a Trinity*) were widely praised and influenced Wesleyan thought.[55] Towards the end of the century, as Tory and orthodox Anglican interests converged in maintaining the existing religious and political orders, Jones became a pamphlet ally of Edmund Burke.[56] His experiments were financed by the Tory Earl of Bute, who paid the bills he ran up at the instrument shop of Adams, himself the expositor of some Hutchinsonian ideas.[57]

Various authors subsequently expressed Hutchinsonian opinions. In Scotland, his views were widespread amongst the episcopalian clergy until well into the nineteenth century. Because of this Scottish constituency, aspects of Hutchinsonian natural philosophy reached English audiences through the second and third editions of the *Encyclopædia Britannica*.[58] Several English writers voiced similar criticisms of Newton's approach to natural philosophy, although they mostly did not subscribe wholeheartedly to Hutchinsonian doctrines.[59] Jones was viewed as the theological leader of the Hackney Phalanx, the London-based network of Tory High Church dispensers of charity, whose members only slowly relinquished their Hutchinsonian leanings during the nineteenth century. For instance, as late as 1835, William Kirby reiterated the Hutchinsonian identification of the physical and the spiritual domains in his seventh Bridgewater Treatise.[60]

Further hostility towards Newton's natural philosophy was articulated by members of the Behmenist movement. Small sects of these visionary followers of the German mystic Jakob Boehme thrived quietly during the eighteenth century. Important for Swedenborgians and other mystically inclined circles, their spiritualist ideology underpinned the imagery of poets such as William Blake and Shelley.[61] The pietist group centered on William Law, the eminent nonjuring theologian, formed a tightly knit community of High Church Tories and Jacobite sympathizers. These included Samuel Richardson, author of *Clarissa* and printer to the Royal Society, and John Freke, a surgeon who wrote about electrical and magnetic topics.[62] Although they vilified each other in public, the Hutchinsonians and the Behmenists shared a wide range of political and cultural attitudes, including their opposition to Newton's natural philosophy. In pietist communities like the one at Bristol, the key criterion cementing allegiances between men from these and other religious backgrounds was their belief in the Trinity and revelation.[63]

Identifying a broadly Tory High Church group of critics might seem to unite disparate groups engaged in local debates, but these people did share a generally consistent set of philosophical, political, and religious orientations.

For advocates of revelation, just as the sun controlled the earth, so too God dispensed knowledge through His sacred texts and ordained the best form of government. Deciphering God's figurative language yielded premillennial interpretations relating social unrest and natural disasters to apocalyptic biblical prophecies. Many Tory Anglicans played important roles in the opposition to the French Revolution, favoring Filmerian patriarchalism over Lockean contractual models of change by human agency.[64]

High Church Trinitarians judged "the weak and glimmering Light of *Reason*" to be an inappropriate tool for learning about God's universe.[65] As they challenged the power of human reason to learn about God and His world, they declared argument by divine analogy to be more fruitful than mathematical methods. They preferred animistic, Neoplatonic views of an interconnected universe to an image of human observers objectively measuring a detached nature. Repudiating Newtonian forces acting through a vacuum, they argued that making power intrinsic to matter limited divine agency and led straight down the road to atheism. The Hutchinsonian and Behmenist universes were operated by impulse on direct contact, not by attraction. These writers often equated their magnetic ethers with solar fire or electricity, viewing them as the terrestrial manifestations of divine omnipotence.

Many High Church Tories accorded higher priority to moral improvement than to controlling nature for material benefit, casting doubt on the rhetorics of progress through natural philosophy. They tended to oppose Mandevillean endorsements of trade and commerce, linking the spread of luxury with corruption and skeptical infidelity. Those of a pietist inclination judged a gentleman's private study to be a more fitting setting than a public stage for philosophical inquiry.[66] Questioning the ambitions of those natural philosophers who sought to establish themselves as privileged holders of specialized knowledge, they stressed the importance of direct individual experience of God's universally accessible world.

The Enlightenment encyclopedists advertised the importance of drawing new boundaries of knowledge, but modern historians need to remap the entire terrain of intellectual life in eighteenth-century England. Individuals and small coteries of eighteenth-century writers opposed projects inspired by Newton's philosophy for specific reasons. Collectively characterizing these objectors as broadly of a High Church Tory persuasion offers one way of tackling a vital historiographical challenge: mediating the gap between detailed studies and broad analytical themes without, on the one hand, undermining the importance of immediate contexts or, on the other, reverting to an abstract history-of-ideas approach. Using this type of scheme enables issues which seem specific—such as debates about magnetic attraction and Hebrew points—to be meaningfully associated with each other within a broader framework.

Transformations in natural philosophy were rooted in deeper conflicts manifested in situations which at first sight seem unrelated. Examining attitudes towards such changes ties together processes which are superficially unconnected. Exploring the contested construction of a scientific discipline reveals—and demands—a panoramic view. Scientific practitioners came to emphasize control through mensuration and the preeminence of mathematical techniques. Historians too readily depict the Enlightenment period as one of progressive quantification, a sweeping generalization which obscures at least two major issues: the multiple components—such as precision measurement, calculation, enumeration, and abstraction—comprising the modern concept of quantifiation, one foreign to eighteenth-century people themselves; and the sustained resistance to these changes in a wide variety of linked contexts. For example, the power of the fiscal-military state depended on numerical information and arithmetical calculation; various factions claimed that its pervasive growth was eroding their liberty. Oceanic navigators mistrusted the new magnetic instruments imposed by natural philosophers, pronouncing theoretical training to be a poor substitute for on-the-job learning. Similarly, Burke championed the "means taught by experience" for steering the ship of state, and agricultural rioters preferred the well-established moral economy of fair prices to the market economy of commercial free enterprise. Within discourses of civic humanism, arbiters of correct taste argued that political authority is rightly exercised by those who reason abstractly, thus excluding women and the poorer classes. Law denounced the reliance on mathematical reason of debauched natural philosophers but praised the virtue of laborers, whose souls instinctively turned towards God like a compass needle towards a loadstone.[67]

Like Law, diverse protagonists found that magnets, referred to both literally and metaphorically, functioned as a valuable polemical weapon for promoting as well as for attacking arguments about other topics. Unlike electrical and pneumatic apparatus, recently invented by natural philosophers, everyone could witness the behavior of magnets and draw on a shared traditional fund of imagery. Since there was no theoretical consensus, magnetic phenomena provided an adaptable rhetorical resource for defending or criticizing a range of ideas. This versatility means that magnetic symbolism yields valuable clues for constructing a larger landscape of the eighteenth century. This book will paint such a picture by exploring how the practitioners and philosophers surveyed in this opening chapter investigated and represented magnetic nature. In the late 1760s, Georg Lichtenberg, the Göttingen natural philosopher and commentator on London culture, composed a vivid aphorism depicting physiognomic connections: "We can see nothing whatever of the soul unless it is visible in the expression of the countenance. . . . As the magnet arranges iron filings, so the soul arranges around itself the facial features."[68] Similarly, by inspecting the magnetic visage of the eighteenth century, we can gain access to the passions concealed beneath its surface.

CHAPTER TWO

"A Treasure of Hidden Vertues": Marketing Natural Philosophy

WHEN CUSTOMERS like Samuel Pepys visited the shop of Thomas Tuttell, instrument maker to the king, they could purchase a pack of mathematical playing cards. The seven of spades, shown as Figure 2.1, depicted the diverse connotations of magnets, or loadstones. At a shilling a pack, these cards were too expensive for many of the surveyors, navigators, and other practitioners shown using Tuttell's instruments. But for historians, they are valuable both for what they are and for what they show.

As marketable commodities, they provide an early example of the products promising both diversion and improvement which were increasingly attractive to the "polite and commercial people" who epitomized English society in the eighteenth century.[1] Iconographically, Tuttell's trade cards offer a valuable entrée into the world of early-eighteenth-century mathematical practitioners and display contemporary attitudes towards natural philosophy. This particular image is emblematic of the multiple interpretations and utilizations of magnetic phenomena during the eighteenth century. The anchor lying in the foreground symbolizes loadstone's significance for navigation, which natural philosophers continually stressed as the most important reason for learning about magnets. The miners to the right might be digging up natural loadstone but more probably illustrate the value of magnetic compasses for guiding men through underground tunnels.[2] The loadstone on the left is, like the others in this picture, a model designed for use rather than display, shown here being tested for its strength with a heavy iron key. The scene is dominated by Mahomet's tomb, a reference to the frequently cited legend that it was made of iron and was held suspended in midair by a powerful loadstone.[3] The advocates of enlightened reason sought to dismiss such traditional magnetic beliefs as superstition. However, many people, including Romantic poets, animal magnetizers, and vendors of philosophical entertainment, drew on the rich cultural heritage of magnetic symbolism.

Tuttell's epigrammatic "treasure of hidden vertues" encapsulates perceptions of loadstone which were to endure through the century. Loadstones were literally a subterranean treasure, since fine specimens were valued for collectors' cabinets. More metaphorically, magnets also proved a lucrative resource for entrepreneurial natural philosophers to mine. As they designed and marketed magnetic compasses and charts, they boasted about their contribution to Britain's expanding trading empire, thus profiting financially and enhancing their public image. Collaborating with instrument makers, they benefited

Figure 2.1. Thomas Tuttell's Magnetic Playing Card. Thomas Tuttell, instrument maker to the king, sold packs of mathematical playing cards around the beginning of the century. The words and images on this seven of spades represent loadstone's importance for navigation, its commercial significance, and the mysteriousness of its powers. British Museum.

themselves by advertising their magnetic lectures and equipment while contributing to the international reputation of London's skilled craftsmen. "Vertue" was equally laden with symbolic undertones, a word reminiscent of older occult explanations of magnetic phenomena. During the seventeenth century, exponents of the new mechanical philosophy rejected scholastic interpretations that magnetic phenomena are inherently inexplicable. Confidently insisting that their approach would make the insensible intelligible, they incorporated occult qualities into their cosmologies. They used artisans' instruments to measure and hence naturalize the world of spirits and sympathies and to bring it under their control.[4] For much of the eighteenth century, natural philosophers retained the word "virtue" for describing a magnet's concealed power because they could use it as a descriptive term. They could thus avoid any specific ontological commitment, as they frequently confessed their inability to penetrate one of nature's most closely guarded secrets. But "vertue" also carried a moral message, particularly for Shaftesburyan civic humanists. For his pre-Mandevillean iconographic endorsement of the public virtues of commerce, Tuttell selected playing cards, symbolic of the private vice of gambling.

"A treasure of hidden vertues": this phrase, with its hints of concealed financial and epistemological benefits, resonates with major analytical themes of the eighteenth century, such as commercialization, the opposition between vice and virtue, and the fascination with the occult in the face of Enlightenment rationality. It thus provides a useful starting point for exploring how English society changed as enterprising individuals concerned with natural philosophy promoted themselves and their products. As these innovators commoditized their expertise, they shaped the future face of public science. Because they were enmeshed in an intricate cultural web, inspecting their activities fleshes out broader analyses of the profitable alliance of commerce and natural philosophy and deepens our appreciation of the multiple transformations distinguishing this period.

Historians have offered starkly contrasting interpretations of the eighteenth century. Jonathan Clark is the leading exponent of a model based on continuity, in which the monarchy and the aristocracy played key roles in a stable society dominated by politics and religion. Other analysts, such as John Brewer and Paul Langford, stress the importance of bourgeois individualism for effecting change, thus differentiating Britain from contemporary European *ancien régimes*.[5] Whichever their allegiance, none of these writers devotes much attention to topics frequently seen as the domain of historians of science. However, the activities of natural philosophers during this period furnish a clear example of how self-interested individuals from diverse origins wrought changes which contributed towards the construction of an industrialized and professionalized nation. For anyone concerned to understand the emergence of our modern consumer society, in which science plays a dominant role, the commercialization of natural philosophy in the eighteenth century must be a key area of study.

Neil McKendrick's famous interpretation of British commercialization stresses revolutionary changes accompanying the birth and explosive growth of a demand-led consumer society in the second half of the eighteenth century. But, as critics are increasingly pointing out, this analysis is too narrow in its time frame and too simplistic in its scope.[6] Economic expansion and an increased market for material possessions were apparent at least by the middle of the seventeenth century. Transformations were more gradual and reflected the complex interaction of many factors. As just a small example, Tuttell's cards, serving as advertisements as well as didactic entertainment, were published right at the beginning of the century; similar in style to others being produced at the time, they represented a long-standing genre subject to several influences, including the suppliers' insistence on government protection against foreign imports.[7]

The upper echelons of society enjoyed excess purchasing power, but emulation is insufficient to explain the uneven patterns of expanding consumption. While leisure became rapidly commercialized amongst the metropolitan élite, in provincial towns the pace of change was far slower, particularly in the burgeoning industrial centers with no traditional gentry. Demand-side emulation models appeal to late-twentieth-century monetarists. They divert attention from class inequalities by providing a gratifying account of successful progress towards wealth, led by the upper classes and without state intervention. Focusing on the demand for luxuries at the tip of the pyramid evades examination of whether those nearer the bottom were financially able to participate in the national spending spree, as taste and refinement supposedly trickled down. Demand does not arise spontaneously to take advantage of a prepared supply waiting patiently in the wings, but is partially created. Trends in consumer patterns are both a cause of and a consequence of deeper cultural shifts in society.[8]

As the century opened, members of the landed gentry and the expanding mercantile community were profiting from England's increasing wealth. Colley has described how these two groups cooperated to forge a prosperous British nation in a mutually advantageous creative cult of commerce. As one of the world's busiest trading cities, London attracted many possessors of both acquired and inherited wealth. Other readily visible indicators of the flourishing economy included the transformation of the provincial landscape through the renaissance of urban towns as thriving centers of enlightenment.[9]

This growing consumer capitalism entailed many types of changes. As men and women redefined their gendered identities within society, the relationships between their public and private lives altered.[10] While élite boundaries were being gradually extended to accommodate the swelling middle ranks, they were also being strengthened by the consolidating distinction—albeit a blurred one—between polite and popular cultures. Entrepreneurs seeking money and advancement invented new strategies for promoting their products and improving their social status. Thus they contributed towards the transformation of a society based on rigid hierarchy into one in which people could build careers and earn their livings in new kinds of ways.

Natural philosophers were centrally involved in these fundamental changes in English life. Stewart and Golinski have painted a vivid picture of how natural philosophers converted their private experiments into a public, commercialized science by demonstrating their successful domination of nature. Newton's growing band of adherents in post-Revolutionary England legitimated their activities by tying the production of knowledge to the public good. Natural theologians validated the exploitation of nature by asserting that the world had been created by a benevolent God for human benefit. By citing this divine sanction of the financial advantages of invention, they effectively translated commercial activity into holy commandment. As philosophical entrepreneurs marketed their products, they participated in building a materialist society dependent on their expertise. Using various tactics to enlist public support and capture appreciative audiences, they packaged their skills, instruments, and knowledge into sellable commodities competing for enlightened income. In so doing, they benefited financially from the demand which they helped to create and simultaneously levered themselves into a position of power. As they employed self-promotional tactics to carve out profitable niches, they helped to construct and enlarge the polite market for natural philosophy. They promoted Enlightenment ideals of a utilitarian scientific culture, governed by an expert élite but with skills democratically diffused through society for the public benefit. By the early nineteenth century, they had established the framework of a professionalized disciplinary science.[11]

Natural philosophers perceived naval improvements to be a key area for establishing credibility in a commercial community whose wealth depended on maritime imports and exports. Many of them turned their attention to inventing and promoting devices for making ships more seaworthy and navigational techniques more reliable. The Royal Society had always had close links with Christ's Hospital, founded as a mathematical school for training naval officers. At the academies catering for the new educational demands of merchants and manufacturers, lecturers in natural philosophy shared classrooms with teachers of navigation and commerce.[12] Natural philosophers underlined the importance of magnetic knowledge for navigation and honored the inventors of new magnetic devices. The president of the Royal Society advertised the national significance of such work by emphasizing the value of Knight's improved compass to the British people, which would enable them "to increase and promote greatly our foreign trade and commerce, whereby we are provided at home with the fruits, the conveniences, the curiosities and the riches of the most distant climates."[13]

But, particularly in the first half of the century, many people disapproved of this harvesting of nature for financial gain. Civic humanists opposed to Mandevillean consumerism were skeptical about rhetorical claims of progress and criticized natural philosophers for seeking to improve their own status by improving the natural world.[14] Defenders of traditional hierarchy bewailed the spread of luxury, a word whose meaning only later shifted from biblical associations with sin towards refinement. The Commonwealthman Andrew Fletcher, for instance, blamed compasses for promoting decadent lifestyles dependent on expensive imports. In a diagnosis paralleling that of his own ailments, the dietary physician George Cheyne proclaimed that England had declined into malady: "Since our Wealth has increas'd, and our Navigation has been extended, we have ransack'd all the Parts of the *Globe* to bring together its whole Stock of Materials for *Riot, Luxury,* and to provoke *Excess.*"[15] Despite such critiques, as natural philosophers marketed themselves, their products, and their expertise, they successfully established scientific invention and discovery as essential features of polite commercial society.

This chapter analyzes the shifting relationships between natural philosophy and commerce from two complementary angles. The first section examines the career of Gowin Knight, England's most influential magnetic merchant. Knight flourished in the relatively unexplored central decades of the eighteenth century, a period when individual innovators in many fields were seizing the opportunity to market their skills. Knight's self-promotional activities exemplify the commercial initiatives of entrepreneurial philosophers who indelibly altered polite English culture. As they sought personal recognition and financial reward, they contributed towards constructing natural philosophy as a visible and powerful public science. By examining Knight's changing role in intersecting networks, we gain a deeper understanding of the transformations affecting so many aspects of English life. The second half of the chapter describes the developing magnetic market place. New products were designed by a range of entrepreneurs seeking to augment their status by enlarging existing audiences and capturing new ones. As polite communities expanded and defined themselves against the other of the lower ranks of society, natural philosophers were concerned to consolidate their own authority. Struggling to differentiate themselves from magicians and quacks, they policed the repertoire and advertising techniques of purveyors of polite rational entertainment. By exploring such demarcation disputes, we broaden our insight into how the self-legitimation of élite experts fashioned class and gender distinctions in this commercial society.

The Rise of Gowin Knight (1713–1772)

The social barriers of metropolitan culture did not correspond to modern academic faculties. Natural philosophers were engaged in many activities besides those which would nowadays be called scientific. At the Royal Society around the middle of the century, aristocratic experimenters mingled with instrument makers, artists, statesmen, admirals, schoolteachers, and physicians. They were united by mutual self-interest in an extended patronage system fostering individual enterprise. The changing relationships amongst these Fellows and between natural philosophers and their audiences cannot be isolated from transformations taking place throughout society. New attitudes and new initiatives at the Royal Society were both a consequence of, and contributed towards, the forging of an increasingly polite and commercial British nation.

Historians of science have defined and divided the eighteenth century in a variety of ways. For England, the central years from about 1740 to 1770 remain an underresearched period. Falling beyond the reach of the Newton industry and before either the northern burgeoning of innovation or the metropolitan expansion of the Banksian empire, this middle third of the eighteenth century comprises its own small "valley of darkness."[16] Between them, Stewart and Golinski provide convincing interpretations of an extended eighteenth century stretching from 1660 to 1820, during which natural philosophers converted their activities into a public scientific culture. But the ostensible ten-year gap from 1750 to 1760 which falls between the stated scope of their studies is effectively far longer. Stewart's narrative of the world of projection and intrigue in the coffeehouses of Exchange Alley rarely mentions events beyond the mid 1730s; Golinski sites the birth of public chemistry in Enlightenment Scotland, essentially delaying his journey southwards until the early 1770s before chronicling Joseph Priestley's impact on English audiences.[17]

Several historians have demonstrated the value of focusing on particular entrepreneurs to provide tangible evidence of how individual activities effected the commercial transformations of eighteenth-century life. For instance, McKendrick's lively analyses of George Packwood's shaving advertisements and Josiah Wedgwood's marketing strategies have acquired classic status. Historians of science and medicine have made some parallel studies. These include James Ferguson, the Scottish autodidact who built up a flourishing London instrument shop, and Paul (?) Chamberlen, operator of an effective publicity machine for promoting such medical remedies as sympathetic necklaces for reducing the pain of teething. The printing house run by Richardson, another self-made man, visibly lay at the generative hub of a vast capitalist network. As he produced the publications of governmental and intellectual institutions, Richardson wrote novels—notably *Clarissa*—which functioned as organizing forces constructing the bourgeois public sphere.[18]

Knight's own rise to eminence resembled that of many of his contemporaries. His close friend the society artist Benjamin Wilson portrayed him as a gentlemanly author and philosophical inventor, flanked by his redesigned compass and his theoretical text: an appropriate choice as the first director of the British Museum (Figure 2.2). The son of an impoverished provincial clergyman, Knight rose through ability, patronage, and commercial opportunism to dine with nobility and be discussed by the king.[19] His career as an upwardly mobile entrepreneur and administrator lasted for about thirty years. Although his life was singular in its details, its broad features were shared by many eighteenth-century men. As they sought to improve their own positions, they established new ways of earning money and gaining status through skill and initiative rather than relying on patronage. They thus contributed to the

Figure 2.2. Gowin Knight. This 1751 etching of Knight as a learned inventor is by his close friend Benjamin Wilson, artist and electrical experimenter. Courtesy of the Royal Society.

commercial transformation of the eighteenth century and the changing nature of organizations like the Royal Society.

For example, Knight's career closely paralleled that of another Fellow of the Royal Society, the artist Arthur Pond. Like Knight, Pond successfully carved out a new form of livelihood during the historically neglected central decades of the century. Through his activities, he helped to transform a London art market based on patronage of French painters into a flourishing commercial and British-based enterprise. Pond and Knight deployed similar tactics of self-promotion and moved in similar social circles. They had prestigious patrons in common and were both members of the exclusive Royal Society dining club.[20] Medical practitioners were also implementing important changes. Enlightenment medicine was a competitive business: physicians vied for patients in a pluralistic milieu shaped by client requirements, and there was no firm boundary between respected practitioners and those denounced as quacks.[21] Many of the Fellows, including Knight himself, were medically trained, and his circle of medical colleagues included William

Hunter, the eminent society surgeon and male midwife. Ignoring traditional career structures based on institutions, Hunter had achieved success by treating medicine as a commercial enterprise.[22] Knight, Pond, and Hunter were entrepreneurs who rose through patronage and merit to attain positions from which they could dispense support to others. These men were inventing strategies of self-advancement as well as marketing new magnetic devices, pictures, and medical therapies. They represent numerous contemporaries who were building career structures outside such existing organizations as the civil service.[23]

Very few studies have been made of the social networks of experimenters at the Royal Society around the middle of the century.[24] Knight is a rewarding example to analyze, because he was engaged in a range of activities at a time when changes being implemented by natural philosophers were resonating throughout society. The Fellows of the Royal Society were seeking to improve their status. They were exerting more stringent controls over the steadily increasing membership and ensuring that their experiments were regularly reported in journals like the *Monthly Review* and the *Gentleman's Magazine*. They advertised the usefulness of experimental research through the new Copley Medal, awarded to Knight and several of his close associates.[25] Knight represented an expanding yet heterogeneous group of people who used natural philosophy to earn their living. Neither traditional instrument makers nor aristocrats, their origins lay between those opposite social poles of the Royal Society. Earlier in the century, experimental projectors like Desaguliers were often bound to aristocratic backers in a symbiotic relationship of dependence. But although these later men did benefit from patronage, they also adopted diverse strategies of marketing and self-promotion to survive and succeed.

Knight's activities exemplify the commercial initiatives of natural philosophers, which affected English society just as much as the more commonly cited industrial innovations.[26] Entrepreneurial philosophers like Knight indelibly altered polite English culture. Through their books and lectures, they marketed rational entertainment as a desirable commodity. They ensured that familiarity with natural philosophy became a badge of culture. Instrument makers consolidated London's international reputation as a center of craft excellence and successfully marketed their instruments as items of domestic display. Consumers purchased books and pictures featuring orreries and air pumps, thrilled to spectacular electrical displays at packed performances, and decorated their houses with thermometers and barometers. Literary authors embellished their texts by portraying philosophical instruments as metaphorical measurers of emotion.[27] As individual natural philosophers such as Knight sought personal recognition and financial reward, they contributed towards constructing natural philosophy as a visible and powerful public science.

Knight's early life resembled that of many who came to hold prominent positions. He was the son of a clergyman, educated at Leeds Free Grammar School like Wilson and their common friend John Smeaton, who became a famous engineer. In 1731, he joined the swelling group of doctors' and clergymen's sons studying at Oxford by winning a scholarship to Magdalen Hall. Four years later, his academic prowess earned him a prestigious scholarship at Magdalen College, relieving him of his duties as servitor to wealthier fellow students. He stayed there until 1741, studying natural philosophy and medicine.[28]

The tuition and examination systems at Oxford were notoriously lax during the eighteenth century: student life revolved around pleasure rather than academic work. A foreign visitor commented disparagingly on the mold covering the library books at Knight's college, where Edward Gibbon judged that he spent "the fourteen months the most idle and unprofitable of my whole life."[29] With only light academic obligations, Knight started investigating the laborious techniques of making artificial magnets: Figure 2.3 illustrates the type which he later marketed. He and his competitors were to develop several different methods, one of which is illustrated in Figure 5.1. These time-consuming operations all entailed a cumulative process of building up the magnetic strength in steel bars by repeatedly stroking them with bars which had been previously magnetized.[30]

Producing artificial magnets was not a new goal. Since the beginning of the century, natural philosophers had investigated the effects of twisting, filing, hitting, heating, and rubbing iron bars and wires of different dimensions. They studied steel producers' experiments with different manufacturing methods and ores.[31] Knight might have read about artificial magnets in the text by the Dutch natural philosopher Pieter van Musschenbroek. He probably knew about the experiments of Servington Savery, a wealthy Devonshire recluse with a lifelong interest in magnets, whose son was a contemporary of Knight's at Oxford. In 1730, a long article by Savery was printed in the *Philosophical Transactions*. He summarized current magnetic knowledge and included detailed instructions for replicating his own experiments on artificial magnets.[32] Operating at a distance, Savery had little control over the fate of his work in London, where his magnets did not come to be well known. Part of his paper had been read out at Royal Society meetings in installments, but it was never completed after the summer break.[33] He gave one of his magnets to the watchmaker William Lovelace, who used it to make smaller magnets for sale, twice visiting the Royal Society to show them off as his own.[34]

In contrast, Knight succeeded by adopting far more direct promotional tactics as soon as he left Oxford. Practicing as a doctor, he took lodgings in London, where he persuaded the president of the Royal Society, Martin Folkes, to visit him and witness his experiments. Evidently impressed, Folkes reported to the Fellows that he had seen Knight's magnets lift heavy iron keys and weights. Knight dramatically displayed his magnetic powers to the Fellows on several occasions. He convinced them that his magnets had lifting powers superior to the Earl of Abercorn's famous terrella, one of the Society's prized possessions.[35] As he deftly manipulated the polarity of pieces of loadstone, Knight illustrated his contribution to the Enlightenment goal of controlling nature. Constantly emphasizing the thoroughness of his research, he also

Figure 2.3. Artificial Magnets. These boxed artificial magnets, now part of the King George III collection, were probably made by Gowin Knight. The steel bars are fifteen inches long, with lines to mark the north poles. The finely crafted box is made of polished mahogany. Knight explained to the Fellows how the magnets' strength could be preserved by storing them separated by a piece of wood, with iron keepers across their ends. Science Museum, London.

stressed its utilitarian value. He showed how to make permanently magnetized compass needles from hard steel and devised a portable case for storing his bars and preventing them from losing their strength (see Figure 2.3).[36]

As he benefited from the patronage of the former president, the eminent physician and botanical collector Sir Hans Sloane, Knight's status at the Royal Society rose rapidly. By 1747, within only three years of his first appearance, Knight had not only been elected a Fellow but had been awarded the prestigious Copley Medal. The following year, he was included in the first restricted membership of the select Thursday evening dining club.[37] Through his spectacular performances, Knight established himself as the Royal Society's expert on magnets, whose colleagues solicited his advice. Recognizing the commercial significance of compasses, he paved the way for his own innovations by condemning the government for irresponsibly economizing on vital instruments of navigation.[38] As will be explored in the next chapter, such pronouncements secured him the influential support of the Royal Society when he came to market his own magnets and compasses to the navy.

Knight continually promoted himself and his inventions, although discreetly tailoring his tactics to match the status he wished to maintain and augment.[39] Artists like Pond and lecturers on natural philosophy were mocked for advertising in the press, while medical practitioners risked accusations of quackery for anything more ostentatious than nailing a small brass plaque to the door. Knight prudently selected gentlemanly techniques of advertisement. To further his medical career and impress his influential visitors, he chose to live in fashionable Lincoln's Inn Fields.[40] In about 1750, he strengthened his ties with the Royal Society by moving virtually next door in the same alley, Crane Court, at the heart of Fleet Street's instrument trade district. He took every opportunity to publicize his prestigious addresses. For example, recounting his experiences of the 1750 earthquake, he included the otherwise gratuitous information that one of his neighbours was the Duke of Newcastle. When he moved, he broadcast the event by arousing public interest in his discovery of an interesting letter by William Warburton which had been left behind.[41]

Using the Royal Society as a promotional platform, Knight ensured that his reputation reached a wider audience. Wherever his name appeared, it was followed by the coveted initials FRS.[42] His talks were published in the *Philosophical Transactions*, to be read by natural philosophers in Britain and abroad. In these articles, Knight boasted about his achievements but concealed his methods. Colleagues excitedly transmitted his polemical messages through correspondence networks linking America, Europe, and England. "Hither to I have wrote only to blot paper," enthused the Philadelphia book dealer Peter Collinson, "but now I tell you some thing new Doc^{r} night a Physition has found the Art of Giveing Such a magnetic power to Steel that the poor old Loadstone is putt quite out of Countenance." Through Collinson and other colleagues, including his close friend the Quaker physician John Fothergill, Knight's work reached American natural philosophers like Franklin.[43] In England, Knight's "very intimate Friend" Baker fielded inquiries about the new magnets.[44] Knight himself was the major channel of communication between the Royal Society and Leiden, where Musschenbroek's lavish praise guaranteed him international fame amongst magnetic cognoscenti. He sent free samples of his bars to such colleagues as René Réaumur in Paris, where his techniques were admired and imitated.[45]

Carefully guarding his methods, Knight used his friends to help him win lucrative individual commissions for making magnets. For instance, as Wilson was sheltering from the rain in Adams' instrument shop, he entertained a fellow customer with tales of Knight's magnetic exploits. Wilson explained how these "delighted him greatly. Now by this introduction, I not only obliged [him], but served my friend the Doctor," who ended up £250 the richer. Knight discreetly targeted the quality end of the market for artifacts of natural philosophy, editing a collection of his articles from the *Philosophical Transactions* into a gentlemanly advertising brochure which he distributed to potential customers. Along with his compasses (Figures 3.1, 3.3, and 5.5), he sold a variety of magnetic products, including three sizes of bars in tailor-made wooden boxes, warding off cheaper imitations with the signed certificate shown in Figure 3.4.[46] Disdaining press advertisements, Knight reached cus-

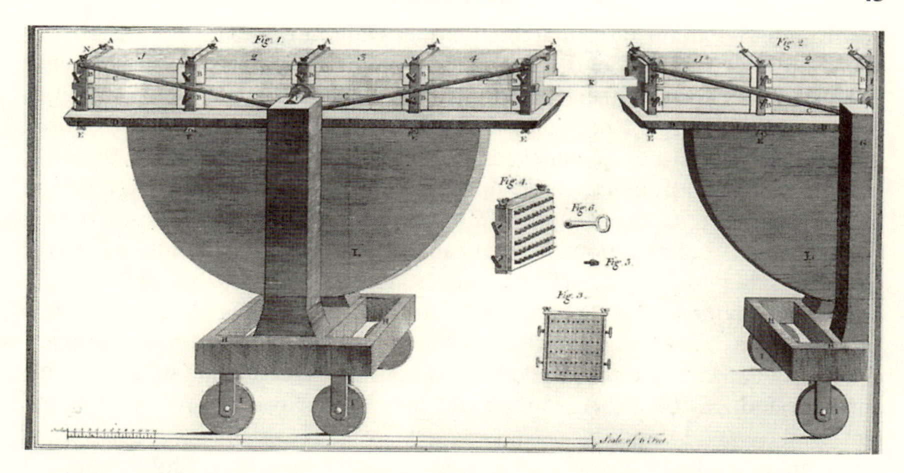

Figure 2.4. Gowin Knight's Machine for Making Artificial Magnets. Knight used a room at the British Museum for displaying his invention, two wheeled magazines each comprising 240 permanent magnets. He could rapidly magnetize a bar (*K*) without going through the laborious stroking procedure. Michael Faraday subsequently adapted part of this device for his own research. *Philosophical Transactions* 66 (1776), facing p. 601. Whipple Library, Cambridge.

tomers outside London through Adams' mail-order catalogues and international dealers. Jean Magellan included Knight's instruments in consignments to the Spanish and Portuguese courts and published instructions in French.[47]

Knight gained further renown from the impressive wheeled apparatus shown in Figure 2.4. Made from two magazines each of 240 permanent magnets, his invention enabled a bar to be rapidly magnetized. After he became head of the British Museum, he insisted that his machine should be assigned a special room. Since his duties included guiding eminent visitors round the exhibits, he could ensure that it became part of their tour.[48] Knight was constantly alert for opportunities to gain free publicity. When his artificial magnets were excluded from a preliminary design for a diploma displaying the Royal Society's achievements, he repeatedly brought the matter up before the Council and succeeded in having a new engraving made. He persuaded colleagues at the Royal Society to praise his compasses in their navigation texts.[49] In advertising himself, Knight contributed to the legitimation of natural philosophers as the authoritative holders of magnetic knowledge.

Knight also sought to take advantage of the expanding market for books on natural philosophy. In 1748, he wrote a theoretical treatise which was completely different from the only other eighteenth-century English book which had so far been devoted to magnetic topics, Whiston's experimental and mathematical confirmation of biblical chronology.[50] Knight may have based his style on that of Bryan Robinson, whose recent book on Newton's ether was already into its second printing. Robinson's renown had attracted Knight's best friend, Benjamin Wilson, to visit him in Dublin, where he was the professor of physics. From around 1740, natural philosophers became increasingly interested in elaborating the etherial model which Newton had proposed in the thirty-first query of the *Opticks*. Robinson had mathematically explored the possibilities of such an elastic ether.[51] Like Robinson's, Knight's dissertation was tailored for natural philosophers reared on Newtonian texts. Laid out in numbered propositions supported by definitions and corollaries, it was a short but dense elaboration of ethers made up from fundamental attractive and repulsive particles. As will be discussed more fully in Chapter 7, Knight aimed to provide a comprehensive cosmology explaining phenomena such as gravity, heat, and light, but the last third of the book was devoted to magnetic activity. He described in great detail how his experimental investigations corroborated his theory of a magnetic fluid made up of mutually repellent particles.

Knight published the first edition at his own expense, investing a couple of hundred pounds in a handsome quarto edition with wide margins. It presumably sold enough copies to motivate John Nourse, who specialized in profiting from such books, to publish a cheaper octavo edition six years later, which was summarized in the *Monthly Review*.[52] Even so, although the book was well known amongst his contemporaries, many of them agreed that Knight "calls Old Discoveries by New Names, and deduces Corollaries till he loses all Sight of his Proposition."[53] Franklin's judgment was probably typical: he thought Knight "the greatest Master of Practical Magnetics that has appear'd in any Age" but, in an oblique reference to the density of Knight's prose, never quite found the "Leisure to peruse his Writings with the Attention necessary to become Master of his Doctrine."[54] For at least six years, Knight nursed plans to tap a wider market with a historically based study, but he eventually cancelled the project because there were insufficient subscriptions.[55]

Knight reinforced his position within the Royal Society. He was elected a member of the Council in 1751 and a couple of months later ran for the post of secretary against the Reverend Thomas Birch, a well-connected Quaker. This appointment was a sensitive issue, because it affected the public image of the Royal Society and was clouded by the controversies which had surrounded Folkes' election and the choice of papers for publication in the *Philosophical Transactions*. With the backing of Lord Northumberland, Knight gained seventy-six votes but was defeated by the more powerful lobby organized by the Lord Chancellor (Hardwicke) for Birch, who won with ninety-one votes.[56]

Knight's status at the Royal Society proved invaluable when he successfully applied for the new post of principal librarian at the British Museum, founded as part of Sloane's bequest. Most of Sloane's executors and the trustees responsible for the Museum's creation were Fellows, including Knight's patrons Northumberland and Folkes, as well as closer colleagues like Baker and Collinson. The Royal Society was intimately involved with constructing this public institution, and Fellows secretly exchanged coded letters on the topic.[57] Knight's two rivals were John Mitchell, also a physician at the Royal Society, and John Hill, a knowledgeable promoter of Linnean botany often marginalized as a Grub Street hack. Hill waged an active campaign of self-promotion, sending off numerous oily letters soliciting patronage and writing long, provocative newspaper articles. Knight rejected such flamboyance. His tactful visit to Birch was rewarded by an invitation to dinner, which enabled him to acquire prestigious references for his restrained letters of application to the Principal Trustees. Hill was notorious for his vicious satires on the monarchy and the Royal Society, and his candidature was not even put forward to the king, the ultimate arbiter for the position.[58]

From his appointment in 1756, Knight remained at the Museum until his death sixteen years later. For £200 a year, he acted as live-in caretaker, responsible for displaying the exhibits and supervising access.[59] Just as the officials at the Royal Academy came to police the construction of public art, Knight and his colleagues monitored the content and presentation of natural philosophical knowledge, thus affecting the future face of public science. Knight participated in framing the regulations governing entry to the Museum, subsequently revised to restrict public facilities still further.[60] He played a key role in the adoption of the Linnean system of classification. Participating in international correspondence networks through associates like Franklin, he contributed to the Museum's overexpenditure as he enlarged the collections and supervised the staff arranging them. While favored visitors dozed off in the stuffy reading rooms, Knight wielded considerable local power in an empire riven by dispute, refusing readers permission to view prints and books and quarrelling with trustees as well as subordinates.[61]

Knight became recognized in polite society as a man of importance, although he was increasingly reputed to be reclusive and ill-tempered. He gained entry into élite medical circles and was known as a distinguished visitor at metropolitan soirées, the intimate colleague of society doctors like John Pringle and Fothergill.[62] Joining richer men like Fothergill, Wilson, and perhaps Samuel Wegg, governor of the Hudson's Bay Company and treasurer of the Royal Society, he gambled in Cornish mining projects. After their collapse, Fothergill paid off Knight's debt of a thousand guineas, probably hoping he would not make public the Quaker doctor's involvement in speculative ventures.[63] Knight participated in Fothergill's new journal intended to help break down the restrictive practices of the Royal College of Physicians, and he probably belonged to the medical club which met informally at the Queen's Arms. He remained a close colleague of Birch, joining him for outings and dining frequently at his house, where he met influential statesmen and aristocrats.[64]

As he negotiated his social climb, Knight achieved a position from which he could wield influence and dispense patronage. He served on the Royal Society Council more than once, rarely missing a meeting and affecting the lives of men who would become more famous than himself. For example, he became involved in diplomatic concealed negotiations to appoint a new president, and he was often present when James Cook was being briefed for the Transit of Venus expeditions. He attended daily in December 1767 when Emmanuel Mendes da Costa, one of his own investment partners, was accused of embezzling the Society's funds.[65] He persuaded Smeaton to come to London as his assistant, where he took him to Royal Society meetings, publicly commended his abilities, and sponsored his application to the Royal Society. Knight backed other applicants, including Birch's protégé James Ferguson, John Michell, John Canton, and John Kidby.[66] He was a member of the Society for the Encouragement of Arts, Manufactures, and Commerce, sitting on various committees assessing which inventions should be rewarded and organizing the selection and viewing procedures for exhibitions.[67]

Only fleeting references to Knight's personal life survive. When he applied for the job at the British Museum, he described himself as a diligent, industrious man who had "spent the greatest part of my Life in the pursuit of natural knowledge. . . . [If appointed I will] render the Museum of Publick Use, & will dedicate the remainder of my Life to that purpose." Other writers testified to this commitment with such conventional terms as "eminent," "knowledgeable," and "distinguished," but they also made less-flattering comments about his reclusiveness and irascibility. This was the man who so antagonized his staff at the British Museum that, according to Thomas Gray, "the keepers have broke off all intercourse with one another, & only lower a silent defiance, as they pass by. Dr Knight has wall'd up the passage to the little-House [toilet], because some of the rest were obliged to pass by one of his windows in the way to it."[68] Gray's remark gives us a rare glimpse of daily reality behind the walls of this new institution. Historians overlook countless men like Knight, who were well known in their day but have left no convenient archive of correspondence. Reconstructing their lives reveals their influential roles, showing us how they contributed to the commercialization of the eighteenth century and the increasing involvement of organizations like the Royal Society in public life.

Magnetic Mystique and Polite Entertainment

Marketing natural philosophy was a competitive business in the eighteenth century. Governmental and commercial organizations had no established procedures for backing inventions, so individual entrepreneurs had to find their own sources of finance. Consumers could choose to spend their money on an enormous range of goods and services. Newspaper advertisers bombarded readers with information about household items, medical therapies, plays, beauty treatments, and much more. Natural philosophers had to convince practitioners like mariners and miners that their products were useful, and persuade polite audiences that their books, lectures, and demonstration devices were entertaining and improving. They created new markets for new types of goods. Like Knight, many entrepreneurial philosophers stressed the élite nature of the information and instruments they were selling. They were keen to differentiate themselves from retailers and popular performers charging lower prices. Through their self-promotional activities, natural philosophers collectively forged an identity as the legitimate purveyors of natural knowledge.

In the first half of the century, maritime practitioners and cabinet collectors were the major purchasers of loadstones and magnetic instruments. Knight's mid-century introduction of reliable artificial magnets diversified and expanded the market. Natural philosophers throughout Europe constructed elaborate precision apparatus. They bought English magnetic instruments and engaged in new projects to measure terrestrial magnetism. Entrepreneurs advertised experimental kits, instruments, and other magnetic merchandise to a wider public, repackaging old experiments and promoting them as enlightening entertainment.

As with electricity, the definition of what should count as acceptably rational recreation was contested. Those seeking public approbation of magnetic science were forced to tread a narrow line. They wanted to differentiate themselves from popular entertainers and conjurors, but at the same time they had to convince polite purchasers that centuries-old experiments with iron filings, compass needles, and magnetic rings were rewarding to witness or perform. They found it relatively easy to denigrate the more flamboyant entertainers as quacks, but they disagreed about how appropriate it was to use elaborate parlor tricks as lures for capturing converts to natural philosophy. Some lecturers and writers avoided this dilemma by sacrificing spectacle to utility, stressing the importance to an island community of learning about magnetic navigation. The danger of leaning too far in this direction lay in being confused with mathematical practitioners. In fashioning their own identity, natural philosophers desired to consolidate public recognition of their élite status as well as of their superior understanding and control of natural forces.

Magnetic goods were commercial commodities throughout the century. Before the 1750s, when artificial magnets became widely available, traders dealt in natural loadstone. The quality, and hence the price, of the loadstone varied widely. Many commentators reiterated versions of the Greek classification into five types, based mainly on the color and the country of origin. The most highly prized were the blue-black varieties like that from Ethiopia, renowned for being consistently powerful and homogeneous. Inferior versions were usually red or black. They were mined in countries throughout the known world, although the ones found in England were of comparatively poor quality. William Petty suggested that, like diamonds and pearls, the price of loadstone should be regulated. Fifty years later, Abercorn provided a book of

Figure 2.5. The Countess of Westmorland's Loadstone. With its casing of a copper coronet, this massive loadstone weighed 171 pounds. The Countess of Westmorland donated it to the Ashmolean Museum at Oxford in 1756 while her Tory husband was the High Steward and three years before his successful campaign to be elected University Chancellor. Courtesy of the Museum of the History of Science, Oxford University.

arithmetical tables, summarized in the *Philosophical Transactions* to reach a wider audience, for calculating the precise monetary worth of a loadstone of any size and strength.[69]

At the end of the seventeenth century, rich collectors of curiosities were condensing nature into a visible compendium, in which they included common objects as well as rare treasures. Like other cabinet pieces, an exceptionally fine mounted loadstone provided a threefold reflection of value. It simultaneously advertised its purchaser's wealth, represented the fascination of natural rarities, and symbolized the monetary cost of labor-intensive craftsmanship.[70] Throughout Europe, individual loadstones were known by the names of their possessors. Large ones were famous for their size, and small ones for their extreme strength. For example, the tiny chip in Newton's ring was renowned because it could support 250 times its own weight, whereas Berkeley's larger loadstone was proportionately far weaker. One of Knight's customers offered him £50 if he could make an artificial magnet rivalling the

Figure 2.6. Loadstones. Natural loadstone was usually shaped into rectangular blocks. Cabinet specimens like the one on the right were mounted in decorative silver casings. A loadstone's magnetic strength was increased by arming it with pieces of iron attached to the poles. These two loadstones have been dipped into a pile of iron filings, still clustering round the iron caps. Science Museum, London.

thirty-pound specimen belonging to the king of Portugal.[71] Proud collectors elicited suitable sentiments of wonder from their visitors. At the emperor's repository in Vienna, Mary Montagu marvelled at the "small piece of loadstone that held up an anchor of steel too heavy for me to lift. This is what I thought most curious in the whole treasure." In Paris, Martin Lister spent hours admiring his English host's expensive magnetic collection, while Celia Fiennes was charmed by the Oxford Ashmolean's Museum's "several Loadstones . . . it is pretty to see how the steele clings or follows it."[72]

Prices were high, up to several hundred pounds changing hands in private purchases or at auctions disguised as social events. Special loadstones became an appropriate gift for seeking favor with royalty or other patrons, and Pepys complained about having to relinquish one of his favorite terrellas to Lord Sandwich.[73] In 1756, the Countess of Westmorland donated a massive loadstone encased in a copper coronet to the Ashmolean. Although she apparently had to be persuaded to reveal her identity as donor, the engraving bearing her name, reproduced as Figure 2.5, must surely have mollified the hostile George II during her Tory husband's successful campaign to be elected the University's chancellor.[74]

Figure 2.6 shows two less spectacular loadstones, both of them typical of the seventeenth or early eighteenth century. Collectors' loadstones varied in design, often being carved into rectangular blocks held in decorative silver mounts, or shaped into spherical terrellas. Many individuals experimented with these magnetic models of the earth: Lord Paisley's collection included a powerful three-inch sphere protected by a green shagreen case, and another enthusiast engraved his six-inch terrella with a global map and meridional lines.[75]

Mathematical instrument sellers competed to market loadstones designed for practical purposes, chiefly navigation. Compass needles were made of poor-quality iron which rapidly lost its strength, so ships carried loadstones for remagnetizing the needles at sea. Loadstones were also used in various medical applications, as well as for assessing the quality of iron ore, testing tools and coins, and retrieving iron filings. Since good loadstone was expensive, astute purchasers were well aware of the risk of being deceived by cheaper alternatives. As shown on the right of John Bennett's trade card in Figure 2.7, mariners used anchors to test the strength of an instrument maker's wares. Although this particular card can be dated to 1760, the loadstone shown here is, like many of the instruments decorating trade cards, stylized and from an earlier period. Far more typical for everyday use were the models portrayed on Tuttell's card, heavy blocks capped with iron feet. In addition, people visited the local ironmongery to buy small unpolished pieces of loadstone for performing simple experiments or entertaining guests with tricks. Village stores pictured loadstones in their advertisements, offering customers "Loadstones contriv'd for children's sports" next to snuffboxes and key rings.[76]

In the first half of the century, the magnetic compasses on sale at mathematical instrument shops had changed little since Gilbert's time. The larger ones were for navigation, but a range of smaller models catered for other purchasers: surveyors marking out boundaries or constructing tunnels, miners searching for seams of iron, natural philosophers, and military engineers aiming their cannon at night. In addition, instrument makers marketed compass dials as a reliable alternative to pocket watches in sunny weather.[77] After the introduction of artificial magnets, instrument makers modified traditional designs to produce instruments specifically tailored to the requirements of distinct communities of users, such as navigators, surveyors, and natural philosophers. These processes of differentiation will be examined in Chapters 3 and 5. The rest of this chapter focuses on how natural philosophers and instrument sellers profited from artificial magnets by persuading polite purchasers to sample magnetic products.

In his book advertising his own techniques for making artificial magnets, the Cambridge theological scholar John Michell hymned their advantages: they were generally cheaper, stronger, and more homogeneous than natural loadstone, making them preferable for navigators and natural philosophers alike; they could easily be constructed in different shapes for carrying out experiments.[78] There was some initial resistance to accepting the artificial substitute: instrument makers advertised both natural and artificial loadstones,

Figure 2.7. John Bennett's Trade Card. John Bennett, a London instrument maker, flourished from 1743 to 1770; a bill on the back of this card is dated 1760. Amongst other navigational instruments shown at the top, Bennett is advertising his loadstones, so strong they can support an anchor. Science Museum, London.

sufferers from gout continued to insist on the best Golconda loadstone, and some artificial magnets were decoratively armed in imitation of the natural product and sold in fishskin cases lined with velvet.[79] Johnson sardonically remarked that "as the highest praise of art is to imitate nature, I hope no man will think the makers of artificial magnets celebrated or reverenced above their deserts." He was satirizing recent rivalries at the Royal Society, but his quip also underlines how artificial magnets fulfilled Baconian prescriptions for unveiling nature's secrets. Whereas for Aristotle art might mirror nature, Francis Bacon taught that imitating nature leads to understanding her concealed operations. Well before the end of the century, artificial magnets had completely displaced natural loadstone, which lost its monetary value and came to be "little esteemed, except as a matter of curiosity."[80]

Although Knight used his magnets for revealing the secrets of nature, he failed to match up to the Baconian ideal of dispelling secrecy. Detractors like Johnson constantly castigated him for being driven by the lust for money rather than by the thirst for knowledge, but he adamantly refused to divulge how he made his exceptionally powerful magnets, "even, as he said, though he should receive in return as many guineas as he could carry." So "these curious and valuable secrets have died with him," although partial accounts were published after his death.[81] Partly because of the mystique surrounding his techniques, Knight remained the acknowledged expert. Taking advantage of the new availability of high-grade steels, experimenters in England and overseas rapidly developed methods of their own, publishing detailed instructions so that soon there were enough to "fill a volume."[82] But some of the commentaries make it clear that replication was not entirely straightforward: some operators made better magnets than others. In addition to secret recipes, Knight's reputation therefore also rested on his particular skills. Just as modern technologists experience problems because instruction manuals omit vital practical details, so too people making artificial magnets needed essential tips like polishing the bars with linseed oil or being fussy about the quality of the steel.[83]

While polemicists preached the virtues of openness, entrepreneurial philosophers were obliged to face the realities of earning a living. Although criticized, William Hunter was prudently secretive about his anatomical innovations, and Knight also felt he needed to protect his inventions. He recognized the commercial value of his magnetic bars and knew that the patent system would afford him little security. As he negotiated a lucrative contract with the Admiralty, he became increasingly concerned about competitors.[84] Many natural philosophers deplored this marketing strategy because it conflicted with their rhetoric of accumulating magnetic expertise for public benefit. When Edmund Stone translated a French text on mathematical instruments, he took the opportunity to comment liberally on English inventions. A former gardener, Stone used horticultural imagery to castigate Knight's commercialization of natural philosophy: "The Plants and Trees of the Gardens, of the Arts and Sciences, cultivated by the Dung of Ambition, and nourished with the Waters of Interest, are very subject to be blasted by the Whirls of Error, and sometimes stunted by the Weeds of Imposition."[85] Writers repeatedly criticized Knight's secrecy in private letters, in texts on natural philosophy and navigation, and in books for polite audiences.[86]

Knight's reports stimulated research all over Europe. In addition to published texts, surviving correspondence indicates that men of leisure searched arduously for methods of their own. William Arderon, an enthusiastic naturalist from Norfolk, articulated their experience. "Hath Mr Knights, yet discovered the Secret, how he adds to the Magnetick Vertue of Natural Loadstones, if he have, pray inform me," he asked of Baker, the Fellow who edited his exuberant prose before publication in the *Philosophical Transactions*. "I can't think what it must be; unless it be done by immerging them in some Chimical

Liquor or by friction, as the Artificial Magnets are made in which Amusement I spent all my Vacant Hours for Several Months some Years ago."[87]

Knight's most important immediate rivals were Canton and Michell. By 1747, Canton, still an unknown schoolmaster, had developed his own method of making artificial magnets, and Knight encouragingly signed his entry certificate to the Royal Society. When Canton revealed his techniques to the Royal Society, Knight was furious. However, the Fellows praised Canton's openness both publicly and privately, and the following year awarded him the Copley Medal.[88] Meanwhile, Michell had been carrying out his own research. He decided to aim for a more popular market through an article in the *Gentleman's Magazine* and by publishing a detailed instruction manual, which was rapidly reprinted in England and translated into French.[89] Knight, Canton, and Michell moved in the same circles, visiting each other's homes: this controversy generated antagonisms which resonated for most of the century. All three protagonists leant on their friends for free publicity, so the numerous accounts of artificial magnets in popular magazines, books, and encyclopedias generally gave preferential coverage to just one of them. The bitter dispute over priority between Michell and Canton was reopened after Canton's death, when his son solicited letters of support from Priestley and other allies.[90] The issue was seen as so disruptive that after Knight died, Wilson decided he could turn it to his own advantage: he disclosed some of Knight's methods to the Royal Society as a peace offering to Joseph Banks after the violent dissensions of the mid 1770s. Authors continued to vaunt the reputation of their particular allies. For example, Erasmus Darwin made Michell central to his celebration of steel as the foundation of Britain's navigational, agricultural, and military achievements:

> Last MICHELL's hands with touch of potent charm
> The polish'd rods with powers magnetic arm;
> With points directed to the polar stars
> In one long line extend the temper'd bars;
> Then thrice and thrice with steady eye he guides,
> And o'er the adhesive train the magnet slides:
> The obedient Steel with living instinct moves,
> And veers for ever to the pole it loves.
> Hail adamantine STEEL! magnetic Lord!
> King of the prow, the plowshare and the sword![91]

Knight's innovation was important because it initiated the transition from natural loadstone to permanent steel magnets, which transformed magnetic experimental practices, particularly on the Continent. For instance, Coulomb could not have constructed his famous torsion balance without long, slender, and powerful artificial magnets.[92] In England, Knight and his emulators effectively revolutionized the magnetic market: craftsmen could make instruments which were far more precise and reliable than earlier ones, and also produce equipment for the burgeoning interest in rational entertainment.

Philosophical entrepreneurs faced heavy competition. London lecturers in natural philosophy were losing students to medical courses. In pleasure-laden provincial spas, potential audiences defected to

> admire the conjurer,
> Fond of deceits and cups and balls,
> While King reads lectures to the walls.[93]

To compete with such enticing conjuring entertainers, natural philosophers devised entrancing performances of their own, using apparatus like air pumps, electrical machines, orreries, and solar microscopes. Like chemical lecturers, they employed different strategies and cultural vocabularies to attract different audiences.[94] Magnetic merchants faced the further challenge that magnetic bars and iron filings could provide only limited fascination. They had to find additional means of expanding their markets. Some of them turned to navigation, advertising improved instruments or delivering didactic lectures about the importance of being familiar with the principles underlying the operation of compasses. Others chose more dramatic routes, developing novel medical therapies or constructing elaborate experimental kits and tricks for performances in public theatres or private gatherings. As they sought personal profit through promoting their magnetic wares, these entrepreneurs expanded the coverage of magnetic phenomena in books and lectures. In a symbiotic relationship of mutual financial benefit, natural philosophers, publishers, and instrument makers cooperated to market magnetic products as they competed for polite audiences. For example, Adams—Knight's retailing agent—wrote extensively about magnetic experiments and may have encouraged Cavallo as he struggled to complete the only English book devoted to magnetism in the second half of the eighteenth century.[95]

Barbara Stafford has recently provided marvellous examples of the continuity between conjuring and natural philosophy, but she clings to false distinctions between their practitioners. Roy Porter has convincingly demonstrated that medical therapists of this period were ranged along a spectrum as they deployed diverse advertising techniques to compete for patients' patronage in a commercialized pluralist milieu.[96] A similar situation prevailed amongst promoters of natural philosophy. Knight and many of his colleagues at the Royal Society were engaged in the same types of self-promotional, commercializing activity as lecturers and popular entertainers who toured the provinces or sought the purchasing power of élite metropolitan audiences. Categories such as orthodox and quack, and professional and amateur, are anachronistic when applied to the eighteenth century. There were no fixed boundaries distinguishing Knight from other lecturers on natural philosophy or from conjurors who employed concealed magnetic devices to spice up their shows. Men like Knight who wished to establish themselves as creditable guardians of legitimized magnetic knowledge strove to establish such distinctions as part of their rhetorical campaign to enhance the prestige of natural philosophy.

Knight performed like a prestidigitator at the Royal Society, dramatically displaying how he could almost magically alter the polarity of pieces of loadstone, and vividly using stage props like keys and swords to demonstrate the power of his new magnets. The president's own report reflects the theatricality of Knight's performances: he related how at home Knight would disappear into his study with pieces of loadstone and emerge triumphant a few minutes later to show off "The Effects of an Art Mr. *Knight* is Master of." Although Knight would not have performed so flamboyantly on a public stage, he certainly behaved as though aggressive marketing techniques would not demean his status as a distinguished natural philosopher. He sold three sizes of steel magnetic bars ranging in price from 2½ guineas a pair to 10 guineas for a pair 15 inches long in a crafted wooden case (Figure 2.3). He also developed a process of heating powdered iron with linseed oil, a traditional conserver of magnetic strength, to make small but powerful permanent magnets.[97] The next chapter will describe how he also made and sold compasses. Knight's comportment as a performer and a merchant exemplifies his participation in a shared yet variegated sphere of rational entertainment.

Knight was tapping the quality end of the market. Rival instrument makers provided a selection of cheaper steel magnets, as bars as well as the new horseshoes. Some of them sold six-inch bars wrapped in black paper which cost only a few pennies.[98] To make their magnetic products more appealing to polite consumers of natural philosophy, dealers sold small boxed experimental kits at various prices. Advertising literature became increasingly sophisticated. Whereas Adams' father had offered an unpriced collection grouped together unenticingly as iron filings, wires, and balls, a decade later Knight was charging a guinea for similar items described as an "experimental apparatus." Later in the century, instrument makers marketed more grandiose versions—such as the one illustrated in Figure 2.8—for between 5 and 7 guineas.[99]

In the first half of the century, magnets had received only a passing mention in lectures on natural philosophy.[100] As artificial magnets became widely available and the polite market for all aspects of the study of nature expanded, natural philosophers included magnetic experiments as a standard, but briefly covered, topic in most courses of natural philosophy at educational establishments. In the provinces, academies successfully applied for grants and bequests to buy magnetic equipment, and experiments were conducted in the local Literary and Philosophical societies.[101]

Travelling lecturers sought to enlarge their audiences by taking advantage of the new artificial magnets. Thomas Peat of Nottingham, for instance, spiced up his advertisements with promises of "*curious Experiments* with the *Artificial Magnets*." James Arden wrote in from Derby for further details as soon as he heard about Canton's experiments, explaining that "as it's a good part of my Employment to read Lectures in Experimental Philosophy nothing at any time can give me more Pleasure than new and curious Experiments." It is difficult to reconstruct the activities of itinerant lecturers, but probably—

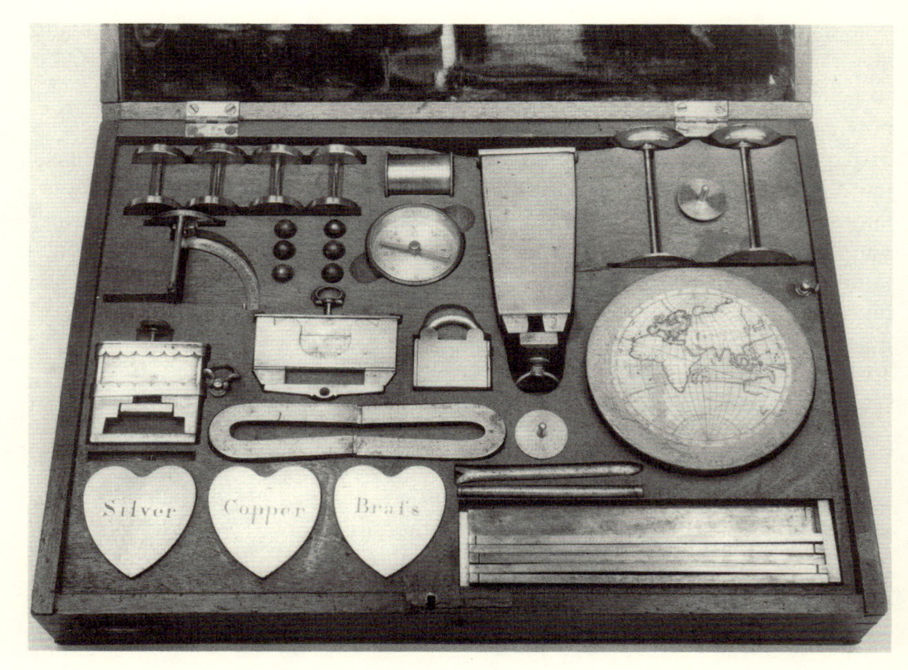

Figure 2.8. George Adams' Magnetic Cabinet. Portable kits of rational entertainment were advertised during the 1790s and contained virtually everything a home experimenter or a travelling lecturer might require. The two horseshoe magnets could support Platonic chains of the small iron balls near the top left. The importance of magnetic studies for international travel is demonstrated by the prominent map (pasted on a wooden box containing a bar magnet), the directional compass, and the dip circle (upper left). The three hearts in the lower left belonged to a popular trick, which entailed concealing small magnets so that a compass needle could apparently pick out a particular metal. The padlock was operated magnetically. Science Museum, London.

like Desgauliers's former student Stephen Demainbray—many of them gradually introduced magnetic topics into their repertoires. Lectures could also serve as a showcase for merchandise. Ferguson and Benjamin Martin were targeting similar London audiences in the 1770s, but only Martin chose to describe magnets, since these were related to the navigational products he was marketing.[102]

As they competed for audiences, some innovators went beyond what more austere practitioners declared to be the legitimate boundary of natural philosophy. During the 1780s, two of the most successful London performers were Gustavus Katterfelto and James Graham, caricatured in Figure 2.9 as warring quacks surrounded by exotic experimental paraphernalia, including the "Leyden Vial Charg'd with Load Stones Aromatic Spices. &c. &c." on the left. Both these men catapulted to fashionable fame for a couple of seasons, performing in front of aristocrats and royalty, before their popularity declined

Figure 2.9. James Graham and Gustavus Katterfelto. Rival natural philosophers and medical practitioners successfully marginalized these two entrepreneurs as quacks. James Graham was advertising the therapeutic value of magnetic beds and chairs similar to those being recommended in Paris by the Société royale de médicine. Wellcome Institute Library, London.

and they resorted to touring the provinces. Like medical practitioners who came to be marginalized as quacks by establishment physicians, they combined the language of theatrical performance with that of natural philosophy. They successfully appealed to the vogue for the mysterious and the spectacular, and simultaneously appropriated respectability by aligning themselves with Enlightenment values of rationality.[103] Elite natural philosophers strove to establish an unambivalent distinction between themselves and these entertainers, who presented themselves as natural philosophers and included magnetic equipment in their shows.

Katterfelto developed diverse self-marketing techniques. He often advertised his courses of natural philosophy lectures by bracketing himself with Newton and Franklin, but he also aligned himself with Masonic and occult practices. Most famous for his solar microscope and black cat, he used picturesque language to package standard magnetic fare as exciting wonders. As well as displaying allegedly splendid loadstones and compasses, he demonstrated "[a] Magnetical CLOCK which has surprised most of the free masons in Europe," and Which probably incorporated some concealed operating mechanism. He also claimed to have invented "A Magnetical APPARATUS Which will take a copy off in five minutes time." He enterprisingly used an artificial magnet to enthral audiences by lifting his own daughter to the ceiling by a steel helmet strapped to her head.[104]

Like his rival Katterfelto, Graham's initial campaign rested heavily on the mystical allure of magnets. He claimed he could restore life through "ærial, ætherial, magnetic, and electric influences," and he is shown astride a large prime conductor, a standard piece of electrical equipment. His metropolitan success as a sexual therapist centered on his celebrated celestial bed, the star attraction of his Temple of Health in Adelphi Terrace. For £50 a night, he promised that a couple would benefit from the stimulating effects of the circulating effluvia emitted by the fifteen hundredweight of artificial magnets installed beneath the bed. Graham also used extravagant language to describe his other therapeutic equipment, including magnetic chairs and thrones, artificial magnets, and loadstones. He offered tours at half a crown for the less wealthy.[105]

In newspaper accounts, books, and satirical cartoons, other medical practitioners, natural philosophers, and social commentators continually ridiculed these two magnetic performers in order to exclude them from the legitimate centers of knowledge production. A more marginal case was Temple Henry Croker, an Oxford graduate who wrote and lectured about his magnetic theories. Croker's checkered career included appropriating Italian translations, coauthoring a three-volume dictionary, inventing a compass, becoming bankrupt, and emigrating to the West Indies. He commissioned Jeremiah Sisson, an eminent London instrument maker, to build the experimental device shown in Figure 2.10. He claimed this was a perpetual-motion machine, in which a bar magnet was constantly rotated by the earth's magnetic forces.[106] Croker was sufficiently well respected to be invited twice to present his ideas at the

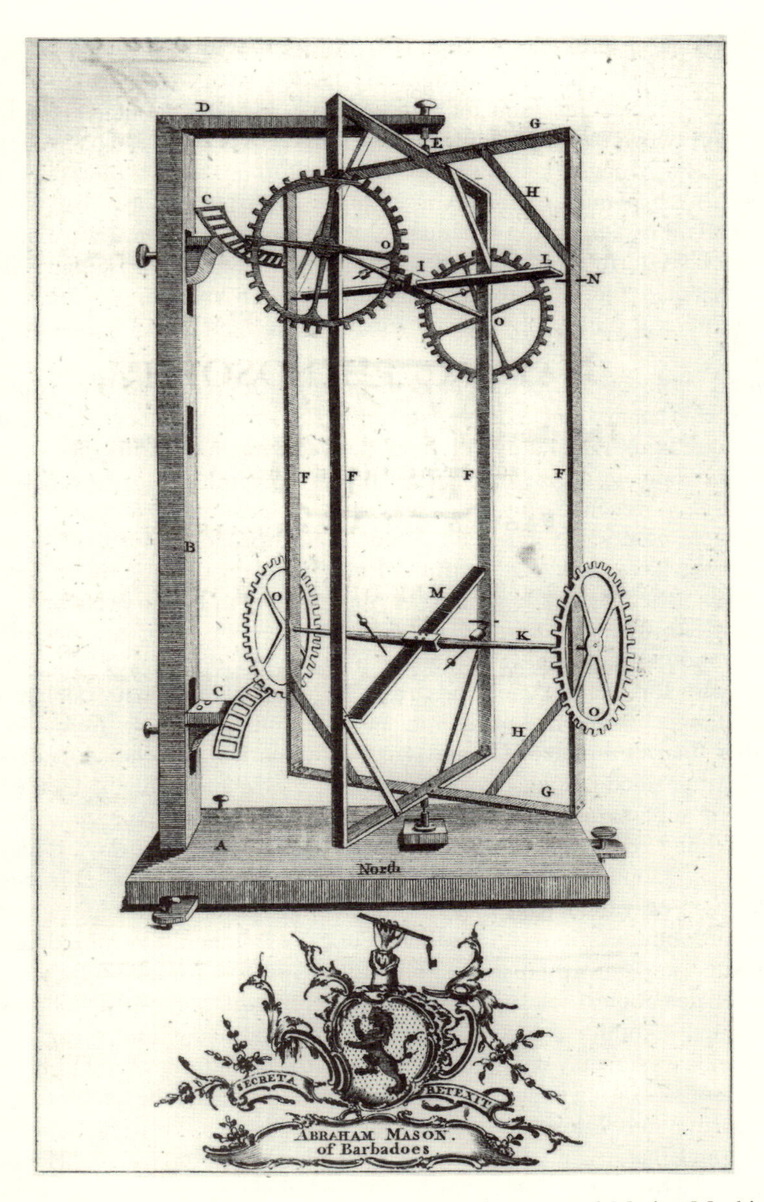

Figure 2.10. Temple Henry Croker's Magnetic Perpetual-Motion Machine. Jeremiah Sisson built this machine for Croker, who twice presented his ideas at the Royal Society. Croker claimed that the apparatus operated perpetually in Barbados, where the earth's strong horizontal magnetic power constantly flipped over the bar magnet (*M*). Frontispiece of Croker, *Experimental Magnetism*. By permission of the British Library (shelfmark 538.g.46).

Royal Society and to join Birch, Knight, and other Fellows for dinner at the Mitre. But he also adopted less-gentlemanly tactics, such as advertising his public lectures in the press and making scathing remarks about Knight in his book.[107] A reviewer reproved him for attacking Knight, well established as the Royal Society's magnetic expert. However, he forbore from criticizing his presentation of perpetual motion, whose impossibility was not universally accepted. Several designs for magnetic perpetual-motion machines were proposed in England during the eighteenth century, but natural philosophers increasingly rejected their feasibility. Perpetual-motion machines were excluded from natural philosophy but remained in the repertoire of travelling showmen.[108]

As natural philosophers contributed to making Graham and Katterfelto symbols of charlatanry and cheap exhibitionism, their territory was also threatened by conjurers. The associations between magic and magnets had always been strong. In the second half of the seventeenth century, Gilbert's Latin *De Magnete* reached élite audiences, but for those who read only English there were plenty of references to the magical properties of loadstone. Many historians have concurred with the Enlightenment rhetoric that magic had largely vanished by the beginning of the eighteenth century. However, contrary to Keith Thomas' verdict, its decline was only apparent: magical beliefs did not quietly disappear in the face of an urban ideology of self-help paving the way for technological progress. Rational philosophers declared that superstitious customs had been eradicated, but older practices were concealed rather than eliminated. Compilers of dictionaries and encyclopedias carefully distinguished amongst various types of magic and retained detailed discussions of supernatural phenomena such as amulets, sympathetic powders, and divining rods.[109] Astrological beliefs were perpetuated as a three-pronged subject adopted by different social strata, while traditional attitudes towards corpses prevailed well into the nineteenth century. But any public expression of confidence in religious therapeutics or witchcraft was hindered by ecclesiastical and political disputes. The Royal Society officially refused to enter into debates about such topics, although behind the scenes the Fellows avidly collected accounts of second sight and conversations with the dead.[110]

Magnetic phenomena lay dangerously close to the boundary natural philosophers were consolidating to separate magic, superstition, and the occult from rational experimental philosophy. Daniel Defoe classed them as tricks of artificial or rational magic, "all Impositions upon the Sight or Hearing of the People. . . . had a Man in those days of Invention found out a *Loadstone*, what Wonders might not he have performed by it? . . . Will any Man believe but he that first shewed these unaccountable things, would have passed for a Magician, a Dealer with the *Devil*, nay, or rather for a real *Devil* in human Shape?" Like the Newtonian expositor Henry Pemberton, Defoe used magnetic behavior to mock earlier gullibility. Such polemicists propagated messages of progressive natural philosophy successfully conquering superstitious belief in magical powers.[111]

English natural philosophers displayed ambivalent attitudes towards magical magnetic effects. They openly admitted their inability to explain the "mysterious phænomenon" of magnetic attraction, frequently declaring that of all God's powers "none excite more astonishment than those of Magnetism." Darwin, in his poetic celebration of modern progressive achievement quoted above, portrayed Michell as a spectacular performer with almost mystical powers. He played on polyvalent words such as "touch" and "charm": people frequently wrote about touching a compass needle with a loadstone to magnetize it, but the term also carried overtones of the healing power associated with the king's touch, as well as erotic connotations which intensified during the vogue for animal magnetism in the 1780s (see Figure 7.6).[112]

Natural philosophers ruled that magical performances were acceptable as amusements or recruitment devices only when based on a complicitous understanding of alternative explanations. Richard Edgeworth, the educator and engineer, learnt how to replicate several magnetic turns performed by the famous French conjuror Comus, enormously successful during his visits to London in the 1760s. The show Edgeworth put on with his aristocratic friend Francis Delaval was designed for prestigious visitors, including Knight and some colleagues from the Royal Society. Edgeworth's magnetic legerdemain impressed Darwin, a fellow member of the Lunar Society, who enlivened his dinner parties by sending an imitation magnetic spider scuttling across a silver salver.[113] But in the public arena, sober natural philosophers condemned such lighthearted displays. As they sought to distinguish themselves from popular magicians, battle was symbolically engaged on the stage of the Haymarket theatre. The conjuror Phillip Breslaw found that his magnetic "learned little swan," magically swimming round a bowl to pick out letters in response to spoken questions, suddenly refused to cooperate. From the auditorium Delaval was wielding a strong magnet: the Enlightened philosopher was overpowering the tawdry performer.[114]

Natural philosophers rhetorically legitimated themselves, defining new public sciences like chemistry and magnetism by contrasting them with magical and superstitious practices. Chemical lecturers carefully distanced themselves from showy competitors. During controversies about chemical measuring techniques, Magellan polemically denounced Cavallo (who was then London's leading magnetic expert) as a conjuror: "Is not this similar to the tricks of Jones, Comus, Breslaw and Katterfelto, who make things appear what in reality they are not?"[115] But for magnetic displays, natural philosophers found it difficult to establish clearly the boundary between deception and rational entertainment. The more relaxed attitude of French educators confused matters. In the magnetic section of a Jesuit professor's book, a teacher told his eager young pupil, "It is permitted to Philosophers themselves to toy, when the Business is only to unravel Truth." The book's English translation was widely recommended, and its educational magnetic boat was illustrated in a popular journal, but élite English natural philosophers never incorporated similar entertainments in their own texts. When Jean-Jacques

Rousseau described how Emile was fascinated at a fair by a magician's duck swimming round a bowl of water, he was obviously familiar with Abbé Jean Nollet's performances with a metallic swan, or M. Servière's magnetic frog, which glided to a number showing the correct time of day. He made this episode central to Emile's philosophical and moral education. English writers viewed such entertainment quite differently. Thomas Day, Edgeworth's collaborator in implementing Rousseau's educational prescriptions, used the same event in his own moralizing children's story. But for Day, the Lunar Society advocate of industrious progress, the duck remained a toy which imparted a single message: that magnetic compasses were vital for navigation and trade.[116] These didactic birds swam from the academic syllabuses of French natural philosophers into the stage repertoires of London conjurers like Breslaw, and—to the consternation of some natural philosophers—into the parlor performances of polite families.

Until interrupted by Delaval's powerful magnet, Breslaw entranced audiences with nightly performances at the Haymarket of his version of the French sagacious swan.[117] An earlier performer on the same stage had been Priestley's protégé Adam Walker, an itinerant lecturer from Manchester whose natural philosophy, flavored by Behmenist ideas, strongly influenced Shelley. Walker's magnetic lectures also came to feature the swimming swan and similar tricks designed to entertain and enlarge his audience.[118] Walker represented a sizeable constituency of instrument makers, lecturers, and writers who found that repackaging old devices into elaborate illusions was an effective way of interesting audiences in magnetic phenomena. These magnetic tricks were qualitatively different from displays illustrating other branches of natural philosophy, since no pretence was made that they exposed or explored the powers of nature. Critics objected that by appropriating conjurers' techniques to willfully deceive their spectators, such performers were transgressing the bounds of polite philosophy.

The major source for these entertainments was William Hooper's four-volume *Rational Recreations*, first published in 1774, which contained detailed illustrated instructions for fifty-seven magnetic recreations. As well as being demonstrated by lecturers, these were widely reproduced in books and encyclopedias. Dignified as "entertaining experiments," Hooper's recreations made up almost a third of the entry on magnetism in the 1797 *Encyclopædia Britannica*. Figure 2.11 shows one of the two whole pages of plates devoted to magnetic tricks. Instrument makers continually enlarged their selection, advertising Hooper's sagacious swan and other games for prices ranging from a few shillings to several guineas.[119] Polite philosophers could entertain their families and friends with dexterous painters, communicative crowns, and enchanted ewers, or magnetically spell out improving Latin mottoes extolling the virtues of virgins.

A century earlier, virtuosi had overtly used small pieces of loadstone for diversions like stopping their visitors' watches. But now, as philosophical magicians wielded wands with a concealed magnetic core, these tricks cele-

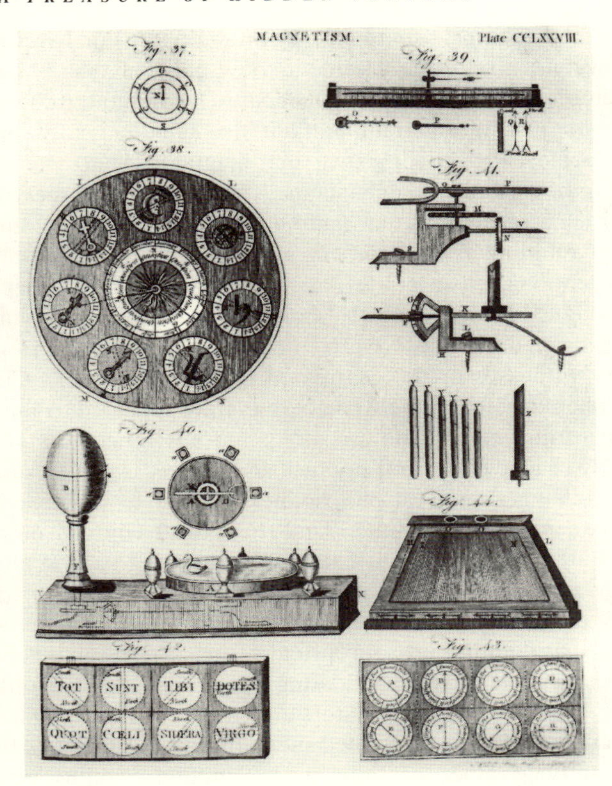

Figure 2.11. Magnetic Rational Recreations. Magnetic illusions like these were widely advertised and illustrated towards the end of the eighteenth century. With their improving moral messages, these parlor tricks demonstrated the performer's conjuring expertise, not the powers of nature. Figures 39–41 show the elaborate geared mechanism needed to make the sagacious swan swim round its bowl. *Encyclopædia Britannica* (1797), vol. 10, facing p. 448. Whipple Library, Cambridge.

brated human expertise rather than natural wonders. Whereas mid-century browsers at an instrument shop could buy simple enamelled swans and frogs to float in water, the elaborate geared mechanism of the sagacious swan required the skills of a craftsman for its construction (Figures 39–41 in Figure 2.11).[120]

One of the most successful entertainments was the magnetic planetarium (Figures 37–8 in Figure 2.11), decorated with zodiacal signs to impart a mystical aura. When the magnetic entertainer set the central pointer to a question, the seven smaller needles rotated to pick out an appropriate response. Hooper recommended an elaborate scheme which enabled a magnetic magician to make the pointers indicate a questioner's month of birth, "which will give the business an air of mystery."[121] The planetarium was an updated version of the magnetic oracle designed 150 years earlier by Athanasius Kircher, a central tower surrounded by small rotating decorated magnetic globes. For Kircher, astrological symbols were central to his Hermetic view of a universe linked directly to God by Platonic chains of magnetic sympathy (see Figure 6.1). He designed such devices not just for entertainment but also as polemical weapons in philosophical debates. By showing that such magnetic devices could work only with the help of concealed mechanisms, Kircher had been seeking to disprove the Copernican contention that natural magnetic forces account for the Earth's diurnal rotation. Enlightenment philosophers converted another of his eristic inventions into a polite parlor trick. Whereas Kircher's illustration of a cryptological machine clearly shows the system of ropes and gears linking two magnetic dials, these were intentionally concealed from late-eighteenth-century witnesses of devices renamed "The Magician's Circles." Natural philosophers reinforced polite values as they appealed to family purchasers by suggesting appropriate questions and answers: "Are you pleased with matrimony? I love quite well my husband."[122]

As social structures changed, natural philosophers sought to maintain a careful distinction between polite consumers and those who enjoyed more popular forms of entertainment. At the same time, transforming natural philosophy into a publicly accredited and accessible science necessitated broadening the constituency of its participants. The potential conflict in achieving these two goals meant that defining an appropriate format for the magnetic aspects of natural philosophy entailed consolidating distinctions amongst different sectors of society. Hooper and his imitators aimed their texts on natural philosophy at wealthier families. They justified including magnetic tricks by stressing the need to captivate rather than alienate overindulged youths, who would thus be encouraged to improve themselves and become "incessantly rapt with joys of which the groveling herd have no conception." But their detractors thought otherwise. Priestley's colleague William Bewley accused Hooper of teaching a pupil "to become rather a conjurer than a philosopher." The Edgeworths singled him out as a narrow-minded purveyor of magical deception, decrying his use of questions on the nature of love as he led students away from the true path of learning through experiment.[123]

Educators in England's rapidly expanding northern urban centers praised the approach of lecturers like Walker, who "conceived a system of education more adapted to a Town of Trade than the Monkish system still continued in our Public Schools." Walker enlarged the audience for natural philosophy by making his tuition appropriate for women and the less formally educated. Although he explicitly distanced himself from theatrical enchanters, his magnetic lectures incorporated instructions for making a sagacious swan and the divining circles illustrated in the diagram at the lower right of Figure 4.9. Following the success of his London lecture course, Walker was invited to import his new didactic techniques into the public schools; but not everyone approved of his popularizing strategies. His fellow performers at the Haymarket had recounted the tale of Mahomet's coffin illustrated on Tuttell's card. Perhaps influenced directly by them, Walker showed his students how to simulate the effect by using a horsehair. An opponent of this popular approach sneered at Walker's "superficial style" and, in a series of selective extracts, reprinted the coffin illusion to denigrate the educational content of Walker's lecture course.[124] But despite such criticisms, instructions for magnetic parlor tricks marketed as rational recreation continued to be included in nineteenth-century encyclopedia entries on magnetism. They were also offered as light relief for those who persevered to the end of weightier expositions of magnetic phenomena.[125]

To become recognized as the legitimate purveyors of magnetic knowledge, natural philosophers had to ensure that lecturers were performing correctly. Strategically consolidating public acknowledgment of their élite status, they fashioned their identity as experts by demonstrating their superior understanding and control of natural forces. The activities of magnetic entrepreneurs exemplify transformations taking place in many fields. As individuals struggled to earn their living and gain prestige, they learnt from other competitors for polite purchasing power how to promote themselves and attract a larger audience for the new science they were collectively constructing. They established magnetism as a public science, an essential component of polite culture dispensed only by accredited suppliers of natural knowledge. These innovators carved out a new type of career while advertising their valuable contribution to the growing commercial wealth of a polite society. Like their contemporaries dealing in other commodities, they contributed to the foundation of specialized institutions whose members were paid for their work. As organizations like the Royal Society levered themselves into powerful positions, scientific practitioners established their role as influential participants in the development of a flourishing industrialized and polite nation.

CHAPTER THREE

The Direction of Invention: Setting a New Course for Compasses

ENGLAND'S NAVAL strength was vital for protection against rival European powers, for increasing her trade, and for defending and acquiring overseas territory. But seafaring still seemed extremely dangerous. Johnson quipped that "being in a ship is being in a jail, with the chance of being drowned," and shipwrecks featured prominently in contemporary art and literature. One particularly popular poem was the sailor William Falconer's epic *The Shipwreck*, in which he dramatically narrated his own narrow escape from death. In a work resembling Darwin's *Economy of Vegetation*, Falconer transposed his personal experiences to classical Greece, including such copious notes on nautical terminology that naval educators recommended *The Shipwreck* for training their novices.[1] As Falconer so vividly portrayed, stormy weather frequently caused enormous losses of men and cargo; other hazards included pirates, enemy ships, and tropical diseases. Poor navigation was also responsible for many disasters. Scurvy decimated Anson's crew as he spent months searching for a Pacific island, and ships sometimes landed hundreds of miles from their intended destination. Nearer home, Sir Cloudesley Shovell's navigational blunder in 1707 became legendary: aiming for the Bristol Channel, he ran aground on the Scilly Isles, losing four ships and about two thousand men.[2]

Entrepreneurial natural philosophers recognized that inventions improving naval safety promised a double reward of money and prestige. They put forward numerous suggestions of varying practicability, but implementing even the more realistic proposals entailed protracted negotiations. For example, although copper sheathing demonstrably protected ships' timber bottoms from rotting, controversies about costs and benefits persisted for around seventy years. Other naval innovations, now often celebrated as technological advances, met with similar resistance before being accepted. The most famous is John Harrison's chronometer, designed to determine a ship's longitude by keeping time accurately at sea; further examples include John Hadley's quadrant for measuring the angular altitude of stars, Stephen Hales' ventilators for circulating fresh air inside ships and institutional buildings, and the new medical approaches to scurvy.[3] This chapter examines how Knight marketed his magnetic compasses to the Royal Navy.

Since the middle of the seventeenth century, naval reformers had been increasingly insistent about the importance of ascertaining a ship's longitude. Spurred on by Shovell's disaster, the Board of Longitude was established in

1714 to administer governmental rewards on a sliding scale, up to a maximum of £20,000, for anyone discovering a method of determining longitude at sea to an accuracy of thirty miles. The search for the longitude became one of the leitmotifs of the century as numerous inventors proposed schemes of varying feasibility and clarity. Satirists mocked extravagant suggestions: many of them reiterated Swift's verdict that it was "[a] Thing as improbable as the Philosopher's Stone, or perpetual Motion." But lured by such an enormous reward, philosophical projectors submitted diverse ingenious plans, often soliciting contacts at the Royal Society to promote their interests. Commercial and philosophical interests corresponded, since resolving the longitude problem would simultaneously benefit the inventor, stimulate international trade and exploration, and demonstrate publicly the value of experimental research.[4]

Several projectors submitted magnetic schemes for approval (see the Appendix). They mostly relied on long-established observations indicating that the earth's magnetic power varies both spatially and temporally. Natural philosophers trying to establish a regular pattern for these fluctuations used two measures of terrestrial magnetism, the variation and the dip. The variation at any particular place and time was the angle between geographic North and the direction a compass needle was pointing. The dip, discovered later and experimentally far harder to ascertain reliably, was the angle between the horizontal and a magnetic needle free to rotate in a vertical meridional plane. Mariners had acknowledged the problems presented by variation at least since the time of Christopher Columbus. They realized that the value of compasses and charts was limited, since they did not know exactly where a compass needle was pointing: its precise orientation depended on the ship's location on the surface of the terraqueous globe.[5] Natural philosophers—particularly those with an eye on the longitude money—hoped to convert this magnetic problem to their advantage; if they could accurately map the terrestrial patterns of magnetic variation, then measuring its value at any location would yield the longitude.[6]

During the first half of the eighteenth century, the compasses used on ocean-going vessels differed little from those of a hundred years earlier. Ships carried several compasses, stored in different places and with their own specific uses. Steering compasses were kept in cupboards called binnacles. These were divided into three compartments, a central one for a nighttime candle, and one on either side containing a compass, so that the helmsman could change position yet easily take a reading. Warships had two binnacles, one for the steersman and one for the conning officer directing the course. Officers had compasses with reversed faces, sometimes mounted in a crown, hung from the ceilings of their cabins so that they could check on the ship's course without going on deck.[7] Some ships also carried azimuth compasses for finding the magnetic variation. These were often of the type shown at the top center of John Bennett's trade card (Figure 2.7), which dates from 1669. Mounted on a portable stand, azimuth compasses were essentially steering

Figure 3.1. Gowin Knight's Steering Compass. Eminent navigators like James Cook initially welcomed Knight's compasses, which were first demonstrated to the Royal Society in 1750. Despite criticisms, they remained official British naval issue until well into the nineteenth century; they were renowned throughout Europe. National Maritime Museum, London.

compasses with some sort of sighting mechanism—such as a string—for taking bearings from the sun or the stars. Finding the magnetic variation entailed measuring the angle between the compass needle and the shadow cast by the string. This was relatively straightforward in fine weather but more difficult at night or in a storm. There were several methods for transforming this reading into the magnetic variation, involving varying degrees of trigonometrical ability. In the second half of the century, azimuth compasses became standard equipment for the Royal Navy and the larger commercial shipping organizations.[8]

Knight substantially modified the traditional design of steering and azimuth compasses, winning a small award from the Board of Longitude (see Figures 3.1 and 3.3). As he marketed his magnets and different versions of his compasses—variants of the one shown in Figure 3.1—his most important customer was the Royal Navy. His compasses became official issue and were amply described in books for mariners and natural philosophers well into the nineteenth century. Compiling the century's major navigational encyclopedia, the poetic seaman Falconer included a typical tribute establishing Knight's priority: "To remedy these inconveniences, the learned Dr Knight was induced to contrive a new sea-compass, which is now used aboard all our vessels of war."[9]

Naval historians continue to portray Knight as the heroic inventor of "the first scientific compass."[10] Historians of science, on the other hand, rarely mention Knight's practical achievements: they use a Newtonian framework for analyzing his magnetic theories.[11] But this artificial dissection into theoretical progress and congratulatory technological advance obscures the interdependence of experiment and theory and their common entrenchment in a wider cultural context. Simplistic narratives of invention conceal the complex processes underlying the introduction of new artifacts.[12] Knight's theoretical commitment went hand in hand with his research into compasses, both of them swayed by financial considerations. The navy did not adopt Knight's compasses because they were indubitably superior; on the contrary, many users criticized their performance. Knight's marketing success was the outcome of numerous negotiations between different groups.

In Barry's painting (Frontispiece), Father Thames' compass occupies a central position, reflecting its potency as a multivalent symbol. In versified equivalents of Barry's celebration of progress, poets such as Joseph Cottle envisaged the "Goddess of Science" commanding her sylphs to describe those inventions which, like printing, had enlightened "Mankind, then immersed in the darkest ignorance." For these armchair polemicists, the introduction of the magnetic compass had vanquished forever the perils of international voyages:

> Hail glorious gift! design'd the world to bless,
> Transcended only by the teeming Press.
> The fearless pilot led by thee shall brave
> The turbid fury of the Atlantic wave

For natural theologians, compasses acted as reminders of "the wisdom of God, and his wonderful Direction and Rule over all things." Religious writers welcomed this divine endorsement of British commercial imperialization: "And a sea-compass . . . is the means of conveying into our harbours the rarities and riches of the universe . . . the choice productions, and the peculiar treasures, of every nation under heaven. . . . London becomes a moot of nations."[13] Because England was a maritime nation, authors with idealized conceptions of compasses liberally sprinkled their texts with nautical metaphors. For instance, Bernard Mandeville commented on "all those at the Helm of Affairs," sarcastically remarking "that the national Interest is the Compass that all Statesmen steer by." More typical was this edifying couplet on a tombstone in Glasgow cathedral:

> Our life's a flying shadow, God is the pole,
> The needle pointing to him is our soul.[14]

But for Falconer, the maritime poet who had experienced the horrors of an oceanic tempest, compasses provided no magic antidote to danger. He regarded them as just one of the aids available for supplementing human skills when disaster seemed imminent, artifacts which were ultimately fallible. Few land-based poets would have understood his reference to using an azimuth compass for measuring the sun's altitude:

> The pilots now their rules of art apply,
> The mystic needle devious aim to try.

The compass, plac'd to catch the rising ray,
The quadrant's shadows studious they survey!
Along the arch the gradual index slides,
While Phoebus down the vertic circle glides . . .
But here, alas! his science nought avails!
Art droops unequal, and experience fails[15]

These examples indicate both the ubiquity of compass imagery and its diversity. They highlight the leverage to be gained from studying the controversies surrounding Knight's compasses. These disputes did involve problems specific to his instruments, but they were embedded in wider cultural disparities. Thus they provide a means of investigating how local controversies are related to broader patterns of change, and of displaying how material objects can act as interfaces between diverse groups. Knight modified his compasses in response to social pressures—such as users' complaints, economic restraints, and production considerations. Reciprocally, members of interacting communities involved in this protracted episode changed alliances and constructed new roles, thus transforming social structures.

Knight's compasses were viewed from varying perspectives by navigators, instrument makers, dockyard workers, naval administrators, natural philosophers, and competing inventors.[16] People from these groups selected contrasting criteria for assessing Knight's compasses and held differing views of how they would impinge on their own lives. Knight had little direct contact with the maritime community, and he produced devices more suitable for the philosopher's private study than the deck of a storm-lashed ship. Navigators were experts in their own environment but insufficiently trained in the use of these new and not necessarily appropriate instruments. They were disappointed in their performance and preferred to rely on traditional methods. Desk-bound administrative purchasers were seeking to balance the conflicting demands of winning wars, transporting goods efficiently, and keeping costs down. And in the dockyards, workers were struggling to balance the books and protect their jobs against outside contractors.

Stone articulated one important aspect of the debate, which lasted for several decades. Surely, he suggested, "skilful Mariners . . . who have used Compasses at Sea for many Years, may at least be allowed to be as competent Judges of the Goodness of their Compasses, as the Doctor."[17] Naval compasses were not primarily appraised by the mariners who used them, but by land-based inventors and naval administrators with their own interests in mind. This was one way in which natural philosophers were usurping the authority of maritime men within their own domain of expertise. The introduction of Knight's compasses into the Royal Navy provides one example of how natural philosophers established themselves as the recognized magnetic authorities and how the Royal Society maneuvered itself into a powerful position.

This chapter is a case study exploring how Knight, a metropolitan entrepreneurial philosopher, successfully negotiated the bureaucratic structure of the navy so that his compasses were transported by mariners around the world. Focusing on this specific yet typical example serves two valuable functions. Historically, as well as fleshing out our still meager body of knowledge about this period, it enables us to explore eighteenth-century attitudes towards invention and reveal the social dynamics underlying change. Many naval inventors and practitioners had only limited contact with each other, dealing separately with an administrative interface. Agents relied on different types of knowledge and were interested in attaining their own immediate objectives. Analytically, this episode illustrates how sociological approaches have transformed our understanding of technical innovation. Material objects are bound in a seamless web knotted by political, social, and economic considerations. Instruments have traditionally been the marginalized preserve of specialized scholars, but historians of science, technology, and medicine are increasingly placing them center stage in their studies of experimental practices. Transcending conventional distinctions between theory and experiment, or science and technology, or science, culture, and society, instruments—like other material items—offer all historians a powerful tool for studying shifting relationships between communities as they fashion themselves against each other. Because compasses were symbolically laden, studying local disputes about Knight's compasses informs us about tensions resonating throughout society.

This analysis draws on sociological models to illustrate how attitudes towards innovation changed during the eighteenth century and how natural philosophers and naval experts redefined themselves against each other. The first section reconstructs differing attitudes towards navigation, making them more accessible to modern readers by concentrating on three communities which contrasted particularly strongly: gentlemanly inventors, navigators, and naval administrators. The second section presents Knight's activities as an example of mid-century promotional enterprise, logging the reception of his compasses as he launched them into the competitive maritime market. Finally, the discussion examines how the questions raised during the debate over Knight's compasses affected participants in different ways: by the end of the century, the relationships amongst navigators, inventors, and administrators, and how they perceived compasses, had altered. Examining compasses as conceptually unstable objects enables us to explore the fluid boundaries between shifting communities.

The Boundaries of Conflict

Virtually everyone agreed that navigational techniques should be improved, although motivations differed. Mariners obviously had a strong interest in ensuring that they reached their destinations safely. Shipowners were concerned about their investments in ships and cargo, while naval administrators were answerable to the Treasury.[18] Government war strategists sought tactical advantages over the enemy, inventors sought financial gain, and natural philosophers were eager to demonstrate the public value of their research. But holding objectives in common does not imply unproblematic cooperation when interested communities hold contrasting short-term aims. Knight took advantage of this shared yet diverse commitment to change: he successfully marketed his redesigned steering and azimuth compasses in the face of criticism and competition.

Retrospective accounts tend to favor the vantage point of the strongest rhetorician or of the group whose view eventually prevailed. Sociologists have developed new methodologies for presenting more balanced versions, which allow the actions of other agents—possibly never stated in print—to emerge more clearly. Such models can provide useful guides for analyzing complex historical situations in new ways. One of the most influential has been Bruno Latour's network model, in which he symmetrically imparts agency to both human and nonhuman actors meshed competitively as they enroll allies to further their own interests. Adopting Knight's point of view, his marketing success with the Royal Navy could be recounted as a Latourian legend of Machiavellian manipulation. But choosing this privileged vantage point would obscure the translations being simultaneously effected by other elements in the extended network within which he was bound, thus effectively returning us to an old-fashioned heroic narrative.[19]

The construction and use of Knight's compasses were outcomes of numerous negotiations, not all of them participated in by Knight. For example, although he used his influence as a Fellow of the Royal Society to ward off competing inventors, he may never have appreciated the existence of his concealed rivals, the naval clerks who were executing some nifty accountancy to safeguard their colleagues in the dockyards. For them, Knight's innovations represented not a desirable improvement in navigation but an unwelcome threat to their livelihoods. Another useful way of viewing compasses is as boundary objects lying at the intersection of disparate social worlds. Their function and value were perceived differently by the various agents involved in the debates surrounding Knight's inventions.[20] Or, borrowing anthropological terminology, we might consider compasses as traded objects bearing various conceptual interpretations. Dissimilar groups can exchange items by agreeing on their local value within trading zones, yet attribute different meanings to them in other cultural contexts. Armchair inventors like Knight viewed compasses as vitally important high-precision instruments, while experienced navigators saw them as unreliable aids to traditional methods. Some French natural philosophers used compass behavior to substantiate neo-Cartesian theories, whereas English expositors denigrated such explanatory projects; they declared compasses to be exploratory tools for displaying the patterns of terrestrial magnetism. Instrument makers assembled compasses from multiple components and literally traded these labor-intensive finished products with Admiralty accountants, who recorded them as ciphers in ledger columns along with food, sails, and timber. By circulating amongst communities, material artifacts subject to differing interpretations can act as bridges between them.[21]

Such sociological models offer stimulating ways of thinking about this episode. One major risk of adhering to them too closely is oversimplifying protracted negotiations between interacting and shifting groups about instruments which were themselves being modified. If frozen in a mid-century snapshot, community boundaries would be blurred: Anson, for example, was a naval administrator yet also an experienced mariner and a Fellow of the Royal Society; the term "instrument makers" embraces distinguished Fellows of the Royal Society as well as waterside craftsmen who have left few historical traces. And by the end of the century the picture would seem very different: a new class of professional compass inventors was emerging, and a disciplinary navy was actively training its officers in natural philosophy. On the other hand, a simpler picture is helpful for exposing important dynamics of this case and for appreciating attitudes and experiences which differed markedly from our own. So this analysis will focus primarily, but not exclusively, on three groups whose views and characteristics contrasted strongly: navigators were concerned with reliability and ease of use; inventive natural philosophers saw accuracy as being of prime importance; and naval administrators wanted to achieve the maximum number of advantages at the minimum cost.[22]

The roles of these groups were unequal in ways which contradict assumptions based on modern values. Navigators were the traditional guardians of magnetic expertise, the community with great practical knowledge using compasses at sea. Particularly in the seventeenth century, natural philosophers often turned to maritime practitioners for advice about using magnetic instruments and for information about terrestrial magnetism. Maritime men were highly esteemed, and the navy enjoyed a far more prestigious role than the army. The personal rewards for oceanic voyagers could be enormous. Anson, for example, gained £60,000 overnight by capturing an enemy convoy: he returned home a rich national hero who soon became effective head of the Admiralty. A contemporary portrait conventionally shows the tools of his trade—including a compass—but also displays a large cornucopia pouring out gold coins.[23]

Towards the end of the seventeenth century, the government embarked on a lengthy and costly program of naval expansion. Pepys introduced extensive reforms to reduce the division between experienced mariners who had risen through the ranks—known as tarpaulins—and well-born commanders, often less competent. Gentlemen were encouraged to enroll in an officer corps, in which promotion increasingly depended on achievement rather than patronage. However, antagonism remained between the upper-class officers, formally trained and examined, and the masters, who had largely learnt their navigational techniques at sea.[24] Members of the seafaring community were physically and culturally separated from contemporary society. Isolated at sea for long periods of time, mariners developed their own social and practical tactics for survival and resented the intrusion of natural philosophers. Halley was esteemed at the Royal Society, but when he took command on an oceanic voyage to measure magnetic variation, the crew mutinied because of his ineptitude. Halley grudgingly admitted that "perhaps I have not the whole Sea Dictionary so perfect" as his experienced lieutenant, Edward Harrison. He plaintively reported to the Admiralty that Harrison "was pleased so grossly to affront me, as to tell me before my Officers and Seamen on Deck, and afterwards owned it under his own hand, that I was not only uncapable to take charge of the Pink, but even of the Longboat." But a naval court-martial exonerated Harrison and the rebellious mariners.[25] Experienced nautical men slated Nevil Maskelyne's *Nautical Almanack* for using incomprehensible terminology and requiring "too nice observations, and too long calculations, to be performed without errors." They mistrusted Harrison's chronometers as mechanical devices which were bound to go wrong, and mocked Franklin's schemes based on outdated charts of the Gulf Stream.[26]

Professionalization may seem an anachronistic term for the early eighteenth century, but, like contemporary physicians, lawyers, and clergymen, naval officers shared some of the characteristics associated with professionals: they possessed a corporate identity as members of an organization, they were salaried for fulfilling recognized roles, and they could expect to be promoted for performing well. In contrast, apart from a few salaried experimental curators, natural philosophers gained no direct financial benefit from their affiliation with the Royal Society. Individuals lacking independent means were forced to rely on their own initiative for earning money. It was only towards the end of the century that practitioners of the new public sciences became able to embark on paid careers. In 1799, Humphry Davy could declare that "philosophy, chemistry and medicine are my profession," because enterprising men during the eighteenth century had collectively established remunerative activities in those fields. Similarly, a new category of professional inventors was being constructed, men who took advantage of the strengthening protection afforded by the patent system to earn their living by marketing their inventions.[27]

But Knight was practicing in an earlier period. He belonged to the substantial heterogeneous group of mid-century gentlemanly inventors, or practical philosophers, who came from diverse backgrounds and adopted a variety of tactics to promote themselves and their products. Because seafaring was so central to Britain's expanding economic, territorial, and military might, these entrepreneurs perceived navigation as a key area for marketing their wares and for demonstrating the value of the philosophical approach. Widely satirized, many of them were out of touch with the needs of practitioners. For example, Whiston designed a magnetic apparatus for measuring longitude whose installation entailed cutting a two-foot hole in the deck; he seemed genuinely surprised that naval commanders were reluctant to embark on such carpentry, let alone engage in complex experiments with his fragile and cumbersome equipment. Michell declared that his tract on artificial magnets was for "Artificers and Seamen," but he buried practical instructions in a morass of experimental details. The president of the Royal Society became wary about recommending the Fellows' inventions, because "the Lords of the Admiralty are so often apply'd to about Schemes absolutely impracticable."[28]

To tap the Royal Navy, the largest market for maritime inventions, gentlemanly inventors had to negotiate with naval administrators. Like other government departments, the naval administration expanded and became increasingly bureaucratic during the eighteenth century. By far the world's largest organization, the Royal Navy had a bipartite management structure. Operational decisions were made in Whitehall by the Admiralty Board, which included civilian politicians as well as retired commanders. Critics castigated members of the Admiralty Board for their ignorance of life at sea beyond the English coast:

> *Would not their Ladies act the farce as well?*
> For are there not, who ne'er Saltwater ey'd
> Except the Scarborough, or the Margate tide?
> The GREAT LOBLOLLY has indeed at large
> Expos'd his person in an open barge;
> From Portsmouth to Spithead has plough'd the main,
> And from Spithead to Portsmouth back again.[29]

Routine administration, including balancing the accounts, was the responsibility of the older and larger Navy Board, a group of former officers liaising directly with crews and dockyards. The Navy Board resented the administrative superiority of the Admiralty Board, whose members were in close contact with government and City merchants. Their financial support was essential to the navy's operations, and, even during wartime, priority often had to be given to shipping cargo. Treasury restrictions forced the naval administration into constant economizing. Subject to both governmental and public scrutiny, naval administrators had to juggle the conflicting demands of winning wars, keeping costs within budget, coping with reluctant sailors and rebellious dockworkers, and transporting goods. Corruption was rife—naval clerks were said to increase their salaries tenfold by accepting bribes. In *The Adventures of Roderick Random*, his fictionalized account of life at sea, Tobias Smollett made frequent allegations of naval malpractice.[30]

Navigators, practical philosophers, and naval administrators held contrasting attitudes towards navigation in general and towards compasses in particular. In printed texts, navigators increasingly campaigned for better charts, compasses, and other instruments, accusing administrators of parsimoniously supplying inaccurate and defective equipment. Harrison, Halley's mutinous lieutenant, described steering compasses—similar to that shown in Figure 3.2—which were "Old and Rusty, and good for nothing, except to throw overboard, &c. Five Hundred such Compasses, I believe, at this present writing, may be found belonging to the *Navy, &c*".[31]

However, navigators' letters reveal that many of them were far more concerned with the problems of acquiring food and men, and getting vital repairs

Figure 3.2. A Traditional Steering Compass. This French compass, made by Joseph Roux of Marseilles in around 1775, is very similar to ones being used a hundred years earlier and still carried by English merchant ships throughout the eighteenth century. The needle is fixed beneath the card, which has a decorative rose labelling the points after the Italian names for winds (for instance, *P* is for Penete, the western wind). Compasses were cased in round wooden bowls with fixed glass which could not easily be replaced when, as here, it cracked. The position of the lubber line marked on the inside of the bowl shows that the card has been rotated through 20° to compensate for the magnetic variation. Science Museum, London.

carried out, than with the state of their compasses. The official inquiry into the Shovell disaster found that only 4 out of 112 compasses belonging to the squadron had been properly maintained, suggesting a vicious circle of malfunction and lack of interest. Many mariners used cheap, poorly maintained compasses and had received little practical instruction in how to use them effectively. They continued to rely on the traditional three L's of navigation: lead, latitude, and lookout. Practical navigation texts explained "that a learner will steer a ship to a greater nicety by a mark-a-head, than a good helmsman can do without a mark by the compass." The objective was to arrive safely, not to pinpoint one's position on a chart which was probably inaccurate. So in unknown or dangerous waters, ships followed expensive local pilot boats; for oceanic voyages, official advice remained the time-honored method of crossing on a particular line of latitude.[32] Navigators only sporadically recorded magnetic variation in ships' logs but carefully noted hourly measurements of the distance travelled and the depth of the water. Towards the end of the century, logbooks were compiled from printed sheets, sold commercially, rather than being ruled by hand. Some of these did include a space for two daily measurements of variation, but navigators generally left it blank.[33]

Harrison was one mariner who actively campaigned for measuring magnetic variation to improve naval safety. He published his strongly worded recommendations because "I Discoursed this Art with some Fellows of the Royal Society whom I found too much aiming at their peculiar Advantage. . . . *England* does not know how many losses hath happened for want of a better knowledge in the *Variation* of the Compass." More typically, navigators found azimuth compasses awkward to use and resented the calculations involved. They judged it pointless to provide vernier scales accurate to tiny fractions of a degree, since "there is no such thing as Preciseness to be expected from any Mathematical Sea Instrument whatever, as most of them are liable to Error from the Motion of the Ship."[34] Maritime practitioners developed methods for finding the variation which were easier and less prone to trigonometrical errors, whose consequences could be fatal. Some of them disdained elaborate sighting mechanisms, taking approximate bearings simply by viewing distant objects along their hands—hence one master's nickname (in 1800) of "Chop the Binnacle." Mariners continued to correct for local variation by rotating the card relative to the needle, perpetuating a much older practice which led to problems when compasses designed for one part of the world were used on oceanic voyages (see Figure 3.2).[35] Presumably corresponding with maritime demand, publishers frequently reprinted seventeenth-century compass rectifiers, two rotating concentric dials used in conjunction with astronomical tables. Another way of avoiding an azimuth compass and its trigonometrical niceties was to suspend a plumb line over the center of a steering compass: by noting the reading of the shortest shadow during the day, navigators could easily find the variation.[36]

Isolated in their "wooden world" and battling for survival, men at sea retained their own traditional beliefs.[37] Many maintained that the direction indicated by a compass needle depended on the loadstone with which it was touched. Even in the nineteenth century, some mariners insisted that "iron will not attract the needle of a compass, provided the iron is covered with wood or puttied up." Ships often banned garlic and onions because they allegedly affected compass readings.[38] Nautical men deployed their magnetic navigational aids in the ways they found most appropriate, ones not envisaged by land-based inventors. So for coastal navigation, they used azimuth compasses not for finding the variation but for steering by known landmarks. To compensate for winds and currents, navigators needed to know the leeway, the angle between the ship's keel and its actual course. They found this by lining up azimuth compasses with the wake of the ship or by attaching compasses to the ship's rails and pinning to their centers lines dragging a piece of wood in the water.[39]

But for natural philosophers, compasses represented completely different resources. Researchers into terrestrial magnetism viewed them as precision aids for laying bare God's natural laws. Entrepreneurs like Knight sought the financial rewards and augmented prestige to be gained by successfully marketing improved versions to navigators. By developing instruments that were more reliable than earlier models, he envisaged lucrative sales from the large naval market and hoped to enhance his reputation as a magnetic expert. Natural philosophers held one of the main criteria of a good compass to be its accuracy. Ensconced at the Royal Society or in their studies, they valued precise measurements made with delicate instruments. From this protected vantage point, they criticized sailors' insensitive compasses with heavy cards and blunt pivots, not realizing that they were more stable during stormy weather. Even practical philosophers claiming maritime expertise scoffed at the practice of keeping two compasses in the binnacle, ignoring the convenience to the helmsman of having one on either side.[40]

Naval administrators, on the other hand, viewed compasses as yet another of the thousands of commodities competing for limited amounts of money. At Admiralty committee meetings, recommendations for expenditure were frequently accompanied by an indication of where the money was to be raised, often by "the Sale of decay'd Naval Stores." Such sales provided ample opportunities for private profiteering. Since each ship carried several compasses, altering specifications for the entire navy represented a significant financial decision. Administrators often failed to implement recommendations for improving compass design or compensated for them financially by reducing the number of compasses allowed per ship. Critics frequently accused the naval authorities of endangering crews' safety through false economies. At Deptford dockyard, where compass supply was centralized, there was a chronic shortage of new compasses and a backlog of damaged ones awaiting repair. Restive workers in the dockyards made the situation worse. Since no single person or department was responsible for supervising equipment like compasses, supply and maintenance depended on local initiatives and influences. Administrators had no established procedures for assessing the worth of new inventions, so tests which seemed expensive and time-consuming were skimped.[41]

Intentionally schematic, this outline of three contrasting groups involved with compasses during the central decades of the eighteenth century clarifies the complexity of interactions involved. Compasses were enmeshed in an extended web linking not only users, inventors, and purchasers but also other commodities and communities. People held conflicting opinions not just on the importance of compasses but on how they should be made and used; they attached different values to modifying their design and performance. Because protagonists with varying types of magnetic expertise disagreed about criteria of superior performance, they had to negotiate what should count as an improvement. Knight's marketing success depended on social status, patronage, and perhaps bribery; the navy adopted his compasses as a consequence of numerous local agreements reached amongst sets of people whose interests clashed. Innovation is, of course, often characterized by dispute. One reward of analyzing this particular episode is that it illustrates one way in which the concept of progressive naval reform was constructed. As natural philosophers successfully implemented their own recommendations within the navy, they concealed the opinions of former experts, the experienced mariners who regarded instruments like Knight's compasses as inappropriate solutions to their problems.

Knight and the Market for Magnetic Navigation

Knight marketed his magnetic bars and compasses direct to the Royal Navy. Through this activity, he furthered his own career as a magnetic entrepreneur and also helped to establish a public image of the Royal Society as a valuable institution. Knight fulfilled one major Baconian prescription for natural philosophers: in his lecture demonstrations, he sought to exhibit and hence explain magnetic forces of nature. Like Hooke, an earlier expert on magnetic compasses, Knight adapted the navigational instruments of mathematical practitioners for revealing and exploring magnetic phenomena.[42] But unlike Hooke, he adroitly used his experiments for commercial self-promotion. As he boasted to the Fellows about his magnetic expertise, the only message remembered by some members of his audience was the parsimonious irresponsibility of naval administrators: "[I]t will cost only about 2s.6d. more to buy a tolerable good [compass]. So that the Lives and Fortunes of thousands are every Day hazarded for such a trifling Consideration."[43] Silencing potential competitors like Canton, Knight negotiated lucrative contracts to supply the navy with magnets and compasses. He was the first compass inventor to take out a protective patent.[44] He promoted his own interests by simultaneously publicizing his membership of the Royal Society and the improvements he was making to maritime safety. In advertising himself, Knight contributed to the legitimation of natural philosophers as magnetic innovators and experts.

The Royal Society had long welcomed accounts of lightning storms as evidence of links amongst meteorological, magnetic, and electrical phenomena.[45] Knight was already well established as the Society's magnetic expert when he was asked to examine a compass damaged by a freak storm at sea. Knight criticized the compass at length, emphasizing the benefits of a philosophical approach to solving nautical problems. The following year, he explained to the Fellows how experiments underpinned the construction of his own compasses. He demonstrated a new steering compass and an azimuth compass designed in collaboration with Smeaton (Figure 3.3).[46] Like other natural philosophers and fine London craftsmen, they were concerned to produce precision instruments made from high-quality materials.

Knight elaborated how their modifications improved traditional designs (like the compass illustrated in Figure 3.2), particularly his major changes in the casing and in the nature and suspension of the needle. Naval compasses were generally set in wooden cases with detachable bottoms so that the needle could be remagnetized. But under maritime conditions, the wood tended to

Figure 3.3. Gowin Knight and John Smeaton's Azimuth Compass. Knight introduced major changes into the design of navigational compasses, principally in the casing and the needle. He placed the compass in a high-quality brass bowl suspended inside a wooden box made with brass screws. His needle was a flat steel bar balanced on a fine point. The azimuth version developed by Smeaton included a stringed sighting mechanism and a finely graduated brass measuring ring. *Philosophical Transactions* 46 (1750), facing p. 515. Whipple Library, Cambridge.

crack and warp, so the bottom fell out, the iron needle rusted, and the card became waterlogged. Knight reported his astonishment on finding that iron nails were used in the outer protective casings (the square wooden box in his illustration). He insisted that brass screws, more expensive but nonmagnetic, were essential.[47] Knight enclosed his compasses in brass held in the outer wooden case by an adjustable suspension system made of rings (the gimballing) to compensate for a ship's motion.

Knight was scathing about the needles he tested, although he probably was unaware that the cards may have been rotated intentionally to compensate for variation (see Figure 3.2). The compasses on many merchant ships did not have slender needles but relied on soft iron wire bent by hand into a rough lozenge fastened underneath the card. He found that these lozenges frequently had systematic errors of several degrees and that the iron soon lost its magnetic strength. More expensively equipped ships carried compasses with waisted needles made from tempered steel. Knight was primarily concerned to improve the accuracy of compasses, and he used magnetic sand to explore the patterns around needles of different shapes.[48] He concluded that a better response would be obtained from a narrow rectangular needle, easier to make symmetrical and with fewer corners introducing irregularities. He used his powerful bars to construct needles from hard steel which would not rust and would remain magnetized far longer. To avoid piercing the needle, which would introduce further problems of asymmetry and stray poles, he balanced it above the card using a point and cap. He experimented with different materials to reduce friction in the needle's suspension. Knight was investing his own money in his research, and he knew that future customers would be concerned about cost. So he fashioned the cap out of cheap ivory, with a small central piece of expensive agate, and made the point from ordinary sewing needles, which could be cheaply replaced when they became blunt. He used light paper for the card, with a supportive brass ring round the edge. The protective glass of many compasses was sealed in with putty, but Knight's was mounted in a hinged lid so that the card could be changed or cleaned and the needle retouched more easily.

Smeaton, enticed down to London as Knight's assistant after a failed romance, described the azimuth version of the compass, the one illustrated on the left of Figure 3.3. An observer used the narrow sights for aligning it with a distant object and then recorded the position of the shadow cast on the face by a horizontal string. The diagram at the top right shows two radially adjusted weights which enabled dip to be compensated for and allowed the compass to be used in both hemispheres. Smeaton explained that the card's brass ring was finely divided so that it could be read to less than half a degree. He did not point out that the instrument's operation depended on manually rotating the whole compass, an operation hard to perform to an accuracy of greater than one degree on land, let alone at sea. "Of all the Instruments now commonly used at Sea," wrote a contemporary mariner, "I do not know one from its Construction so clumsy, or in the Use of which People are more imposed on, than that which goes by the Name of the Azimuth Compass."[49]

Knight had been selling magnetic bars to individual captains for several years when, in 1751, he decided to venture into a far larger market, the Royal Navy. He charged high prices, which he justified by advertising his compasses as high-quality precision instruments made by skilled craftsmen from the best materials. Figure 3.4 shows the signed certificate accompanying his compasses. When writing and editing navigational books, his colleagues at the Royal Society reinforced Knight's own publicity campaign: "These Compasses . . . are made by *George Adams*, Mathematical Instrument-maker to His Royal Highness the *Prince of Wales*, and before they pass out of his hands, are examined and attested by the said Doctor *Knight*, whose certificate is fixed up to the Cover of the Box; without which they are not to be depended on."[50]

As Table 3.1 shows, cheap mariners' compasses were on sale for around 5 shillings, and the Ordnance Office was paying only £1.8.0 for its best steering compasses. At the beginning of the century, the navy had been paying £1.0.0 for a steering compass and £5.0.0 for an azimuth compass. But Knight charged the navy £2.5.0 and £3.13.6, while customers buying them from Adams, the only authorized retailer, paid even more. Many potential purchasers were deterred by this expense, although the Liverpool dockmaster urged the benefits: "Where there is so many lives, and so much property depending

I hereby certify
that I have examined this Compass,
and that it appears to be rightly
Constructed
Gowin Knight

Figure 3.4. Gowin Knight's Certificate. To justify his high prices and ward off competitors, Knight advertised that he would personally inspect his compasses before they left the workshop. He warned purchasers that magnetic equipment lacking his signed certificate might be shoddily constructed. Science Museum, London.

on good Compasses, I have been surprized and vexed to hear some people begrudge the price of Dr. *Knight's* improved steering and azimuth Compasses, which I thought, when I bought one of each, not only deserved the price, but the inventor the thanks of the public as a trading nation and a maritime power for so great an improvement in that important instrument."[51]

Knight's position at the Royal Society gave him an immediate entrée to the lucrative naval market. Several Fellows, including his patron Folkes, were Commissioners of Longitude.[52] Even more significantly, Anson—also a Fellow—was now the effective head of the Admiralty. Anson was keen to implement reforms throughout the naval service, and the introduction to the hugely popular account of his own world voyage had included a plea for the government to finance the collection of magnetic data. Anson visited Knight's home to witness his magnetic experiments, and he ordered the Navy Board to examine Knight's bars and compasses. Knight and Smeaton met experts from the Navy Board and the dockyards several times and convincingly demonstrated the value of their instruments.[53]

The Admiralty Board ordered trials to be carried out on five ships, but, ever economizing, the Navy Board bought only three sets of bars, three steering compasses, and one azimuth compass.[54] An attempt was made to instruct the navigators how to use the new instruments, but only three captains obeyed orders to make the journey to London, at least one of them regarding the trip as a welcome holiday. However, even that minimal contact resulted in the practical suggestion of supplying a spare card.[55] Only one captain bothered to comment on the instruments' performance at sea. He judged "the im-

TABLE 3.1
Eighteenth-Century Compass Prices

	Steering Compass	*Azimuth Compasses*
Navy in 1705[a]	£1.0.0 (brass box)	£5.0.0
Ordnance Office[b]	£1.8.0 (brass box)	
	£0.7.6 (wood box)	
	£0.10.0 (hanging)	
Mariner's Compass with Lozenge Needle		
ca. 1750[c]	£0.5.0	
ca. 1761[d]	£0.4.6	
Knight's Compasses		
1751 Navy Price[e]	£2.5.0	£3.13.6
Adams' Shop[f]	£2.12.6	£5.15.6
Henry Pyefinch[g]		£5.5.0
Benjamin Martin[h]		£3.13.6 and £5.5.0
Edward Nairne[i]	£2.12.6	£5.15.6
Kenneth McCulloch[j]	£1.11.6 to £2.12.6	£9.9.0 to £12.12.0
W. and S. Jones[k]	£0.10.6 to £4.4.0	£5.5.0 to £12.12.0
George Adams[l]	£3.13.6	£5.15.6 and £10.10.0
Ralph Walker[m]		£15.15.0

[a] May, "Naval Compasses in 1707," p. 405.
[b] Millburn, "Office of Ordnance," p. 261.
[c] G. Knight, "Account of the Mariners Compass," p. 117.
[d] Crawforth, "Evidence from Trade Cards," p. 507.
[e] Letter from Navy Board to Admiralty Board of 27 March 1751: PRO: ADM106/2185, p. 431.
[f] G. Adams, "Catalogue," in *Celestial and Terrestrial Globes*, p. 157.
[g] Calvert, *Scientific Trade Cards*, no. 38.
[h] Millburn, *Retailer of the Sciences*, pp. 53, 41.
[i] Wheatland, *Apparatus of Science*, pp. 155–60.
[j] McCulloch, *New Improved Sea Compasses*, p. 17.
[k] W. and S. Jones, Catalogues in 1795 and 1800 editions of Cavallo, *Treatise on Magnetism*.
[l] G. Adams, *Geometrical and Graphical Essays*, p. 490.
[m] RGO 14/17, fols. 5–6.

proved Compasses to be preferable to the other Compasses now in use at all times, except in Stormy Weather."[56] A couple of years later, Knight's compasses were sent for further testing on two ships going to the East Indies, but no report seems to have been made. The Admiralty also financed an experimental voyage for Knight and Smeaton off the English coast. The captain recorded that in light winds the new compass worked better than the old ones, but Knight abandoned his tests when a storm blew up: "Fresh gales with strong Squalls and hails pm filled 15 punch[s] with Salt water Opened a Cask of Beef at 6am weighed & Came to Sail lost the Logg & three Lines trying Experiments."[57]

Despite the paucity of the evidence provided by this perfunctory evaluation program, with its prescient hints of problems in bad weather, Knight managed to convince key administrators of the value of his new products. In 1752, Anson eventually convened enough Commissioners of Longitude to award Knight £300, money to be raised by selling off old equipment. Later that year, the Admiralty Board ordered the Navy Board to supply all ships being fitted for foreign service with one of Knight's compasses.[58] Knight was initially an external supplier to the Admiralty Board, but in 1758 a rival compass maker's death conveniently coincided with an acute shortage of compasses during the war, and he obtained a contract with the Navy Board.[59] However, for at least the next ten years there was a three-way tussle amongst Knight (seeking to maximize his profit), the dockworkers (creatively comparing the charges of outside contractors with the cost of their own labor to justify extra work for themselves), and the Navy Board (constantly trying to economize).[60]

All the navigators who reported on Knight's artificial magnets agreed that they were greatly superior to natural loadstone. George Rodney—not yet a famous admiral—praised them for charming all the *Rainbow*'s compasses into agreement; Mr. Wager, from Deptford dockyard, confirmed that magnets which had already seen several years' use abroad outperformed his own loadstone. Nevertheless, naval economies and failures to disseminate information meant that these new magnets were not immediately adopted by the entire maritime community. Even at the end of the century, manuals still recommended remagnetizing needles with a knife, and a master about to set off for South Africa had heard that magnets might be useful but wondered where to buy one.[61]

Initially, many maritime practitioners were also enthusiastic about Knight's new compasses and asked the Admiralty Board to be supplied with them. One captain "found by experience that they are by much the best Compasses I ever have met with, at Sea," and this endorsement spread round the naval community.[62] By 1766, his reputation was sufficiently great for a ship's master to welcome Knight's latest inventions: "As Boath is Invented by the famos Docter Knight I make no doubt of their answering what they are Intended for."[63] Knight was present at most of the Royal Society meetings briefing Cook for his first voyage and may well have persuaded him to write to the Admiralty secretary: "Doctor Knight hath got an Azimuth Compass of an Improv'd con[s]truction which may prove to be of more general use than the old ones; please to move my Lords Commissioners of the Admiralty to order the Endeavour Bark under my command to be supplyed with it."[64]

However, two problems hindered universal adoption of Knight's compasses: supply and performance. Enthusiastic navigators asked the Navy Board to supply a Knight compass, but only ships going abroad were allowed to carry these expensive new instruments, which had to be returned into store as soon as the voyage was completed.[65] Augustus Keppel gave a vivid account of how his faulty compasses led him dangerously off course "at four oClock on a very dirty hazey Morning," pleading, "The very great Error I found in the

Torbay's Compasses . . . puts me upon Solliciting the Board to order her to be Supplyed with One of Doctor Knights." The Navy Board turned down his request because he was not on foreign service, but he persuaded the Admiralty Board to override this decision.[66] Dockyards ran out, and the Navy Board was faced with the conflicting demands of navigators requesting a Knight compass and accountancy clerks in the docks, perturbed by the cost of repairs and a surplus of 1268 older compasses.[67]

The high initial demand for Knight's compasses suggests that many navigators found them to be superior to earlier models. But when navigators who had managed to obtain Knight compasses tried them out at sea, their comments rapidly became less favorable. Knight had designed an instrument that was accurate and sensitive but that performed badly in the unstable conditions of an oceanic voyage. Rodney complained that they were "impossible to steer by," and Falconer's *Dictionary* reported that "experience sufficiently proves, and truth obliges us to remark, that the methods he has taken to ballance the card with more accuracy than had been formerly attempted, have rendered it by far too delicate to encounter the shocks of a tempestuous sea."[68] After Cook lost his treasured old compass overboard (similar to the one shown in Figure 3.2), he asked for an identical replacement, explaining that "Doctor Knights Stearing Compas's from their quick motion are found to be of very little use on Board small Vessels at sea." As for the azimuth versions, Cook judged them "by far too Complex Instrument ever to be of general use at Sea" and averred that they were "universally allowed to be defective at Sea."[69]

In 1766, Knight patented a revised design intended to make his compasses perform better in rough weather. Designed by a natural philosopher, his stabilizing mechanisms were incorporated in devices for demonstrating Newtonian mechanics in public lecture rooms. But they proved of little value for delicate compasses in ships on the high seas.[70] Just by looking at the new model, one experienced mariner correctly predicted the problems he would encounter in his forthcoming voyage to Tahiti:

> The New Compass is a very fine Instrument for observing the Variation when the Ships in smooth water. It is likeways very good for observing the variation ashoar, but will not answer at Sea in bad weather when the Ship has a quick motion. It then runs round and never stands Steedy. This Defect I think is owing to the new method of hanging the Card, the Soket wheir the card traverses is too wide, and the Card is a great deal too weighty which keeps the Compass in a Continual Motion at Sea.[71]

Although Knight continued to implement minor alterations, this metropolitan inventor was never able to make his instruments appropriate for maritime users.[72]

But in addition to these defects specific to Knight's compass, navigators found that *all* compasses gave inconsistent readings. Cook complained that "[t]his Variableness in Magnetick needles I have many times and in many places expieranced both a shore and on board of Ships and I do not remember of ever finding two needles that would agree exactly together at one and the same time and place." He drew up comparative tables and speculated with his assistants about whether the sun's position was affecting the readings.[73]

Knight and Cook held different types of magnetic expertise. Just as Knight had very limited practical experience of using compasses at sea, so navigators did not appreciate the magnetic environment for which Knight had designed his instruments. Although some individual seamen had been stressing the adverse affects of iron on compass performance since the sixteenth century, few navigation texts discussed everyday details of compass use. Practical philosophers like Knight were shocked to discover that thrifty carpenters used cheap iron nails for making compass casings. Navigators were so oblivious of the effects of iron on compasses that the binnacle was used by Cook for storing the keys to his leg irons, by William Bligh for his pistols, and by soldiers as a convenient prop for their muskets. The binnacles were often not lined up properly with the ship, and they also were frequently made with iron nails rather than the far more expensive brass ones.[74] Compasses were "committed to the care of the boatswain . . . who, of course, pays no more attention to [them] than he does to his other stores; and, even when returned to our dockyards, they are stored away like other stores, without any sort of regard being paid to preserve their attraction."[75] Mariners continued to blame their instruments for conflicting or patently erroneous readings. Even in the nineteenth century, navigators were unaware of magnetic information that was common knowledge amongst natural philosophers. One ship's captain declared that several compasses must all be defective: the passengers had to explain that the problem lay in the ship's iron cargo.

So the debates about Knight's compasses had several components. On land, critics were concerned about his marketing strategies. At sea, difficulties arose from the divergence between the perspectives of mariners—whose expertise lay in practical techniques of navigation—and natural philosophers, who were applying to nautical instruments their magnetic knowledge derived from experience gained on terra firma. Knight died in 1772, but in 1778 his compasses became official issue for all the ships in the Royal Navy, and they featured in naval discussions well into the nineteenth century.[76] Evaluations of Knight's compasses were related to other changes characterized by conflict, such as administrative struggles within the navy, the shifting relationships amongst the Royal Society, the Admiralty, and the government, and contested alterations in the instrument trade. Tracing some wider ripples of Knight's naval engagement will reveal some of the mechanisms underlying these transformations.

Compasses and Competition

Admiralty records are sketchy for this period, but Knight was evidently competing with rival petitioners for naval money. His contemporaries submitted numerous devices, including lifting chairs, pumps, way measurers, quadrants, and timekeepers. Naval administrators strove to allocate money for evaluating these devices, some of which were—like Knight's compasses—eventually adopted.[77] Knight was also constantly vying against some unseen adversaries: the dockworkers at Deptford, who were concerned about the erosion of their jobs through the intervention of this external contractor. They used various tactics to persuade the Navy Board to reemploy the yard's former compass-maker. As well as accusing Knight of overcharging, the Clerk of the Cheque emphasized criticisms of Knight's compasses: "Having frequent Demands from the Out Ports for Steering Compasses of the Old Construction for the use of small Vessels Doctor Knights being so ticklish and Subject to Vibrate it is with the utmost difficulty the true points can be ascertain'd therefore humbly propose as we have many in store to have some of the best of them Repair'd by Sam[l] Saunders the Compassmaker in the Yard upon his former terms." He produced detailed calculations to substantiate his argument that the most economical course would be for their local men to repair the hundreds of old compasses in stock.[78]

In addition, Knight realized that other inventors were criticizing his design to promote models of their own. When these competitors approached the navy direct, Knight benefited from being well established as official compass supplier. Warned in advance by the Admiralty about one rival, he persuaded the Board to delay interviews for several weeks while he perfected some of his own modifications.[79] He may have influenced Anson's treatment of Croker, who had savaged Knight in his *Experimental Magnetism*. In 1761, Croker claimed to have remedied some of the defects in Knight's compasses by turning the rectangular needle on its side and by adding a mirror to the azimuth compass. Although the Admiralty Board had paid for assessing Knight's compasses, Anson fobbed Croker off with "a Chaplainship in the Navy to go out & try it." Unfortunately for Croker, Anson died the following year; writing in 1788, "buried . . . in the Desarts of Deep Bay" (West Indies), Croker despairingly abandoned hope.[80]

Knight was less able to ward off competitors allied with London's instrument trade. Knight used Adams and Magellan as retail agents, but prominent instrument makers soon started to cut into his market by selling high-quality instruments of their own. Canton, Knight's former protégé, helped Edward Nairne secure the contract to replace Harvard College's magnetic instruments after they were destroyed in a fire.[81] Towards the end of the century, instrument makers diversified to cater for specialized groups of users. Maritime practitioners increasingly favored instruments designed specifically for use at sea, which they bought in waterside compass shops rather than in general instrument stores. Navigational innovators were no longer gentlemanly entrepreneurs associated with the Royal Society but were trained instrument makers and nautical experts. These compass inventors regularly patented their designs, recognizing the strengthening protection afforded by the patent system.[82] Whereas Knight had stressed his expertise in natural philosophy, these new specialists included signed affidavits from mariners in their promotional literature. Some of their advertising emphasized the limitations of Knight's compasses, and they introduced modifications tackling the instability of compasses in stormy weather. Marketing quality instruments designed for a discerning clientele, they charged even more for their azimuth compasses than Knight (Table 3.1).[83]

Naval reforms continued to alter the navigators' world. Life on board ship was increasingly disciplined, and training in a naval academy became an essential component of structured careers. Promotion depended on passing examinations in the instruments and techniques of natural philosophy, but magnetic expertise was still unevenly distributed amongst different communities. The Board of Longitude included natural philosophers who provided several cogent reasons why ships' compasses might give inconsistent readings: "by the Iron about a Ship, and that variously, according to the way the Ship's Head stands, and its place upon the Globe, and by the irregularity of the magnetic attraction of the Earth." But such knowledge was not necessarily shared by skilled practitioners: clockmakers did not realize that watches could be affected by carrying a knife in the same pocket, or that longitude chronometers might give misleading readings because of iron on a ship.[84] Similarly, navigators remained perplexed by the conflicting readings of their compasses. They commented on the disparities between observations recorded near land and out at sea. Like Constantine Phipps, the polar explorer, they concluded that, provided you always used the same one, compasses were accurate enough for indicating direction but not for taking precise measurements.[85] In one international episode, Cook angrily denied the very existence of Cape Circumcision, land which French explorers had earlier claimed to locate and identify as part of Australia. Defending Cook to the Royal Society against allegations of poor navigation, the mathematical astronomer William Wales argued that since critics' "reasons are wholly grounded on the variation of the needle," Cook was vindicated—it was well known that even compasses by the same maker could give readings differing by several degrees.[86]

Preparing for his expedition to record the Transit of Venus, Charles Douglas wrote to the Admiralty Board: "[H]aving been informed of the great Utility, of Doctor Knight's lately invented Boat Compasses . . . I beg that you'd be pleased to move their Lordships, to order my being supplied with two thereof." He subsequently reported that they so "far exceeded [his] warmest expectations . . . as to be a safe and unerring Guide to a Boat, either in sailing or rowing." Knight's compasses impressed this experienced mariner so much because they never deviated from the true reading by more than a point. But a point is over eleven degrees, and Knight himself had been concerned to obtain readings of fractions of a degree. Navigators trained in natural philosophy—men like Wales and William Bligh—were perturbed by inconsistencies of only a few degrees.[87]

Despite this growing philosophical insistence on precision, navigators continued to attribute discrepancies to the instruments. Isolated individuals pointed to other explanations: Wales found that the observed magnetic variation depended on the ship's direction, and Ralph Walker, a prominent compass inventor from Jamaica, explained that iron on a ship would affect magnetic instruments. But these perceptions were neither widely disseminated nor acted upon.[88] When Flinders was surveying the Australian coast, he systematically investigated the influence on a ship's compass of its magnetic environment. He concluded that inconsistent readings were due to the magnetic effects of iron, both on the ship—such as guns—and in nearby mineral deposits. But Flinders found it hard to persuade administrators and other navigators of the importance of his research. He undertook a lengthy campaign of experimentation and publication to confute the reactionary arguments of seasoned explorers. Patronized by Banks, Flinders was able to exert greater leverage than individual mariners like the colonial Walker, but his project was nevertheless marginalized.[89]

Naval administrators reacted slowly to innovation, partly because financial restraints limited their purchasing power. As Knight's outdated compass remained official issue until well into the nineteenth century, Falconer's *Dictionary* editor added: "The mechanical parts of our compasses, too, are generally, very indifferently executed; indeed, the low price allowed by the government for supplying the navy with that instrument will not admit of their being made by the best workmen and in the most correct manner." The naval authorities were sometimes obliged to pay the costs of skimping effective testing programs. For instance, their overhasty adoption of Davy's recommendations for protecting copper sheeting resulted in the scheme's expensive failure. The Royal Navy did establish a more systematic program for testing new inventions, but it continued to economize by cutting back on equipment.[90]

Critics with diverse agendas frequently condemned the navy for not buying modern and reliable instruments. When Peter Barlow, mathematical master at Woolwich Military Academy, went to examine the stores in 1824, he exclaimed he "could scarce bring myself to believe that the instruments exhibited to me were those actually employed in His Majesty's vessels. . . . it does appear to me very unaccountable that vessels of such immense value, and the safety of so many valuable lives, should be so endangered by the employment of instruments that would have disgraced the arts as they stood at the beginning of the eighteenth century."[91]

Heroic accounts of discovery and invention conceal conflicts between communities with their own areas of expertise. Attaining broad objectives such as improved navigational safety entails effective organization for ensuring that diverse constituencies collaborate and communicate. Studying Knight's compasses has revealed many administrative problems which affected not only their introduction into the British navy but also their assessment and replacement by more appropriate models. Parallel cases include the Admiralty's resistance to hydrographic reform and to new charting methods, similarly compounded by the ineffective transfer of information between the Royal Society and the maritime community. Such features characterized other fields of activity at the end of the eighteenth century, such as underground exploration, for which different types of expertise about the subterranean world were held by three disparate groups: élite geologists, practicing miners, and wealthy mine owners.[92]

In the nineteenth century, iron ships rendered the problems of magnetic navigation more acute. Now that European wars no longer hindered voyages of exploration and imperial conquest, Edward Sabine and his allies convinced scientists, navigators, and administrators that it was in their mutual interest to cooperate in projects for improving magnetic instrumentation. Continuing Banks' diplomatic maneuvering, the Royal Society had levered itself into an influential position. Seen as experts, practitioners of the new scientific discipline of magnetism could affect government policy and expenditure on navigation.[93]

CHAPTER FOUR

An Attractive Empire: Mapping Terrestrial Magnetism

As LONDON's economy boomed, enterprising managers fostered public enthusiasm for marvels by opening museums displaying automata, freaks, and natural objects brought back by international travellers. When James Cox's celebrated collection was sold off in a controversial private lottery, the advertisers hymned British commercial expansion:

> Thus Britain's white sails shall be kept unfurl'd,
> And our commerce extend, as our thunders are hurl'd,
> Till the Empress of Science is Queen of the World[1]

This flamboyant declaration of British supremacy articulates one of the major themes of this chapter, the symbiotic alliance amongst commerce, natural philosophy, and imperial expansion. After the loss of the American colonies, Britain's international possessions multiplied rapidly as ships carried mercantile explorers to remote parts of the globe. Governmental and mercantile interests corresponded in developing shipping routes, mapping unknown territory, and settling lands to provide profitable markets and raw materials, strategic military sites, and new sources of natural wonders. As the rhetorics of national glory merged with those of commercial gain, natural philosophers ensured that they too were on board for the colonizing voyages of discovery.

Magnetic phenomena and overseas expansion were bound together in a two-way interaction. Explorers and surveyors needed effective magnetic instruments to reach and map new territories. Reciprocally, people compiling up-to-date information about terrestrial magnetism depended on receiving data from all over the world. Natural philosophers and mapmakers recruited maritime travellers as assistants to measure magnetic variation and dip in different places and at different times. In return, they claimed to offer navigators more reliable instruments and charts.

By measuring and recording magnetic patterns in this way, natural philosophers were effectively transforming the world into their laboratory staffed by maritime practitioners. Such global experiments act as levers to alter the balance of forces in society. While the government found it advantageous to advertise the state's contribution to voyages of discovery, natural philosophers benefited from public financial support and maneuvered themselves into a powerful advisory role. They fortified their own status by proclaiming that their knowledge of terrestrial magnetic phenomena was indispensable for successful overseas exploration and settlement. Establishing themselves as experts, they influenced government policy and commercial decisions as well as navigational training and techniques. Conversely, nautical practices molded the collection and representation of magnetic information. Maritime instruments, logbooks, and oceanic charts affected the magnetic observations presented in texts of natural philosophy. Modern meteorology provides a parallel example. This science gained public credibility at the beginning of the twentieth century because it benefited commercial and military aviation projects; but it was shaped by the requirements of aviators, whose cooperation was essential for acquiring atmospheric data.[2]

Lacking central direction, overseas expansion proceeded sporadically and unevenly.[3] The Cox advertisement voices the sentiments of the multiple advocates of commercial imperialism in the face of diverse critics. Civic humanists, for instance, argued against imperial growth, citing the corrupting nature of imported luxuries. Critics maintained that the financial benefits of gaining new markets and sources of natural materials had to be weighed against disadvantages—administrative costs, the dangers of despotism, the immoral nature of the slave trade. Gibbon's monumental *History of the Decline and Fall of the Roman Empire* reformulated in rich documentary detail some familiar warnings of the dangers attending imperial greed. But many natural philosophers successfully allied themselves with those who campaigned for commercial and territorial enlargement. They preached the public benefits of harvesting the natural world, and natural philosophy became harnessed to imperial expansion in a powerful ideology of progress. When William Wordsworth recollected gazing from his college window at Newton's statue, he chose a maritime image for articulating the new public face of abstract science, a Newton "Voyaging through strange seas of Thought, alone." As Wordsworth celebrated scientific achievement, he remained silent about the exploratory oceanic voyages in which the imperial ambitions of natural philosophers had been coupled with slavery and commerce.[4]

Conquering the terraqueous globe had two intimately related aspects: physical domination and intellectual possession. Large sections of the world were scarcely mapped, since Europeans had not ventured far beyond narrow coastal strips of the major land masses, and ships tended to follow set routes across the oceans. European powers competed to colonize new territories for political strategy and commercial advantage. In one of the century's longest didactic tracts of versified natural philosophy, the famous London radical orator Capel Lofft poetically circled the globe cataloguing the devastation wrought by the avarice of commercial imperialism in the guise of exploration:

> Yet powerful it urges: thirst of gold,
> And lust of sway, and fiercer than these fiends,
> Relentless superstition; not alone
> The love of knowledge, and the towering step
> Of Virtue, all divine, have led the way.

> Thus hath the form of this our earthly world
> Been learnt

European explorers and natural philosophers sought to make the world and its peoples their own through encyclopedic naming and systematic ordering. Patriotic travellers collected and classified information about the countries they visited, converting this mass of detail into pleasurably instructive narratives for avid readers back home. Enlightenment philosophers increasingly assumed a role as detached observers of the terraqueous globe as well as of its inhabitants: one naval captain impassively sandwiched his comments on magnetic variation between detailed descriptions of the Hottentots and experiments with a microscope.[5]

Information about terrestrial magnetism was politically and commercially valuable, and this greatly affected the collection and publication of international data. In 1701, Halley published the chart, or marine map, reproduced as Figure 4.1. Presented to the Royal Society after his return from two voyages to measure variation, it portrayed his ideas about the earth's magnetic patterns in the Atlantic Ocean. Halley's theories dominated studies of terrestrial magnetism, both in England and abroad, and versions of his chart were reproduced throughout the century.

Far from being objective representations, maps provide visual evidence of ownership; they reflect the interests and tacit assumptions of their makers. Halley's chart carries concealed texts. The curved lines supposedly link points with the same magnetic variation, but their smooth, continuous form reflects Halley's claim to possess comprehensive knowledge of a natural phenomenon with an intrinsic regularity. The two decorative cartouches are also revealing. The one in the middle of Africa, with its dedication to King William, shows the importance of royal patronage for the pursuit of natural philosophy. It also proclaims the usefulness of Baconian measurement and experiment in acquiring valuable overseas possessions. Conforming with contemporary iconography, three female muses display the azimuth compass and other navigational instruments necessary for these masculine activities. Halley was using this chart as an advertisement not just for himself but also for the practices of the natural philosophers at the Royal Society. By illustrating a family lolling under a geographically inappropriate palm tree in South America, Halley invited armchair travellers to share the fascinated gaze of early explorers and conveyed a moral message about the improving influence of British rule.[6]

To explore the relationships amongst commerce, empire, and magnetic natural philosophy, this chapter is divided into three sections. The first investigates why people studied terrestrial magnetism and how their inquiries were framed by different questions. The second examines the patchy emergence of an organizational framework for amassing international magnetic data as people perceived the immediate local advantages of participating in a global project. Last, the discussion explores how magnetic information was

Figure 4.1. Edmond Halley's 1701 Magnetic Chart of the Atlantic Ocean. The statement inside the decorative cartouche in North America claims: "The Curve Lines which are drawn over the Seas in this Chart, do shew at one View all the places where the **Variation** of the **Compass** is the same; The Numbers to them, shew how many degrees the **Needle** declines either **Eastwards** or **Westwards** from the true **North**; and the Double Line passing near **Bermudas** and the **Cape de Virde** Isles is that where the **Needle** stands true, without **Variation**." The dotted line indicates the route of Halley's second voyage. The straight rhumb lines emanating from the central compass rose conform with older chart conventions for helping navigators to find their bearings. The chart, engraved by John Harris for William Mount and Charles Page of Tower Hill, measures 22½ by 19 inches. British Library.

summarized in tables and maps. Borrowing nautical methods of display, natural philosophers developed new visual techniques for representing and constructing the earth's magnetic nature.

Magnetic Interests

England's wealth and international power had depended on its naval strength since Elizabethan times. As the government poured money into the navy, Britain's maritime supremacy became a dominant theme in patriotic writing and iconography. By the middle of the eighteenth century, the image of Britannia ruling the waves was a publicly recognized symbol glorifying the seapower of a united nation.[7] Nevertheless, sailing remained a very dangerous enterprise: in addition to frequent storms and other hazards, compasses were unreliable, and charts were inaccurate and incomplete. Dutch and Portuguese experts had long recognized that magnetic variation compounded the difficulties of drawing and using charts accurately, and their work helped to convince English people that navigational safety could be improved by collecting more precise information about terrestrial magnetism.[8] Although a pattern of increased collaboration during the eighteenth century may be discerned in retrospect, there was no coordinated project to examine terrestrial magnetism systematically for furthering imperial expansion. People held conflicting opinions about the types of magnetic information which could be useful. As the last chapter made so clear, altering navigational practices entailed confrontation and conciliation. Diverse people involved in the confluence of navigation, trade, and empire—seasoned mariners, politicians, shipowners, factory owners, natural philosophers, naval administrators, explorers—maintained their own vested interests.

Maritime practitioners, as one might expect of experts concerned to reach their destination, showed great interest in navigational tools and techniques. The last chapter described how they often preferred to rely on their own traditional tactics, regarding some innovations of land-based inventors as unsuitable for use at sea. But many representatives of the maritime community did call for better magnetic measurements and for more comprehensive and accurate tables and charts. Mixed motives lay behind their pleas for improved navigational equipment. As a lieutenant with the East India Company, Harrison denounced magnetic ignorance and sought promotion by appealing to his employers' financial sense: "[M]y Honourable *Company Masters*: little do they know, how many Ships have been lost for want of a better knowledge of the *Variation*." Other nautical authors deprecated existing instruments and charts to lend support to rhetorical advertisements of their own inventions.[9]

James Fergusson, an instrument maker from Stepney, averred that "there are so many imperfections, and so many causes of error in the present Art of Navigation, that a mariner scarce knows to which of them an accident is to be imputed." He penned around twenty pages of swingeing critique as a rallying call to enlist maritime support for his British Naval Society, which would stand against the hegemonic Banksian learned empire at the Royal Society.[10] However, other experienced navigators were closely allied with the Royal Society, and a few were even Fellows. Like their colleagues in the new naval academies, they promoted instruments devised by natural philosophers and were interested in procuring government backing for developing new techniques.

Several maritime travellers cooperated with the Royal Society by sending in magnetic observations as they voyaged around the world.[11] Although some of these men appear to have gained little beyond seeing their name in print, for others personal advantages are easier to detect. The naval commander Christopher Middleton, for example, was rewarded for his Canadian magnetic research by being made a Fellow of the Royal Society. Awarded the prestigious Copley Medal, he used the *Philosophical Transactions* as a desirable medium for advertising his new azimuth compass.[12] Joseph Harris' comments on compass design and magnetic variation, compiled during his trips as a naval schoolmaster, were directly related to his activities as an instrument maker. More unexpectedly, the theological writer John Maxwell contributed about forty measurements of magnetic variation from his voyage to South Africa; he belonged to the coffeehouse culture linking instrument makers, projectors, and natural philosophers in financially beneficial ventures.[13] Later in the century, as the Royal Society, merchant companies, and the government collaboratively funded overseas expeditions, navigators were explicitly required to make magnetic measurements.[14]

But far more influential people than ships' captains stood to gain by better navigation. Shipowners, merchants, and naval administrators wanted to negotiate trade routes more safely and use military tactics more effectively. Geographical records, including magnetic information, were not regarded as neutral scientific data but were closely linked with war, exploration, and trade. Petty compiled the first survey of Ireland because the English government intended to pay off thousands of creditors with forfeited land. The world was still so incompletely mapped that fabricated documentation flourished about the Mississippi and the Northwest Passage, and Samuel Taylor Coleridge was accused of spying for the French when he sketched a Somerset river. For the first two-thirds of the century, the Hudson's Bay Company guarded its surveyors' reports as commercial secrets.[15] Britain was frequently at war with France and Spain, and magnetic data became politically important for ensuring supremacy at sea. Anson returned from his voyage around the world a wealthy man and a national hero because he had captured Spanish ships loaded with gold. But he had also seized their secret logbooks, containing valuable magnetic information about the Pacific Ocean. British exploration of the still largely uncharted Pacific was inhibited by the Spanish determination to maintain its domination of the area. The author of the best-selling account of Anson's voyage declared this magnetic data to be "of infinite import to the commercial and sea-faring part of mankind." He urged the government to

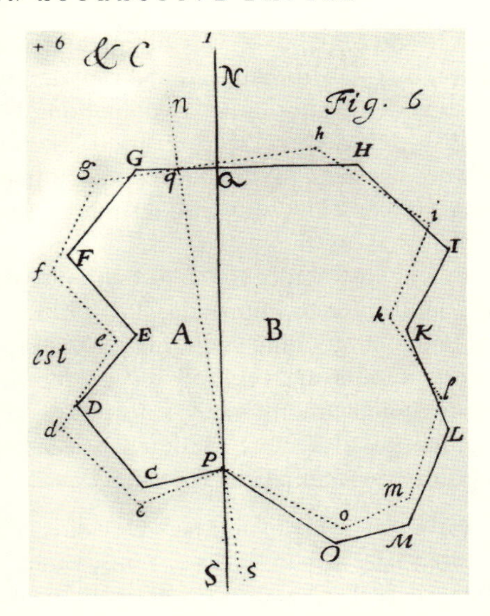

Figure 4.2. The Boundaries of Magnetic Surveys. Propagandists used diagrams like this to explain why magnetic surveys carried out in previously unexplored territory became increasingly inaccurate. Because the magnetic variation is constantly changing, the direction of North indicated by a compass gradually rotates. After several years, the boundaries of a property recorded on a map will no longer correspond to those of a new survey. *Philosophical Transactions* 19 (1697), facing p. 593. Whipple Library, Cambridge.

follow the French example and collect terrestrial measurements in order to prevent Britain's enemies from gaining further tactical advantage.[16]

Understanding terrestrial magnetism was also important for mapping land. Surveyors relied on magnetic instruments in unexplored territories, such as Ireland and the West Indies, where standard equipment was useless because there were no known landmarks. But when boundary disputes arose, critics inevitably debated the trustworthiness of such measurements. Defending Petty, William Molyneux of the Dublin Philosophical Society drew the diagram shown in Figure 4.2 to explain how natural changes of the earth's magnetism would gradually, but increasingly, affect the line between two adjacent properties. His solid and dotted outlines represented magnetic surveys taken at two different times. As the direction of magnetic North rotated from the vertical solid line to the dotted line, so the area covered by a parcel of land would also move round. Molyneux argued that this explained the discrepancies between recent measurements and ones which had been made forty years earlier. By the end of the eighteenth century, this problem was leading to frequent lawsuits in the eastern states of America.[17] Drawing diagrams like Molyneux's, the maritime compass expert Ralph Walker campaigned for surveyors in Jamaica to take account of variation changes. Himself perhaps a victim of land appropriation, he complained that "the present laws now in force in our colonies are very defective, and leave a very wide opening for litigation, and the oppression of the poor, by their ill-disposed and rich neighbours." But the king's surveyor presented an alternative explanation. He maintained that since Jamaican property deeds were accompanied by diagrams indicating that the boundaries were unchanged, the earth's magnetism must be unusually constant near the Equator.[18] Decisions about the earth's magnetic nature were thus linked to territorial arguments.

The Fellows at the Royal Society had always endeavored to demonstrate the contributions their investigations could make to improving navigation, a key commercial and political concern. In the burgeoning economy of Restoration London, they mingled closely with traders making and selling instruments and charts, as well as with navigational practitioners, reformers, and educators.[19] In 1683, during a lecture at the Royal Society, Halley insisted that terrestrial magnetism should be studied more systematically:

> [T]he great utility that a perfect knowledge of the Theory of the Magnetical direction would afford to mankind in general, and especially to those concerned in Sea affairs, seem[s] a sufficient incitement to all Philosophical and Mathematical heads, to take under serious consideration the several *Phænomena*, and to endeavour to reconcile them by some general rule.[20]

Halley was polemically presenting two major themes which underpinned subsequent English research into terrestrial magnetism: its practical benefits and the condensation of diverse phenomena into clear natural laws. But although Halley articulated utility and rationalization as justifications for his programmatic agenda, these broad labels concealed a complex pattern of less immediately apparent objectives held by his successors.

Entrepreneurial authors seeking profit from marketing diverse texts of natural philosophy all stressed the utility of learning about magnetic phenomena. Like lecturers on chemistry, popularizing writers knew that audiences were attracted by guarantees of the usefulness of educational material. Aligning themselves with advocates of commercial expansion, they tendered promises of progress through experimental investigations of the natural world. As training for navigational careers became increasingly academic, natural philosophers sought nautical purchasers for their didactic magnetic information. Some of them published magnetic charts, aiming for financial gain by presenting themselves as recognized authorities patriotically assisting British mariners.[21]

Polemicists advertising the possibilities of progress through the alliance of natural philosophy and commerce cited magnetic discoveries and innovations. Under Robert Walpole's stable administration, merchants constantly but unsuccessfully pressed for war in Europe to reduce competition for lucrative markets and sources of raw materials. Campaigners like Defoe emphasized the financial benefits of philosophical inventions, declaring that "by the discovery of the Magnet and the use of the Compass, Men were particularly qualify'd to visit remote Countries, and make both Discoveries and Improvements also in Trade and Plantation." Around the middle of the century, as the Earl of Chatham embarked on an aggressive policy of commercial expansion, the Royal Society's Copley Medal was twice awarded for magnetic inventions. As the president stressed how this research would benefit Britain's trading economy by facilitating the import of foreign luxuries, he reinforced the Society's role as a disinterested contributor to public welfare. Only recently, he declared, "we have been enabled to make, with security and ease, long voyages by sea. . . . [The Copley Medal is] an honourable distinction from a Society no ways rich . . . but earnestly wishing to become day by day, more and more remarkable for their constant application to promote useful knowledge . . . to the advancement of Science, and to the general benefit of the Publick."[22] As Britain's trading empire expanded, writers repeatedly used magnetic instruments for vaunting the commercial advantages natural philosophy could bring to the nation.

The prize money offered by the Board of Longitude provided a direct incentive for some natural philosophers to undertake extensive magnetic investigations (see the Appendix). Whiston, for example, had been instrumentally involved in the campaign persuading the government to offer the longitude reward in 1714. Frequently mocked by his contemporaries for the impracticality of his schemes as well as his religious beliefs—"poor Mr *W——n* has been so often handled as a *Longitudinarian*, and a *Latitudinarian*"—he started studying magnetic dip only after reading a claim that it provided a method of measuring longitude.[23] Hoping to make his fortune from his magnetic theories, the Welsh physician Zachariah Williams migrated to London. Although he languished in the Charterhouse for nearly twenty years, he did successfully seek Johnson's patronage before dying in poverty.[24] Ralph Walker claimed he had voyaged from Jamaica explicitly to present his compasses to the Longitude Board, and he subsequently composed a theory of terrestrial magnetism. Richard Lovett, a lay clerk at Worcester Cathedral who wrote extensively on electrical medicine, discreetly concealed his request for money on the very last page of a lengthy philosophical and religious analysis of terrestrial magnetism.[25]

Besides utility, the other reason Halley suggested for studying terrestrial magnetism was to unify different phenomena under a single law. Enlightenment natural philosophers repeatedly declared that through understanding the powers of nature they would be able to control them. In a presidential speech, Pringle singled out magnetic research to underline how the Royal Society's Baconian approach contributed to British command of the terraqueous globe: "Let those who doubt, view that Needle, which, untouched by any loadstone, directs the course of the British mariner round the world."[26] Natural philosophers also hoped that rationalizing the earth's magnetic behavior would enable them to interpret other apparently erratic events, particularly the aurora borealis, well known to affect compass needles. Interest in the aurora ran high for much of the century, when there were some exceptionally dramatic appearances over England. Gilbert White and other diarists frequently noted auroral observations and tried to link their incidence with meteorological phenomena. Particularly in France, natural philosophers promised to benefit medicine and agriculture by establishing regular predictive laws for fluctuating quantities like air pressure, temperature, and magnetic variation: "Perhaps the time is not far distant, when we may make such predictions without being deemed visionaries."[27]

In addition to furnishing rhetorical endorsements of progress through harnessing natural forces, some natural philosophers formulated elegant laws to reinforce arguments for providential design. Demonstrating the uniformity and simplicity of nature provided evidence of a divine blueprint and of God as a causal agent. Chapter 6 explores how magnetic theologians incorporated variation in their arguments. Halley, for instance, presented his own influential explanations of the alterations in terrestrial magnetism at a time when he needed to refute charges of atheism. By assuming that God's world functioned with law-governed regularity, men keen to vindicate scriptural truth supplied magnetic verification of the dates of the biblical flood. But other writers began to formulate natural explanations for the temporal changes in magnetic phenomena. Their arguments supported the concept of a world with a history, a world which had been changing since its creation without God's direct intervention.[28]

As natural philosophers participated in Britain's territorial and commercial expansion, they also contributed to European cultural domination. William Derham, clerical author of the century's most popular text on natural theology, welcomed navigational improvements as "a Means to transport our Religion, as well as Name, through all the Nations of the Earth." Numerous authors invoked compasses as evidence of God's munificence towards human beings, His privileged creation: this divine magnetic gift should fill "our hearts with gratitude and adoration for his goodness in placing us over all his other creatures in it."[29] Europeans extended such common sentiments to embrace their superiority over other human societies. British people justified overseas expansion for its commercial rewards but also confidently boasted about the improving benefits imparted by Western civilization. Priestley promoted natural philosophy by linking it with predictions of human progress: "[I]t is nothing but a superior knowledge of the laws of nature, that gives Europeans the advantages they have over the Hottentots. . . . science advancing, as it does, it may be taken for granted, that mankind some centuries hence will be as much superior to us . . . as we are now to the Hottentots." Using a Mandevillean metaphor, a popular journal reinforced Derham's religious imperialism by explaining that compasses enabled "the industrious bees, from the hives of *Europe*" to reach and so instruct "the *Indian* and *African* savages . . . in the knowledge of that supreme Lord and Governor of the universe, of whom, before that, they had such odd and uncouthly confused notions."[30]

So by reiterating Halley's twin objectives of utility and rationalization, polemical writers validated their personal visions of financial and social improvement and hence collectively promoted magnetic research. They claimed that it benefited commercial trade, increased knowledge of God's natural world, and promised moral and material progress to societies throughout the world. Magnetic investigations of the terraqueous globe were firmly entrenched within the expansion of Britain's trading empire.

International Magnetic Measurements

At the end of the seventeenth century, magnetic investigators could read about two types of observations of variation. One category comprised earlier measurements made by English natural philosophers. Authors listed these readings chronologically, attempting to derive laws for predicting future values. The other category included measurements which had been made at various times and places throughout the world, mostly since the Portuguese voyages of the sixteenth century. Analysts endeavored to organize these sporadically accumulated data into orderly patterns reflecting changes in the earth's magnetism. During the eighteenth century, numerous polemicists called for more systematic global projects to compile magnetic information. Halley, for instance, frequently urged Spanish and French cooperation, since "I am in hopes I have laid a sure Foundation for the future Discovery of an Invention, that will be of wonderful Use to Mankind when perfected; I mean that of the Law or Rule by which the said Variations Change, in Appearance regularly, all the World over."[31] However, no administrative framework existed for superintending and financing such investigations. Collective ventures sounded appealing, but conflicting political and commercial interests impeded collaboration. Mounting a research expedition depended largely on personal initiative, although the Royal Society played a key role in facilitating financial support. It was not until the nineteenth century that European governments and scientific societies found international cooperation mutually advantageous.[32]

European communities of natural philosophers developed an increasingly formalized network of communication. They launched several major projects, most famously the two expeditions to measure the Transit of Venus. Viewing themselves as enlightened members of an international Republic of Letters, some experimenters freely exchanged magnetic information, quite often in Latin.[33] For example, Halley corresponded with a German colleague about variation, and well-located Swedish observers shared useful details about the magnetic effects of the aurora borealis. Networks for transmitting magnetic information between London and Russia were established through letters and emissaries. When the new Imperial Academy was opened in St. Petersburg, the Royal Society's secretary immediately sent over a long list of imperious experimental demands, including "the Deflection of the Magnetic Needle. I would like this Observation to be repeated frequently, at different times and for many days."[34]

In print, the Royal Society welcomed the free distribution of magnetic measurements and other geographical data collected by merchants and explorers, "which contribute to the enlarging of the Mind and Empire of Man, too much confin'd to the narrow *spheres* of particular *Countries*." But when put to the test, this idealistic vision failed to materialize. In the middle of the eighteenth century, when two naval educators, William Mountaine and James Dodson, appealed for the Fellows to solicit measurements of variation from their colleagues throughout the world, only five individuals responded.[35]

One problem was the ambiguous nature of voyages of magnetic exploration. Halley criticized the Spanish and French for refusing to provide him with magnetic readings, but his own survey of the English Channel was seen as furnishing cover for secretly observing the French coast. Although he complained about being arrested and fired at during one of his Atlantic expeditions, he seized a small island for Britain when supposedly committed to drawing up magnetic charts of the Atlantic.[36] Because Britain was obviously trying to expand its trading networks and was intermittently at war throughout the century, enemies detected ulterior motives in projects claiming to be devoted to the cause of natural knowledge. So the Spaniards exerted diplomatic pressure to curb Cook's activities, and European governments were reluctant to grant explorers documents of immunity.[37]

International projects faced other obstacles. For example, a worldwide meteorological survey was launched in the 1720s. Like magnetic phenomena, the earth's weather patterns affected shipping and carried religious implications. But the organizers found it difficult to compare readings made with different instruments and to persuade volunteers that taking long series of measurements was worthwhile. Financing and coordinating large-scale series of measurements presented a greater problem in England than in France, where cartography and hydrography benefited from state subsidies.[38] From the outset, the Royal Society decided to enlist the voluntary cooperation of mariners. The first volume of the *Philosophical Transactions* included instructions about a vast range of detailed measurements, including magnetic ones, to be collected on long voyages and submitted in three copies. This ambitious project of marine investigation petered out, partly because sailors were not sufficiently enthusiastic about shouldering this unpaid clerical workload. Until around 1740, a few committed travellers and navigators did send in sporadic reports, but readings were often recorded unsystematically, and there was no established procedure for vetting their reliability or comparability.[39]

In the first half of the century, magnetic research was either personally financed or dependent on patronage. Experiments and expeditions were achieved through individual initiative rather than collaboratively, and were undertaken for a variety of reasons. Halley originally suggested that his voyages would be backed by a merchant friend, but he eventually negotiated a royal subsidy. Whiston persuaded "publick spirited men, and my particular friends" to subscribe almost £500.0.0 towards his investigations into magnetic dip, "by far the greatest sum that was ever put into my hands." He hoped that he would be able to recoup his costs by claiming a reward from the Board of Longitude. Although the government appeared to be offering generous rewards for useful discoveries, the scheme acted as a disguised incentive for people to invest their own resources.[40]

Natural philosophers gradually persuaded commercial and government organizations of the value of their magnetic explorations. In the middle of the century, although Mountaine and Dodson's appeal for international magnetic data in the *Philosophical Transactions* elicited little response from maritime volunteers, it did receive the official sanction both of the Admiralty, which gave them free access to naval logbooks, and of the East India and Hudson's Bay companies. A few years later, Mountaine and Dodson, now both Fellows of the Royal Society, were commissioned by London's major map publishers, William Mount and Thomas Page, to update their chart. Despite this commercial backing, they maintained the stance of altruistic public benefactors: "[N]otwithstanding we have the Satisfaction to receive several Accounts and Testimonials of its Accuracy and Service . . . yet it has never returned the first Charge and Expence. . . . However this has not discouraged Us from making a second Attempt, for the Public Service."[41] During the rest of the century, the Royal Society, the government, and private enterprise increasingly perceived it as beneficial to cooperate in producing and disseminating information about terrestrial magnetism.

In the last third of the century, the government was concerned to consolidate its possessions in the Pacific as insurance against French expansion. At the same time, the major trading companies wanted to protect and increase their shipping routes and preserve access to natural resources in Canada, India, and elsewhere. Because of the close liaison between government and merchants, commercial considerations influenced government policy.[42] The Royal Society, the government, and the major trading companies all had reasons for believing they could gain from voyages of exploration, and a few individuals played key roles in effecting closer collaboration. For example, the governor of the Hudson's Bay Company was Samuel Wegg, treasurer of the Royal Society for over thirty years. Under his direction, the Royal Society gained access to reports on many aspects of Canadian topography and natural life which had formerly been kept secret. In return, the company profited by borrowing magnetic and other instruments and gaining free publicity when experiments were reported in journals like the *Philosophical Transactions* and the *Gentleman's Magazine*.[43]

Joseph Banks played a pivotal role in consolidating the rapport of the Royal Society with commercial and governmental organizations. The cartoon in Figure 4.3, published in 1772 shortly after his return from the *Endeavour* expedition as Cook's botanist, caricatured him as a macaroni straddling the terrestrial poles. Satirists frequently mocked the foreign manners of wealthy

Figure 4.3. Joseph Banks as a Botanic Imperialist. Around the time of Banks' voyage on the *Endeavour* with James Cook, macaronis were frequently ridiculed as foppish young men sporting such extravagant fashions as Banks' ostrich feather and elaborate hairstyle. This caricature was one of a series. "The Simpling Macaroni" showed Daniel Solander, Banks' natural history collaborator: the accompanying rhyme read: "Like Soland-Goose from frozen Zone I wander, / On shallow Bank's grow's fat, Sol*****." British Museum.

gentlemen returning from the European Grand Tour, but Banks had laid claims to the whole world. "Every blockhead" travels to the Continent, he allegedly jeered; "my Grand Tour shall be one round the whole globe." Botany Bay was christened in Banks' honor, and today there are at least five geographical features in Australia and Canada named after him. Six years later, Banks became president of the Royal Society, a position he held for over forty years. Banks strengthened the status of the Royal Society by politically manipulating the structure of its governing body. At the same time, he ensured that its activities became indispensable to the British government, which felt itself under public pressure to participate more decisively in projects of international territorial and commercial expansion.[44] Banks embraced a philosophy of empire that would further British aggrandizement by mining colonial resources through coupling natural knowledge with commercial enterprise. He came to exercise a pervasive influence over a wide range of activities. For example, he demonstrated the contribution the Society could make to the new Ordnance Survey, thus taking over a military responsibility. He gained an increasingly free hand in managing foreign expeditions financed to varying degrees by public and private institutions. By the turn of the century, during preparations to survey the Australian coast, it was Banks who liaised between Flinders, the Admiralty, and the East India Company. He organized salaries and equipment and drew up documents regulating the allocation of proceeds from the voyage.[45]

In the last third of the century, voyages of exploration simultaneously served the interests of several constituencies, notably the Royal Society, the government, and the major trading companies. The most famous were led by Cook, but others included Phipps' search for the North Pole and Bligh's voyage on the *Bounty*. Good magnetic compasses were needed to reach and map new areas. The Royal Society increasingly used government money to provide instruments and instruct travellers how to use them. In exchange, trained navigators and explorers collected vast quantities of terrestrial data, including extensive magnetic measurements.[46] Because such expertise was becoming a route to promotion, naval men had a direct incentive to participate in projects of magnetic mensuration. Voyagers and publishers profited by marketing travel literature and books on natural history, thus fuelling the enormous public interest in these expeditions. When Mary Shelley wrote *Frankenstein*, she confidently assumed that her readers would understand her polar explorer's claim: "[Y]ou cannot contest the inestimable benefit which I shall confer on all mankind to the last generation . . . by ascertaining the secret of the magnet, which, if at all possible, can only be effected by an undertaking such as mine."[47]

By the end of the century, although—apart from western Europe—the interiors of the continents remained largely unexplored, natural philosophers had extensively charted oceanic magnetic patterns. They had built an iron-free geomagnetic laboratory, equipped with up-to-date instruments and research literature, in an isolated spot in Sumatra, then owned by the East India Company. Individual imperial settlers were communicating their magnetic measurements to researchers in England either privately or through the new colonial press. Demonstrating the value of their contribution to the imperial enterprise, natural philosophers had instituted new ways of organizing scientific expeditions and local research projects.[48]

Representation and Possession

Natural philosophers compiled an accumulating fund of magnetic observations and developed new ways of representing terrestrial magnetism. At the end of the seventeenth century, international measurements were mainly collected by maritime voyagers. But as metropolitan investigators miniaturized the world into their notebooks, they transformed the chaos of life at sea into an orderly rendition of magnetic nature. They demonstrated their intellectual possession of terrestrial magnetism by translating the overseas readings of itinerant observers into a convenient format for stationary experimenters. By converting the global vagaries of magnetic variation into a two-dimensional inscription, they stabilized the constantly fluctuating magnetic patterns into a fixed representation. They demystified the world's magnetic behavior by capturing it in black and white on a piece of paper.[49]

Authors commented on the large number of theories which had been proposed to account for the mysterious phenomenon of terrestrial magnetism. By the end of the century, many natural philosophers thought it was fruitless to search for a mathematical solution. They related the cautionary tale of the mathematics teacher Charles Leadbetter, dismissed from his job of updating Halley's chart after he "plunged himself into an inextricable labyrinth. . . . His attempt was frustrated by the opinion he possessed of an uniform and regular mutation."[50] However, they did use tables and maps as visual demonstrations of their ability to condense a confusing array of measurements into comprehensible magnetic knowledge. Natural philosophers marshalled irregular sets of magnetic data into clear displays supporting their accounts of the terraqueous globe.[51]

At the end of the seventeenth century, natural philosophers frequently recounted magnetic measurements, and the conclusions they drew from them, in prose sequences. But gradually they came to present their observations in tables—a general tendency adopted for other types of data, such as mortality figures or meteorological records. The format of ships' logbooks was also being regulated. It became standard practice for navigators to enter daily measurements in ruled columns; later in the century, printers sold special stationery. Like maps, tables can embody investigators' classificatory systems for structuring their world. The dramatic shifts in chemistry towards the end of the eighteenth century provide a good example. Chemists initially compiled synoptic tables displaying as many relations between individual substances as possible, hoping to discover nature's truths by collectively accumulating neutral facts. Following Antoine Lavoisier's innovations, chemistry acquired a new disciplinary status, and stylized tables fashioned nature into his predetermined categories. Similarly, as natural philosophers constructed a new magnetic science, they converted traditional maritime records into authoritative tabular displays encapsulating their theories of terrestrial magnetism.[52]

Figure 4.4 shows the readings of magnetic variation sent in to the Royal Society by a ship's captain in 1721. The measurements have been made sporadically, perhaps corresponding to convenient lulls in activity when someone could spare the time. In contrast, the table reproduced in Figure 4.5 is typical of the more systematic arrangements later preferred by natural philosophers. The dates have been omitted, and the readings are presented for equal inter-

Month and Year.		Latitudes.	Meridian Diſtance.	Longitud.	Variation.
Auguſt 1721.	24th	9° 8′ Sou.	9° 23′ W.	9° 25′ W	2° 13′ E
Ditto	26	11° 12′ S.	10° 46′ W	10° 50′ W	4° 30′ E
Ditto	27	11° 34′ S.	11° 28′ W	11° 41′ W	4° 29′ E
Ditto	28	12° 32′ S	11° 31′ W	11° 43′ W	4° 27′ E
Ditto	31	15° 46′ S	10° 53′ W	11° 6′ W	6° 10′ E
Septemb.	2d.	16° 26′ S	8° 25′ W	8° 30′ W	7° 16′ E
Ditto	5th	18° 45′ S	9° 31′ W	9° 39′ W	6° 17′ E
Ditto	6	19° 47′ S	9° 10′ W	10° 0′ W	8° 6′ E
Ditto	17	28° 43′ S	1° 7′ W	1° 9′ E	5° 53′ E
Ditto	22	31° 33′ S	3° 41′ E	3° 56′ E	4° 10′ E
Ditto	27	33° 30′ S	11° 29′ E	12° 57′ E	0° 11′ W
Ditto	30	32° 40′ S	19° 6′ E	12° 1′ E	3° 0′ W
October	1ſt,	32° 53′ S	21° 18′ E	24° 59′ E	5° 41′ W
Ditto	3	32° 30′ S	25° 33′ E	30° 0′ E	7° 47′ W
Ditto	5	32° 28′ S	30° 37′ E	35° 52′ E	8° 44′ W
Ditto	6	31° 22′ S	31° 40′ E	37° 7′ E	10° 57′ W
Ditto	7	31° 11′ S	32° 4′ E	37° 47′ E	11° 20′ W

Lat.		Long.		Vari.		Lat.		Long.		Vari.	
D.	M.	D.	M.	D.	M.	D.	M.	D.	M.	D.	M.
50	00	12	00	14	00	50	00	18	45	17	00
51	00	12	00	14	15	51	00	18	45	17	1[illegible]
52	00	12	00	14	30	52	00	18	45	17	30
53	00	12	00	14	45	53	00	18	45	17	45
54	00	12	00	15	00	54	00	18	45	18	00
55	00	12	00	15	15	55	00	18	45	18	15
56	00	12	00	15	30	56	00	18	45	18	30
57	00	12	00	15	45	57	00	18	45	18	45
58	00	12	00	16	00	58	00	18	45	19	00
59	00	12	00	16	15	59	00	18	45	19	15
50	00	14	15	15	00	50	00	21	00	18	00
51	00	14	15	15	15	51	00	21	00	18	15
52	00	14	15	15	30	52	00	21	00	18	30
53	00	14	15	15	45	53	00	21	00	18	45
54	00	14	15	16	00	54	00	21	00	19	00
55	00	14	15	16	15	55	00	21	00	19	15
56	00	14	15	16	30	56	00	21	00	19	30
57	00	14	15	16	45	57	00	21	00	19	45
58	00	14	15	17	00	58	00	21	00	20	00
59	00	14	15	17	15	59	00	21	00	20	15
50	00	16	30	16	00	50	00	23	15	19	00
51	00	16	30	16	15	51	00	23	15	19	15
52	00	16	30	16	30	52	00	23	15	19	30
53	00	16	30	16	45	53	00	23	15	19	45
54	00	16	30	17	00	54	00	23	15	20	00
55	00	16	30	17	15	55	00	23	15	20	15
56	00	16	30	17	30	56	00	23	15	20	30
57	00	16	30	17	45	57	00	23	15	20	45
58	00	16	30	18	00	58	00	23	15	21	00
59	00	16	[illegible]	18	15	59	00	2[illegible]	15	21	15

Figure 4.4 (*left*). Captain Cornwall's Measurements of Magnetic Variation. Printed as tables with no accompanying text, around thirty observations were sent in to the Royal Society by Captain Cornwall from his African voyage of 1721. Taken at irregular intervals, these readings show no systematic patterns of variation. *Philosophical Transactions* 32 (1722), p. 55. Whipple Library, Cambridge.

Figure 4.5 (*right*). Captain Middleton's Measurements of Magnetic Variation. Like the table reproduced as Figure 4.4, this set of data was printed with no accompanying text. Christopher Middleton contributed around two hundred observations during his Canadian voyages of 1721 to 1725. In 1727, he was made a Fellow of the Royal Society for his magnetic studies. These readings have apparently been made at regular intervals of latitude and longitude; the magnetic variation also seems to alter smoothly. *Philosophical Transactions* 34 (1726), p. 73. Whipple Library, Cambridge.

vals of latitude and longitude. By focusing exclusively on the data, the observer's shipboard life has been erased, and the earth's characteristics seem objectively displayed. However, the recorded values of the variation increase suspiciously regularly, suggesting that the readings were rounded off to comply with an assumed regularity. Such tables enabled natural philosophers to demonstrate how they disciplined maritime confusion into an orderly world governed by natural laws.

Back in England, experimenters amassed huge sets of magnetic data from their own investigations as well as collating international measurements from maritime travellers. At first, their published reports incorporated lengthy tables listing all the relevant observations. Towards the end of the century, they condensed these comprehensive sets into more easily absorbed digests.[53] Such distillations made it far simpler to understand the information being presented but enabled the author to exert greater editorial control over the data. Some natural philosophers used techniques of visual rhetoric to reinforce their message that the earth's magnetic variables obey regular laws. Impressive tabular displays accord factual authority to estimated values which have been derived theoretically. So some experts printed pages of calculated figures resembling tables of logarithms. The absence of the gaps and contradictions which characterize observed data provided immediate visible testimony of theoretical superiority—and also implied certainty. Other polemical devices included mixing observed and predicted data in the same table or recording figures to a degree of accuracy far beyond that which instruments were capable of achieving.[54]

Modern readers are used to interpreting numerical data plotted in graphs, which show at a glance the pattern of the relationship between two variables. But the impact of visual technologies depends on the tacit skills of both their producers and their consumers. Before about 1820, virtually no English analysts tried to communicate numerical information graphically, even though mathematicians were studying curves, and the appropriate printing techniques were available. In 1823, when Samuel Christie did use a graph to present his results, he attached a long verbal explanation, nowadays superfluous.[55] Unlike experimenters exploring other branches of natural philosophy, magnetic researchers acquired much of their information from the maritime community, whose nautical experts were accustomed to interpreting navigational charts and manipulating trigonometrical expressions. For them, it would seem quite natural to summarize magnetic data visually in maps or spherical representations. As some natural philosophers came to be closely involved with navigational education—for example, at the training academies at Portsmouth and Woolwich—they adopted these maritime formats of representation.

In England during the eighteenth century, two important English global magnetic charts were produced, both in several versions: Halley's and Mountaine and Dodson's.[56] Although their makers claimed that they had been derived empirically, they were affected by theoretical considerations. Mapping the measurements of terrestrial magnetism entailed solving probability problems similar to those involved in analyzing population data: how to derive a representative average of a great number of observations; and how to cope with gaps in the records or with readings disobeying a general pattern. Halley and Dodson were both involved in commercial and political projects using mortality figures to calculate appropriate insurance premiums. Enmeshed in the London culture of projectors, maritime insurers, and government financiers, they developed techniques for coping with large disorderly sets of figures. When they compiled their magnetic charts, they used similar tactics to persuade their data to demonstrate their ideas.[57]

Halley was world-famous for his innovative maps displaying the patterns of natural phenomena such as winds, ocean currents, and terrestrial magnetism.

Nowadays, everyone is familiar with thematic maps picturing, for example, the world's temperature zones or income distributions. But at the beginning of the eighteenth century, the concept was a novel one. The practice of thematic mapping originated with Halley, who seems to have thought in an exceptionally visual way.[58] Petty, for example, was also involved in maritime affairs, map making, and probability but never brought these three spheres of interest together. Joseph Moxon, a prominent instrument maker and map seller who imported Dutch expertise into English cartography, reported that as a boy Halley "was very perfect in the caelestiall Globes in so much . . . that if a star was misplaced in the Globe, he would presently find it." Halley undertook several sea voyages to collect data and introduced new mapping notation indicating the directions and strengths of tides and winds. Atypically for the period, he constructed three-dimensional diagrams of mortality data and drew graphs of pressure as a function of altitude. Apart from a few mineralogical maps, Halley's cartographic initiative was scarcely imitated until the nineteenth century, when analysts started to represent environmental and social phenomena in maps.[59]

Between 1698 and 1700, Halley completed two voyages in the Atlantic. Commanding a ship loaned by King William III, he drew up his own Admiralty instructions to provide the government with useful information about African and South American ports; he also emphasized that "in all the Course of your Voyage, you must be carefull to omit no opportunity of Noteing the variation of the Compasse, of which you are to keep a Register in your Iournall." He collected a total of about 150 observations and, on his return, almost immediately published the magnetic chart of the Atlantic Ocean shown in Figure 4.1. Hung in the Royal Society's meeting rooms, it provided impressive evidence of the Fellows' mastery of maritime matters.[60] Halley's major innovation was to link points of equal variation with smooth curved lines, implying an underlying regularity. Although earlier writers had suggested making magnetic maps, Halley was the first to publish and advertise this particular technique. Natural philosophers referred to these curves as "Halleyan lines" until the term "isogonics" was introduced in the nineteenth century. Halley's magnetic charts provided an important precedent for later cartographers using isogonics to depict other abstract quantities, such as the spatial variation of temperature and pressure. Alexander von Humboldt, for instance, explicitly cited this magnetic inspiration when he introduced isothermals to exhibit his laws of heat. Isogonics are conceptually different from contour lines, which directly represent the observable world. Although contours are usually regarded as a French innovation of the early nineteenth century, John Churchman—a Philadelphia surveyor practicing in England—proposed a contouring technique after producing his magnetic atlas showing isogonic lines of equal variation.[61]

As a natural philosopher ridiculed by his own crew, Halley had to establish credibility amongst the maritime community. At the Royal Society, he was concerned to consolidate his reputation as a magnetic expert by substantiating his theories of the earth's magnetic structure. One immediate priority was to convince the Admiralty that its financial contribution towards his voyages was justified. While still overseas, he sent back letters to his sponsors confidently maintaining that "I have found noe reason to doubt of an exact conformity in the variations of the compass to a generall theory." He promised that his research would "appear soe much for the public benefit as to give their Lord^sps intire satisfaction."[62] Thus he simultaneously promised experimental corroboration of his theories and vindicated the Admiralty's investment of public money in the pursuit of natural philosophy.

Within a year, Halley had produced a second chart of the whole world, shown as Figure 4.6. Halley profited from this chart socially as well as financially. Widely reproduced into the second half of the century, this version made it appear that he had charted the whole world. Swift later articulated contemporary skepticism about traditional maps:

> So Geographers in *Afric-maps*,
> With Savage-Pictures fill their Gaps;
> And o'er unhabitable Downs
> Place Elephants for want of Towns.[63]

But Halley had already given his chart a modern authority by dropping the old-fashioned rhumb lines and eliminating the pictorial cartouches. Instead, a Latin poem celebrated the commercializing initiative of the compass's unknown inventor, the forgotten hero who "linked / Shores . . . and by wind / Brought mutual products to remotest lands."[64] Proclaiming his affinity with the nautical world, Halley attracted maritime purchasers by placing explanatory advertising copy next to the chart. In this, he claimed that his charts could be used both for estimating longitude and for correcting compass readings: when it was more important to determine a ship's direction than its precise location, navigators could estimate the variation from the chart and hence amend the value given by the steering compass.[65]

For at least the next two decades, Halley appealed to his audience of natural philosophers by stressing how his chart confirmed experimentally his theories of terrestrial magnetism. He had postulated four magnetic poles, two on a central magnetic nucleus rotating in liquid inside the earth, and two on the outer crust. His chart hardly provided the unbiased record which people like to view as the outcome of scientific activity. To construct it, Halley had been faced with the limited extent of his observations, particularly in the Pacific Ocean. He drew liberally but selectively on other navigators' readings made with different instruments and at different times, when the magnetic variation would not have been the same. Declaring himself justified in drawing a curve on the basis of only three observations, he candidly admitted extrapolating to cover gaps in the data: "I never observed myself in those Parts; and 'tis from the Accounts of others, and the Analogy of the whole, that in such cases I was forc'd to supply what was wanting." In elaborate circular arguments, he enlisted single readings to substantiate hypothesized tendencies for large tracts

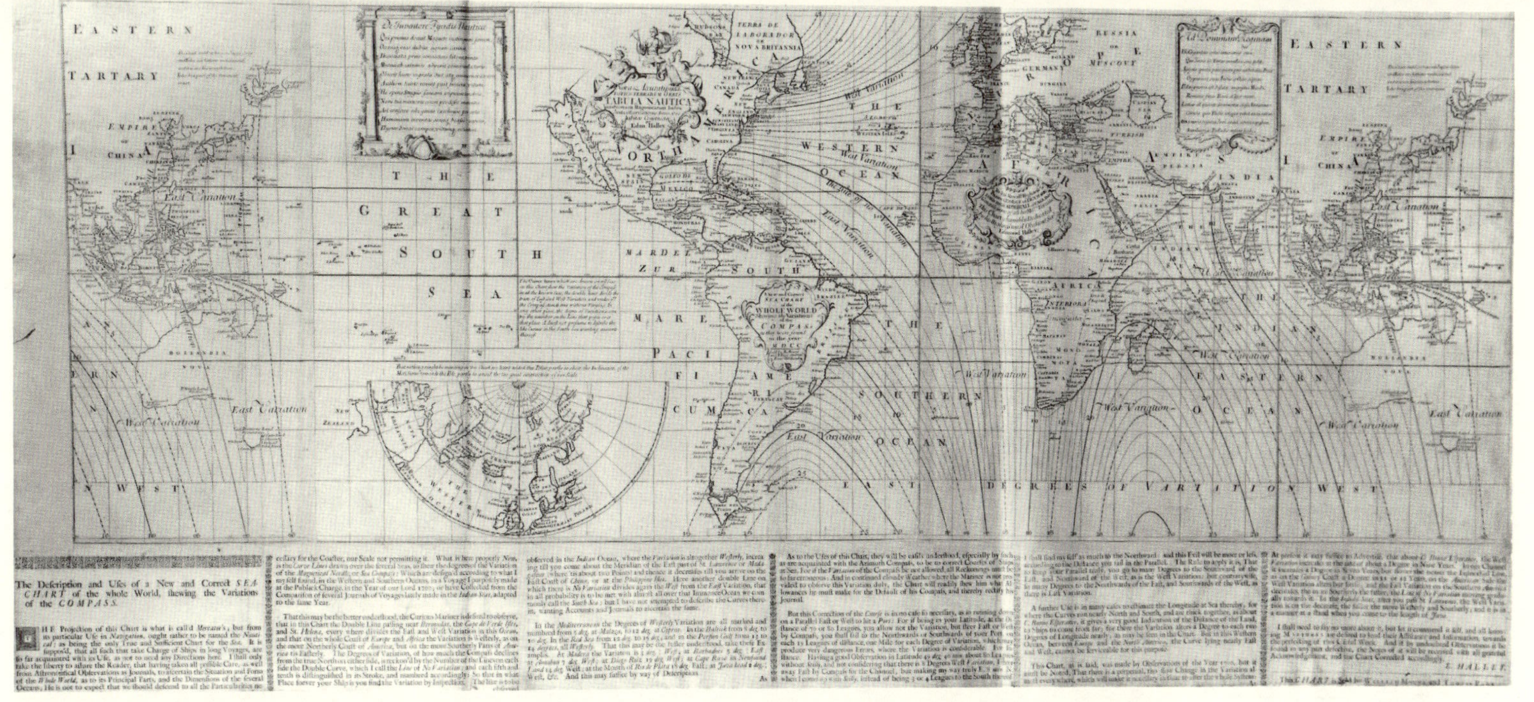

Figure 4.6. Edmond Halley's World Chart of Magnetic Variation. The chart's title inside the South American cartouche reads "A New and Correct Sea Chart of the Whole World Shewing the Variations of the Compass as they were found in the year MDCC." The same title in Latin is printed inside the North American cartouche. The African cartouche contains a dedication to Prince George of Denmark, Lord High Admiral of England: he was the consort of Queen Anne, lauded in a Latin verse in the European decorative box. The box in the Great South Sea contains the Latin poem praising the heroic inventor of the compass. The chart, engraved by John Harris for William Mount and Charles Page of Tower Hill, measures 20 by 57 inches. Royal Geographical Society, London.

of unexplored ocean. Halley vigorously countered criticisms—particularly French ones—by citing new measurements which confirmed his own theories and finding various excuses for dismissing the ones which could not be accommodated. Conveniently retrieving earlier readings of his own, he emphasized the likelihood that other people had made mistakes: "I am willing to believe this to be an Error of the Press . . . or rather the Memory of M. *Couplet*, who, it seems, lost all his Papers by Shipwreck in his Return."[66]

Halley's chart had no competitors until the middle of the century, when Mountaine and Dodson were commissioned to modernize it. They immediately stressed the poor quality of their data: ships' journals were poorly kept, readings were inconsistent, and they lacked observations for huge areas of the world, particularly over land. Although they scoffed at Halley's tactic of "reconstructing a Sett of Lines by Analogy," they themselves adopted a probabilistic approach based on preconceived ideas. When observations seemed to conflict, they devised an elimination procedure based on conformity with the majority. For places with few confirmed observations, they drew broken lines rather than leave blank patches.[67]

Maritime practitioners denigrated such magnetic charts as inadequate navigational aids, complaining that, though "laid down from very uncertain observations, and the rest from the mere imagination of the draughtsmen, they are not withstanding held forth as unerring guides . . . but experience quickly informs us they are only pictures, the creatures of the brain."[68] Critics pointed out that the charts were inherently of little value for finding longitude in one of the most important trading areas, the North Atlantic, where the lines of equal variation run approximately east-west. Additional problems stemmed from financial limitations on naval expenditure. Oceanic charts were produced privately in a competitive business dominated by one publishing company, Mount and Page. Even Royal Navy commanders had to buy their own, which were frequently poorly printed and based on old surveys. They were reproduced from costly copper plates engraved by hand, which became increasingly worn with repeated use. Paper was also expensive, so charts were printed to an economical scale. Any magnetic chart suffered an additional ageing problem: "[T]he continual change in the variation makes it necessary that it should be renewed at the end of every three or four years. This short time of sale cannot possibly repay the charges of the compiler."[69] Naval entrepreneurs selling their own services were particularly vocal critics, but towards the end of the century there were widespread demands in books and naval encyclopedias for state subsidization of large accurate charts: "The engraving of new plates on every publication of such charts is a charge too great for private persons to incur; and should, like the nautical almanack, be done at the expence of the public."[70]

At the beginning of the nineteenth century, some of the new sciences acquired a visual language from practitioners and field workers. For example, geologists derived their systems of representation from techniques developed by surveyors and mine workers for mapping terrains and sketching land formations. Botanists and meteorologists benefited from pictorial methods generated by landscape artists for discriminating amongst tree species and cloud types.[71] Similarly, the science of magnetism absorbed diagrammatic technologies from its maritime roots. Instead of presenting measurements of terrestrial magnetism in lengthy tables, natural philosophers came to display them in the style of navigation charts. Thomas Young, for example, reproduced in the printed version of his lectures at the Royal Institution two full-page plates of global charts portraying magnetic patterns. He used the diagrams 574–6 shown in Figure 4.7 to summarize visually how magnetic variation alters with time. By juxtaposing images of measurements made in 1700, 1744, and 1790, he made the changes immediately apparent.[72]

Diagrams enabled natural philosophers to perform visualized thought experiments for studying the earth's inaccessible magnetic core. Young used the diagrams 577–8 in Figure 4.7 to illustrate how the lines of dip would appear if a small magnet lay angled at the earth's center, or at a distance from it. John Lorimer, a Scottish physician, depicted the patterns of terrestrial magnetism which would result from different configurations of four Halleyan poles. In Figure 4.8, the top two maps show the lines of equal variation that would result from symmetrical arrangements; he rejected these because they differed markedly from observation. The lower two illustrate the patterns to be expected from an asymmetrical arrangement of the four poles. This was the model he selected, because it was the one which most closely resembled Mountaine and Dodson's chart of magnetic variation. One advantage of this type of visual reasoning was that it did not demand a precise mathematical model of the complex patterns of terrestrial magnetism. Although the lower patterns did not correspond precisely to the mapped data, they were generally of the same form. Lorimer explained that discrepancies could be accounted for by inaccuracies of measurement or by local features such as deposits of magnetic iron ores.[73]

Diagrams can fulfill other experimental functions, not only summarizing data and testing ideas but also acting as an effective format for disseminating theories. By portraying an invisible underlying order, they come to constitute, rather than just represent, the real world. For Halley, the lines of variation marked on sheets of paper embodied his four-pole hypothesis of the earth's magnetic structure. But, surviving independent of its creator, his chart was open to other interpretations. Advertised as an objective record of the natural world, a navigational tool developed with philosophical vision, Halley's chart travelled out from the meeting room of the Royal Society across national and social boundaries. Navigators pored over it in their cabins, natural philosophers perused it in their studies, cartographic publishers debated the profitability of updating it, polite purchasers flicked past it as they browsed their new books for rational entertainment. In Latour's memorable phrase, the chart acted as an "immutable mobile."[74]

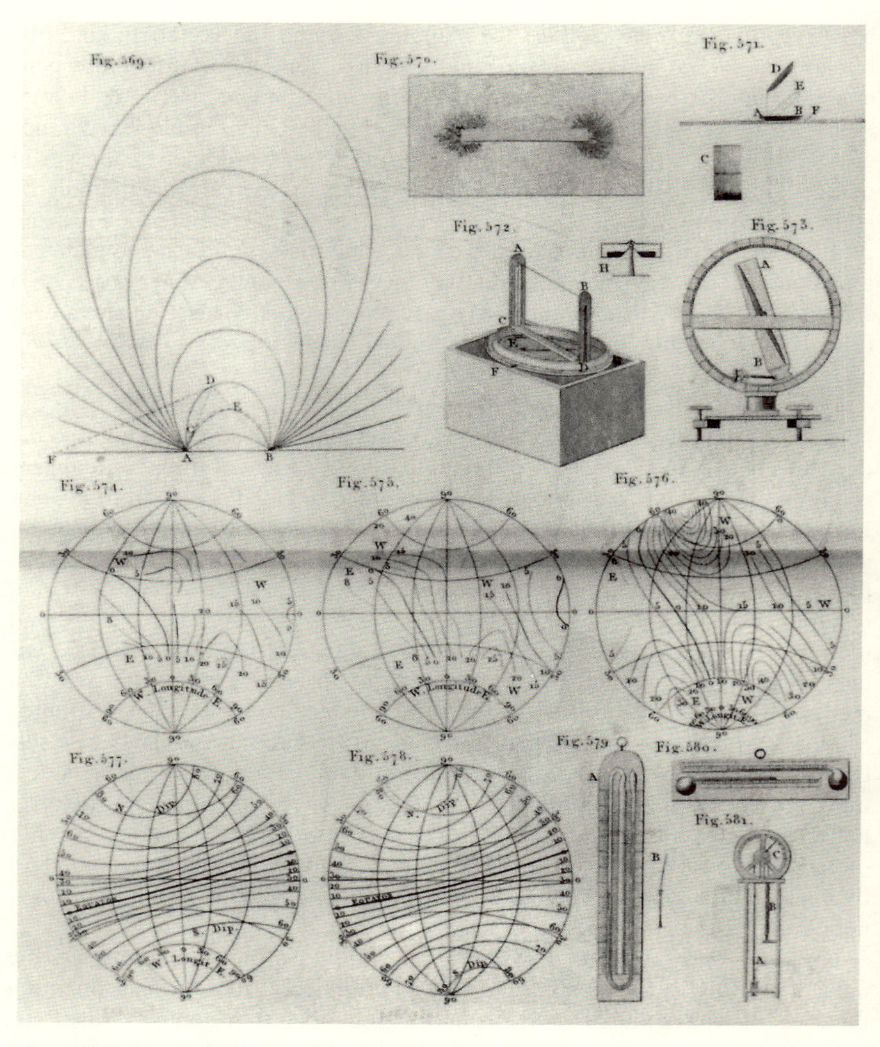

Figure 4.7. Magnetic Illustrations from Thomas Young's *Lectures*. The three images 574–6 compare data on magnetic variation in 1700, 1744, and 1790. They display at a glance how the patterns of terrestrial magnetism alter with time. Young used the two diagrams of dip, 577–8, for displaying the isogonic patterns to be expected from different magnets placed at the earth's center. Diagram 569 shows the geometrical method introduced by John Robison for analyzing patterns of filings round a bar magnet, as in 570. Young, *Course of Lectures*, vol. 1, plate 41. Whipple Library, Cambridge.

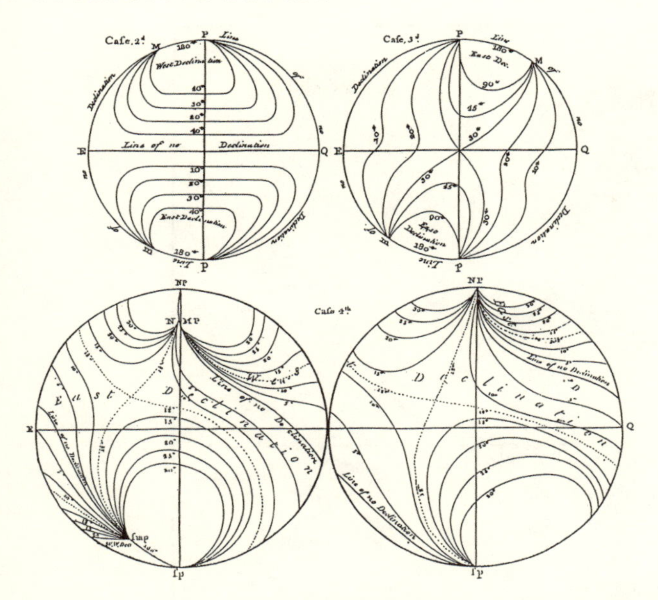

Figure 4.8. John Lorimer's Thought Experiment. The diagram at top left shows the pattern of magnetic variation to be expected if the earth's four magnetic poles were placed symmetrically in the same meridian but on opposite parallels. The one at top right corresponds to their being placed in opposite meridians and on opposite parallels. The bottom two figures show the patterns to be expected from an asymmetrical arrangement, when the poles are in neither the same nor opposite meridians. Based on Lorimer, *Essay on Magnetism*, plates 5 and 6.

Instruments developed in one theoretical framework can function as durable artifacts which natural philosophers subsequently endow with their own constructs. For instance, Franklin revised the meaning of electrical vocabulary to explain how the Leiden jar could operate; rival practitioners adapted Priestley's pneumatic devices in ways he had not envisaged. Similarly, natural philosophers reinterpreted Halley's magnetic charts. Figure 4.9 shows how Adam Walker literally twisted the evidence, converting isogonic abstraction into physical reality. Rejecting Halley's magnetic nucleus, Walker made the sun central to his terrestrial theories based on a universal circulating fluid conducted through the earth by iron. He pictured the earth's conductive core as a serpentine shape directly reflecting the charts' curved variation lines.[75]

In the early nineteenth century, there were several constituencies of magnetic researchers, including international field workers like Humboldt, laboratory experimenters like Faraday, and naval mathematicians like Barlow. Bonded by their common interest, such investigators continued to develop these techniques of representation. Following Humboldt's adaptation of Halleyan magnetic lines, isogonic mapping became an increasingly popular way of displaying the earth's other physical characteristics. As Faraday explored

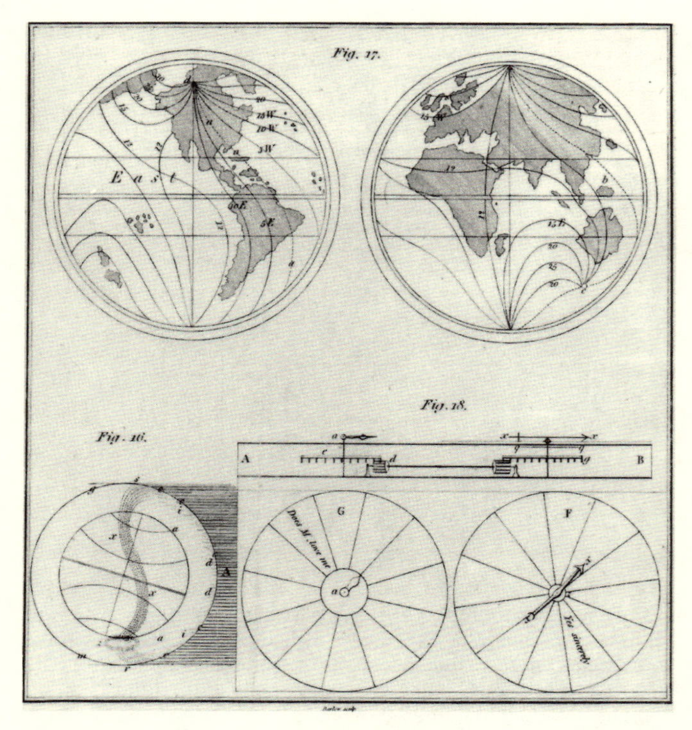

Figure 4.9. Adam Walker's Visualization of the Earth's Internal Magnetic Structure. The top two diagrams show observed patterns of magnetic variation. In the bottom figure on the left, Walker conjectured that the earth's magnetic core curves like a serpentine, reflecting the shape of the mapped isogonic lines. The figure at lower right illustrates Walker's detailed instructions for making divining circles, a deceptive rational entertainment playing on the metaphorical associations between love and magnetic attraction. Walker, *System of Familiar Philosophy*, vol. 1, plate 8, facing p. 68. By permission of the British Library (shelfmark 8706.f.4).

the recently established links between electrical and magnetic effects, he gave new meaning to an older magnetic vocabulary. Building up his novel theoretical vision of fields, he reinterpreted the terminology of curves which Halley and many others had used to describe the patterns of iron filings round magnets and the isogonic lines of variation. Linked by his experimental colleague Barlow to the naval world of the Woolwich Academy, with its maritime tradition of three-dimensional representation and trigonometrical calculation, Faraday extended older visual and verbal descriptive technologies to depict a new space traversed by electromagnetic lines of force.[76]

In addition to this legacy of conceptual approaches to studying and representing the earth's magnetic features, physicists and terrestrial scientists of the nineteenth century also inherited numerous sets of observations, precise measuring instruments, and the basis for an international observational network. Technical innovations transformed magnetic practices: revolutionary methods of paper manufacture and printing enabled charts to be produced more cheaply and in finer detail to a larger scale; new experimental apparatus facilitated more accurate measurements; and faster and safer ships encouraged voyages of exploration. Administrative changes were also vitally important. As Britain's empire rapidly expanded, scientists and cartographers cooperated in geodesic projects with governmental organizations like the navy and the army. In common with other scientific activities, recording terrestrial magnetism was an imperial enterprise.[77]

CHAPTER FIVE

Measuring Power: Patterns in Experimental Natural Philosophy

EIGHTEENTH-CENTURY magnetic experts based their knowledge on everyday experiences. At sea, navigators had acquired their skills not from books but from their daily familiarity with using compasses to guide them round the magnetic globe. On land, natural philosophers also perceived that they inhabited a magnetic world. Examining their built environment, they reported that iron window bars and the crosses on church steeples became magnetic, as did pokers leaning against the fireplace. Visiting workshops, they collected information from craftsmen who found their tools becoming magnetized. Their personal possessions acted as magnetic measuring devices—keys and swords, and also portable compass dials, small boxes with a magnetic needle and a gnomon, for telling the time. Figure 5.1 illustrates the importance of familiar household items: Canton explained that the first step in making artificial magnets involved hanging an iron bar from a poker with a silk thread and stroking it twenty times with the tongs until it could lift a small key.[1]

Encyclopedias and journals included information about magnetic experiments with such readily available objects. They repeated the anecdote recounting how Canton had been inspired by noticing that his fireside poker leant at the same angle as a dip needle. Secrecy, pokers, and stroking came together in a skit on Freemasonry:

> Next for the SECRET of their own wise making,
> HIRAM and BOAZ, and Grand-Master JACHIN!
> Poker and tongs! the sign! the word! the stroke!
> 'Tis all a nothing, and 'tis all a joke

But it was keys rather than pokers which emblematized the knowledge to be gained from magnetic experimentation. In his popular London lectures, Desaguliers demonstrated how a loadstone could "give Power to a great key, to draw away a little key from the Stone it self," an attractive feat undoubtedly replicated by other performers. Well into the century, experimenters continued to use keys as a rough measure of magnetic power. As John Bennett's card illustrates (Figure 2.7), instrument makers marketing their wares to maritime practitioners advertised a loadstone's strength by showing it supporting a heavy anchor. Similarly, in encyclopedias designed for polite audiences, natural philosophers used a key—a domestic item—to symbolize magnetic experimentation.[2] For natural philosophers propagating a Newtonian message, keys delivered additional metaphorical punch. Halley had concluded his Latin eu-

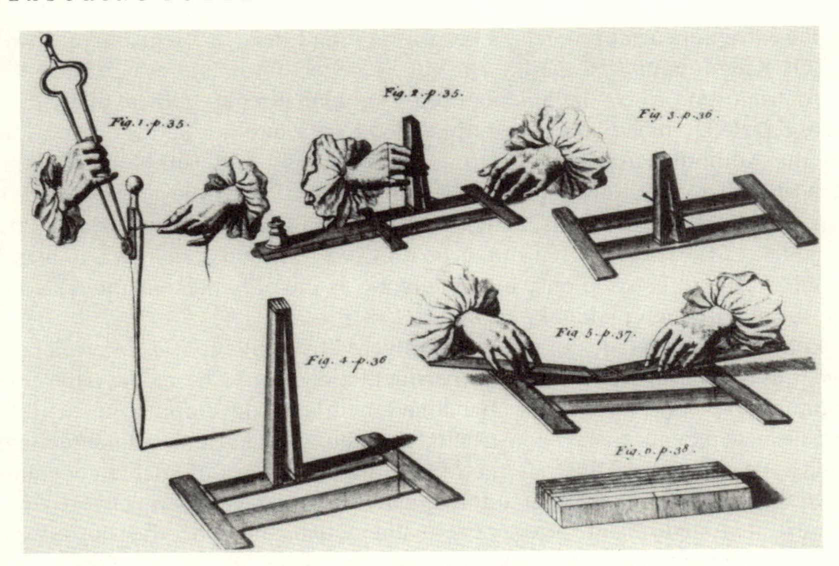

Figure 5.1. John Canton's Method of Making Artificial Magnets. The mathematics teacher John Canton explained his double-touch technique to the Royal Society but declined to perform before such an august assembly. He first magnetized four steel bars with a poker and tongs, then used repeated stroking processes for cumulatively building up their magnetic power. These pictures were widely reproduced in journals, encyclopedias, and texts on natural philosophy. *Philosophical Transactions* 47 (1751), between pp. 34 and 35. Whipple Library, Cambridge.

logy prefacing the *Principia* with an image of Newton unlocking "the hidden treasuries of Truth," articulating the Royal Society's commitment to convert the secrets of nature into public knowledge. Introducing his own magnum opus, Knight declared that studying magnetic effects promised to reinforce Newtonian attraction with repulsion as one of the "Principles [which], as extensive, and as clearly explained, may open a Door to the most secret Mysteries of Nature." Figure 2.10 shows how Abraham Mason incorporated a key in the coat of arms beneath his magnetic perpetual-motion machine; the motto is *Secreta retexit*—Let the secrets be unravelled.[3]

But not everyone agreed that experimental philosophers were the best explorers to venture into this hidden magnetic nature, inevitably gendered as feminine. Johnson, for instance, commented sardonically on Knight's and Canton's activities. Writing in the guise of Hermeticus, an aged philosopher dedicated to retrieving ancient wisdom, he mocked "the wonders every day produced by the pokers of magnetism" and questioned "the value and usefulness of natural philosophy . . . in this age of enquiry and experiment." He also voiced reservations through the mouth of his protégé Williams, the elderly physician whose magnetic longitude scheme had been rejected by Newton and other men associated with the Royal Society. Characteristically scathing about entrepreneurial projects, Johnson regretted that "the legislative Power of this Kingdom invited the Industry of Searchers into Nature by a large Reward proposed. . . . By the Splendor of this golden Encouragement many Eyes were dazzled, which Nature never intended to pry into her Secrets." Petitioning the Admiralty, Johnson—writing as Willliams—flatteringly averred that "the only test of my tables, and of the system on which they are formed, is experience. Mathematicians, mere mathematicians, are apt to be misled by the prejudices of theory. . . . If I might be allowed to propose my own judges, I should desire to be tried only by navigators, as the only persons interested in the success of such undertakings."[4]

Echoing the messages of progress delivered by contemporary rhetoricians, one way of narrating magnetic experimentation would be to describe how natural philosophers gradually abandoned their keys, developing increasingly precise instruments as they constructed a mathematical science of magnetism. Johnson's comments illustrate the misleading simplicity of such an account: natural philosophers were not automatically regarded as experts but strove to acquire that status; and introducing mathematical methods was not always viewed as guaranteeing improvement. Writing from the vantage point of a world dominated by international mathematized sciences, it is tempting to trace a clear path leading to the present from the multitude of past events. When historians choose to accord mathematics a definitive role as the rational framework of science, they inevitably contrast qualitative approaches unfavorably with the forward thrust of sophisticated quantitative techniques. In retrospect, operations such as classifying, abstracting, calculating, enumerating, and measuring can conveniently be swept under a single Whiggish umbrella of quantification.[5] However, Enlightenment writers and practitioners perceived them as separate processes, ones neither solely associated with natural philosophy nor necessarily esteemed as indicators of progress. Many natural philosophers did come to place greater emphasis on systematic ordering, precision measurement, and mathematical techniques, but people questioned their value in various contexts: such changes were not inevitable accompaniments of an increasingly powerful description of nature. This hostility was not merely a reactionary resistance to novelty but was embedded in deep-rooted political, religious, and economic concerns. Specific contests about local practices revolved around rival claims to authority. Their resolution entailed legitimating one value system at the expense of another: in other words, quantification demands a social history.[6]

Seduced by the image of Newton as the heroic founder of modern science, historians have depicted the eighteenth century as the consolidation of his fruitful union of mathematics and natural philosophy. But written in Latin, the *Principia* was theoretically challenging, and the accessible suggestions of the English *Opticks* proved more effective at stimulating investigative research. As far as numerical data collection was concerned, natural philosophers were influenced less by Newton than by Hooke. Eager to establish certainty by replacing fallible human perceptions with objective, repeatable measurements, they followed Hooke's insistence on discovery through experiment, continuing his policy of converting the traditional tools of mathematical practitioners into accurate instruments of mensuration.[7] But in England, this emphasis on precision measurement was not accompanied by the adoption of mathematical techniques. Newton and his exegetes did seek to draw magnetic attraction into the Newtonian fold by deriving a simple force law, but this project petered out. Flicking through natural philosophy texts from the end of the century immediately reveals the strong contrast in approaches between England and continental Europe—unlike French texts, English ones are virtually devoid of algebraic formulae.

As Desaguliers urged the Newtonian troops to rout the Gothic vandals fired by "warm Imaginations," he laid out his plan of campaign. Formulating the goals of many English experimental philosophers, he aimed to show

> that those Properties of Bodies, such as Gravity, Attractions, and Repulsions . . . do really exist, and are made by Experiments and Observations the Objects of our Senses. These Properties produce Effects, according to settled Laws, always acting in the same Manner under the same Circumstances: And, tho' the Causes of those Causes are not known; since we do not reason about these hidden Causes, it is plain that we reject occult Qualities, instead of submitting them in our Philosophy, as the Cartesians always object to us.

Over fifty years later, Cavallo delivered a similar message, informing the readers of his *Treatise on Magnetism* that he had allotted by far the shortest section to theoretical suppositions: he intended to impart "the knowledge of certain facts."[8] However, establishing those facts was problematic, as magnetic phenomena proved difficult to pin down. Experimenters discovered that a loadstone would perform differently on separate occasions, so they could give no accurate measure of how much weight it could support or from what distance it would attract a piece of iron. Maritime travellers reported no clearly discernible regularity in the way a compass needle's orientation depended on longitude. Comparing earlier measurements with their own, natural philosophers concluded that the earth's magnetism varied slowly with time, as well as fluctuating daily, in complex and unpredictable patterns. Faced with a plethora of confusing observations, they concentrated on repeating old experiments with increasingly delicate apparatus. Following Desaguliers's prescription of confirming "settled Laws," they systematized their results into a coherent body of information which they could promote as scientific magnetic knowledge.

Magnetic experiments lacked the spectacular appeal of demonstrations using electrical or pneumatic apparatus. The route to understanding magnetic activity seemed to lie in patient and meticulous observation rather than in dramatic discovery. Interest inevitably flagged: three years after revising his book on artificial magnets, Michell apologized guiltily, "You will perhaps expect now that I should give some account of Experiments made with a Needle I talked to you of some ten months ago, but I am so far from having made any experiments with it yet that I have not so much as finished the Instrument."[9] Other people were more enthusiastic, although not even Knight maintained a lifelong commitment. Some investigators embarked on lengthy projects of mensuration which yielded more accurate information about previously known effects. As well as collating international measurements, they explored the local shifting patterns of terrestrial magnetism. Researchers also analyzed the magnetic behavior of different materials, constructed sophisticated measuring instruments, and developed techniques for improving artificial magnets. By focusing on practical applications—notably navigation—they promoted natural philosophy as a valuable public enterprise contributing to England's commercial improvement.

Johnson was not alone in his mistrust of the increasing authority accorded to experimenters associated with the Royal Society. Seasoned maritime practitioners resented the erosion of their traditional methods of navigation: they accused natural philosophers of imposing precise instruments unsuitable for life at sea and claimed that mathematical expertise was an inappropriate substitute for the skills acquired through experience. The transformations which naval historians celebrate as making navigation scientific not only meant altering traditional practices but also entailed power conflicts between different communities of practitioners. Similar contests were being waged in other walks of life. For instance, skilled physicians preferred to rely on diagnoses informed by the intuition of experience rather than use the new instruments introduced by natural philosophers. Excisemen, on the other hand, became increasingly powerful partly because they did adopt precise measuring equipment for vetting imports. Locally detested, they were also decried as symbols of the state's expanding pervasive reach. Defending the British constitution, Burke argued, "The legislators who framed the antient republics knew that their business was too arduous to be accomplished with no better apparatus than the metaphysics of an under-graduate, and the mathematics and arithmetic of an exciseman." In his rhetorical outbursts against the political innovations of French revolutionaries, Burke frequently cited their policies of surveying land and counting people as further evidence of threats to individual liberty. Profiting from the metropolitan insurance business, experienced financiers concerned to set realistic premium levels ignored the attempts of theoretical classical probabilists to express human judgment and good sense by mathematical formulae; similarly, jurists clung to traditional criteria of adjudication.[10]

Rhetoricians of progress during the eighteenth century admired projects to abstract general laws from individual observations—the goal articulated so clearly by Desaguliers and exemplified by the increasing codification of magnetic experimental knowledge. Enlightenment spokesmen of polite culture claimed that political authority belonged to people capable of producing abstract ideas from the raw data of experience. The clergyman George Campbell, for instance, compared the advancement towards "knowledge of general truths" with the ascension of "people, who, from a low and confined bottom, where the view is confined to a few acres, gradually ascend a lofty peak or promontory." From their lofty vantage point, public men enjoyed a panoramic perspective denied to those less privileged beings, the laborers of limited vision whose individuality became obscured in the larger landscape. Being able to appreciate the view from the top of the hill implied being also at the summit of the social order. Such criteria excluded women, working people, and the uneducated from the republic of taste. Similarly, by demonstrating their ability to derive clear laws from the confusion of the world about them, natural philosophers underpinned their bids for authority. Just as gentlemanly artists distinguished their profession from the trade of mechanical painters, so too natural philosophers stressed their superiority over mathematical practitioners and instrument makers.[11]

This chapter explores various aspects of the multiple transformations collectively labelled quantification by examining magnetic experimentation during the eighteenth century. Through considering these changes as the outcomes of contests for authority, it reveals how natural philosophers augmented their power within society and established themselves as the skilled producers of magnetic knowledge. The first section explores some relationships between mathematics and mensuration, investigating why English magnetic researchers concentrated on collecting more accurate measurements rather than verifying mathematical laws. The second section discusses how experimenters focused their attention on deriving practical advantages from their research. As they redesigned magnetic instruments, found ways of making stronger artificial magnets, and decided which materials were magnetic, they reinforced their public image as commercial benefactors. Thirdly, the chapter describes how natural philosophers refined older observations and repackaged them into systematized magnetic facts. They promoted their texts as treasuries of authoritative magnetic knowledge based on experiments performed with the precision tools of élite specialists.

Magnetic Measurement and Mathematics

As London's instrument trade boomed, the Fellows of the Royal Society chauvinistically congratulated themselves on "the singular success with which this age and nation has introduced a mathematical precision, hitherto unheard of, into the construction of philosophical instruments."[12] But although they valued precise measurement, numerical data collection, and arithmetical and trigonometrical calculations, they generally spurned the algebraic approach which Continental philosophers were adopting. Like "physics" and "science," the word "mathematical" has to be carefully interpreted, because its meaning was rather different from its present one. For magnetic research, the distinctions between the different types of mathematics and their practitioners were particularly relevant. Such subjects as navigation, surveying, and astronomy were traditionally known as mixed mathematics, which also embraced the more recent commercial arithmetic. Emphasis was placed on results rather than on understanding, and instruction books were crammed with arithmetical or trigonometrical rules to be learnt by heart. Such practices were distinct from the activities of the pure or abstract mathematicians, often from a higher social bracket, who were developing new theories in algebra, geometry, and calculus. In addition, philomaths, many of them schoolteachers, contributed to the numerous mathematical journals analyzing standard problems in geometry, Newtonian calculus, and astronomy.

These distinctions played an important role in the dissensions at the Royal Society in 1783–4, when hostile dissidents viciously attacked Banks, recently elected president. Amongst other criticisms, they accused him of packing the Council with men ignorant of mathematics. Contemporary polemicists reported these rows "as a struggle of the men of science against the Maccaronis of the Society," while Banks himself spluttered "that howsoever respectable mathematics as a science might be it by no means can pretend to monopolize the praise due to learning it is indeed little more than a tool with which other sciences are hewd into form." But modern facile accounts of a confrontation between Banksian virtuosi and progressive mathematical scientists erroneously rely on a stable diachronic meaning of "mathematics," and conceal the complex ramifications of these debates. Unlike gentlemanly natural philosophers, many eighteenth-century mathematicians were men who worked for their living: attitudes towards mathematics were politically and socially shaped.[13]

Traditionally, mathematicians and natural philosophers were engaged in contrasting activities and had different objectives. One distinction concerned formulating results: Galileo famously averred that the book of nature is written in the language of mathematics, but Robert Boyle retained the philosophical view, arguing against expressing experimental results mathematically. Aristotelian natural philosophers were trying to understand nature by building causal models, whereas mathematicians were more interested in measuring: formulae were descriptive, not explanatory, while practitioners at the less theoretical end of the mathematical spectrum used their instruments as tools for solving problems. Social status was another important distinguishing characteristic, although the boundaries differed from modern ones. Craftsmen were socially inferior to gentlemanly philosophers, but even at the universities, mathematics was regarded as less prestigious than natural philosophy. These long-standing distinctions affected people's conduct during the eighteenth century.[14]

Constructing magnetism as a science dependent on precision measurement entailed redefining not only the practices of various communities but also the relationships amongst them. Magnetic behavior seemed so erratic that English natural philosophers often reiterated Halley's sentiments that theoretical difficulties "are secrets as yet utterly unknown to Mankind; and are reserved for the Industry of future ages."[15] While insisting that methodical investigation based on precise mensuration was the best strategy for reducing magnetic phenomena to an ordered system of knowledge, they sought to differentiate themselves from mathematical practitioners. To establish themselves as élite purveyors of magnetic knowledge, they widened their separation from men such as instrument makers and navigators with whom they were in close contact and whose techniques they were utilizing. Elite book reviewers commented contemptuously on "mere English mathematicians. . . . however skilled such writers as these may be in the theory, or expert in the practice, of mechanics, yet, when they take upon themselves magisterially to decide upon philosophy in general, they should be checked with a *ne sutor ultra crepidam* [Pliny the Elder: Let the cobbler stick to his last]."[16] In Canton's explanatory illustrations (Figure 5.1), the frilly cuffs and the smooth hands, with their delicate gestures, emphasize that the repetitive strokes of magnetization could constitute an appropriate occupation for a gentleman: manual labor need not imply getting his hands dirty even when using the fireside instruments of a servant.

Experimental replication was extremely difficult, since natural philosophers found that the measured strength of magnets not only depended on their size, shape, and the ore or metal from which they were made, but also varied with time. They explained away results that did not match up to expectation by blaming the weather or "all the Inconveniencies and Disadvantages of a crouded Room." They stressed the importance of individual skill and sent detailed instructions: "Place the piece of Iron upon the palm of your Left Hand . . ." ; "Fixing the poker upright between the knees . . . a piece of sewing silk, which must be pulled tight with the left hand . . ." It was well known that, like loadstone, artificial magnets gradually lost their strength, so rival manufacturers could never be completely confident in their claims of superiority.[17]

Merchants demanded more accurate criteria than keys or anchors for determining a magnet's strength. Instrument sellers listed bars by their length: presumably customers used price as some guide to quality. In 1729, Lord Abercorn produced a set of tables relating the monetary value of a piece of loadstone to its "Character of Goodness." The calculations entailed considerable arithmetical persistence, and the lack of subsequent references suggests that his book was little used. Nevertheless, like Abercorn, people often described a loadstone's strength by the weight it could lift as a multiple of its own weight.[18] Natural philosophers favored two methods for assessing magnetic power. Some of them followed Hooke's technique of suspending a magnet from one pan of a balance and finding the weight which exactly counterbalanced how strongly it attracted another magnet or a piece of iron placed at varying distances beneath it. Halley introduced the deflection method: the angle through which a loadstone turned a needle was recorded for different distances between the loadstone and the needle.[19]

In the first half of the century, Newton and some of his proponents tried to measure magnetic strength accurately for deriving a mathematical law

Table 5.1
Measurements of Magnetic Force up to 1750

Date	*Name*	*Law Derived*	*Method*	*Comments*
1666 & 1680/1	Hooke	None	Balance	Unpublished before 175(some results lost[a]
1687	Halley	None	Deflection	Commented on Hooke's experiments[b]
1713	Newton	Inverse cube	Unspecified	May have used Taylor's results[c]
*1712–*1721	Taylor and Hauksbee	None	Deflection	Ratio altered with distance[d]
*1719	Whiston	Sesquiduplicate (inverse $r^{5/2}$)	Deflection	Tried an oscillation method[e]
1722	Stephens	Inverse cube disproved	Timed movement of needle	Refused publication[f]
*1725–*1745	Musschen-broek	Various/none	Balance	Found several different relationships but no single law[g]
*1734	Desaguliers	Nearly as inverse cube and a quarter	Unspecified	Cited Taylor's results as evidence[h]
*1739	Helsham	Inverse square	Balance	Gave only two measurements[i]
1742	Calandrini	Inverse cube	Deflection	Obscurely published[j]
*1747	Martin	Sesquiduplicate	Balance	Referred to Newton and Helsham[k]
*1750	Michell	Inverse square	Unspecified	Gave no data[l]

Note: Accounts of experiments marked with an asterisk (*) were published in English.

[a] Palter, "Early Measurements," pp. 544–5.

[b] Palter, "Early Measurements," pp. 545–6; de Pater, *Newtoniaans Natuuronderzoeker*, pp. 122–4.

[c] Palter, "Early Measurements," pp. 546–8; Home, "Introduction," pp. 168–9; de Pater, *Newtoniaan Natuuronderzoeker*, pp. 124–6.

[d] Hauksbee, "Account of Experiments"; B. Taylor, "Account of Experiment" and "Extract of Letter."

[e] Whiston, *Longitude and Latitude*, pp. 8–25.

[f] Heilbron, *Physics at the Royal Society*, pp. 75–6.

[g] Musschenbroek, "De Viribus Magnetibus," *Elements of Natural Philosophy*, pp. 205–8, and elsewher

[h] Desaguliers, *Course of Experimental Philosophy*, pp. 17, 30.

[i] Helsham, *Lectures*, pp. 19–20.

[j] Palter, "Early Measurements," pp. 551–8; de Pater, *Newtoniaans Natuuronderzoeker*, pp. 179–81.

[k] Martin, *Philosophia Britannica*, pp. 38n-39n.

[l] Michell, *Artificial Magnets*, p. 19.

describing how it diminished with distance (see Table 5.1). Newton himself had been concerned to refute the magnetic structure of the universe proposed by Gilbert and elaborated by Johannes Kepler. As he constructed his own cosmology, Newton supplanted magnetic attraction by gravity. Revising his opinions about the sun's magnetic control of cometary orbits, he engaged in a lengthy written debate with the Astronomer Royal, John Flamsteed.[20] Gaining mastery over these astronomical visitors was not just of theoretical significance: successfully predicting their arrival helped the new mechanical philosophers displace the authority of traditional astrologers. Deploying comets as weapons against a Cartesian interplanetary ether, Newton and Halley also converted them into essential guarantors of the stability of the solar system and the scriptural history of the earth. In 1685, Halley was still telling his international colleagues that gravitational and magnetic attraction did not seem so very different. But the following year, as part of his prepublication promotion of Newton's *Principia*, he scathingly remarked, "Some think to Illustrate this *Descent* of *Heavy Bodies*, by comparing it with the Vertue of the *Loadstone*; but . . . this Comparison avails no more than to explain *ignotum per aeque ignotum* [the unknown by the equally unknown]"—precisely the accusation that would be levelled against gravitational attraction by many of Newton's critics.[21]

When it suited his arguments, Newton bracketed magnetic and gravitational attraction, but in other places he insisted they were different. Unlike gravity, he declared, magnetic attraction neither depends on the quantity of matter nor follows an inverse-square law. In the first edition of the *Principia*, he claimed that magnetic force diminishes more rapidly than the second power of the distance, and his supporters attempted to garner quantitative experimental support.[22] As Halley steered the *Principia* through the press, he commented disparagingly on Hooke's experiments, intended to find similarities in the decrease of gravitational and magnetic forces with distance. Halley carried out several experiments of his own but declined to derive any force law.[23] In 1712, working on revisions for the *Principia*, Newton suggested a project to measure magnetic power, which, he hinted heavily, "he believed would be nearer the Cubes than the Squares." Brook Taylor and Hauksbee, enrolled as Newton's Royal Society assistants, did carry out experiments, but their results were inconclusive, and publication was delayed for several years. Despite this, in the new edition of 1713, Newton confidently distinguished gravitational and magnetic attraction:

> The power of gravity is of a different nature from the power of magnetism; for the magnetic attraction is not as the matter attracted. Some bodies are attracted more by the magnet; others less; most bodies not at all. The power of magnetism in one and the same body may be increased and diminished; and is sometimes far stronger, for the quantity of matter, than the power of gravity; and in receding from the magnet decreases not as the square but almost as the cube of the distance, as nearly as I could judge from some rude observations.[24]

But many of Newton's successors, less interested in perusing the *Principia* closely than in consolidating a Newtonian orthodoxy, ignored this pronouncement (see Table 5.1). Guided by their conviction that a simple law existed, some of them extrapolated readings or proclaimed an inverse-square law with virtually no supporting evidence to back their claims.[25] The Royal Society placed stringent requirements on hopeful contributors. When the Fellow William Stephens sent in results from Dublin which apparently contradicted Newton's inverse-cube relationship, James Jurin suppressed the account, allegedly in "regard for your Reputation." He made numerous detailed criticisms of Stephens' experimental protocol: "You do not say, whether your stone was arm'd or naked, nor whether you measured your distances from y[e] Centre, or surface of y[e] Stone, or from y[e] surface of y[e] armour." Musschenbroek performed by far the most extensive investigations. His varying results, widely disseminated in England, seemed to preclude the possibility of formulating any single law, and English natural philosophers like Desaguliers echoed his recommendations of theoretical neutrality.[26]

From the middle of the century, English natural philosophers abandoned the search for a magnetic-force law. Showing little interest in such investigations being continued on the Continent, they focused on packaging their magnetic expertise for public consumption. They concentrated on topics which were both profitable and demonstrably useful, such as improving navigational safety by recording terrestrial magnetism, constructing better instruments as a contribution to Britain's international trade and reputation, and advising steel manufacturers on production methods. Engaged in Kuhnian projects of normal scientific activity, magnetic researchers carefully reexamined existing observations and systematized their results into paradigmatic exemplars which they could advertise as solid knowledge. They repeated experiments with increasing care and precision, investigated conflicting results, and explored the usefulness of theoretical analogies.[27] Relying on accurate mensuration to clarify uncertain points, they recorded the temporal and spatial patterns of terrestrial magnetism and experimented on magnetic bars. This second category of investigations included ascertaining if iron changed in weight after being magnetized; exploring the relationship amongst the weight, dimensions, and strength of a magnet; examining whether the relative attractive powers of a magnet's two poles depended on which hemisphere it was in; finding under what circumstances repulsion could change to attraction; comparing the efficacy of different techniques for making artificial magnets; and examining to what extent arming a magnet increased its strength.

Although they promoted their sensitive instruments for collecting accurate numerical data, natural philosophers did not introduce mathematical methods into their accounts. This rejection stemmed from a variety of origins. For authors trying to market their texts to polite readers, formulae seemed likely to diminish a book's appeal. Cavallo explicitly excluded mathematical demonstrations from his *Treatise on Magnetism* because they "were incompatible with the genius of the generality of readers." Similarly, for didactic purposes,

Young intentionally reversed the English preference for analytical methods (arguing from effects to causes) to the synthetic approach of proceeding from general principles. But simplicity of explanation is itself too simple an explanation: the philomaths' contributions to journals like the *Ladies' Diary* had reportedly resulted in "no inconsiderable degree of useful mathematical knowledge being more widely diffused in England than in any other country in the world," yet the articles on magnetic topics appearing in the *Philosophical Transactions* and other specialized works were no more mathematical than those written for public consumption.[28]

Mathematizing natural philosophy was bound up with cultural issues, some of which were specific to England. Religious allegiances underpinned nationalistic expressions of an aversion to the mathematical techniques being developed in France. In his influential *Light of Nature*, favored by the prominent natural theologian William Paley, Abraham Tucker explained that just as experiment was preferable to hypothesis for learning about the natural world, so too everyday experience should provide the basis for moral knowledge. He articulated English distrust of the French tactic of "drawing us aside from the plain road of common sence into the wilds of abstraction."[29]

Diverse educators cited John Locke's endorsement of mathematical thought as a technique for training the mind. So teachers from puritanical backgrounds fostered the mental discipline needed to acquire mathematical facility but warned against the risk of taking too much pride in expertise. This ideology affected students of mathematics and navigation in the dissenting academies, which stressed the utilitarian value of their syllabuses. At Cambridge, as the tripos became entrenched as a quantified adjudicator of merit, mathematical subjects replaced the logic still being taught at Oxford to encourage mental dexterity. But substantial numbers of lecturers as well as students objected to this transition, protesting that mathematics took time away from classical studies, precluded an understanding of the basis of natural philosophy, and led people to cavil at revelation. Instructors continued to emphasize the value of mathematics for practical projects and for strengthening the rational faculties. These attitudes pervaded nineteenth-century teaching at both Cambridge and the newly founded London University. Unlike in France, where mathematics was a specialized pusuit of an educated élite, English reformers viewed mathematics as playing an important role in a broad liberal education. The difference strongly affected the adoption of French mathematical techniques in England.[30]

One obvious reason for avoiding mathematics is lack of ability. The chemist Bryan Higgins ambitiously suggested that a substance's magnetic power could be computed by compounding the forces of all the polar atoms it contained, but decided to leave this task for someone else. Similarly, Wilson explained to Aepinus that his mathematical treatment of magnetic and electrical phenomena was received unfavorably in England because "[t]he introducing of algebra in experimental philosophy, is very much laid aside with us, as few people understand it; and those who do, rather chose [*sic*] to avoid that close kind of attention."[31] But Wilson's specification of algebra, rather than mathematics, hints at the complexity of cultural attitudes shaping the mathematization of English magnetic texts. Different practices now bracketed together as mathematical—such as enumeration, geometrical proof, algebraic manipulation, and arithmetical calculation—were each laden with its own religious, social, and political connotations. Appreciating such distinctions is essential for understanding how natural philosophers formulated the new science of magnetism at the turn of the century.

Consider, for instance, John Robison, the Edinburgh professor of physics whose activities provide a singularly vivid example of how the new sciences served political and religious interests. Himself the author of a vehement tract blaming the French Revolution on the machinations of concealed Masonic conspirators, Robison was also responsible for the revised articles on natural philosophy in the 1801 supplement to the third edition of the *Encyclopædia Britannica*. Under Scottish direction, this supplement was dedicated to counteracting "the seeds of Anarchy and Atheism" allegedly disseminated by "that pestiferous work" the French *Encyclopédie*. Although Robison criticized English natural philosophers who shared Wilson's willful ignorance of Aepinus' *Tentamen*, his insistence that they adopt mathematical methods was not a blanket endorsement. Trained in the Scottish commonsense school of philosophy, he vaunted his own geometric analysis of the forces between pairs of magnets and pieces of iron, the approach further developed by Faraday's circle in the nineteenth century. Like other Tory–High Church men of this period, he rejected French algebraic methods, which writers often described pejoratively with chauvinistic attributes similar to those used for denigrating the French themselves—flowery, superficial, ostentatious. Robison used magnetic examples to advertise his treatment of dynamics, further castigating French mathematicians by reminding his readers of the sordid commercial roots of algebra and arithmetic: French algebraic manipulation, he sneered, resembled "the occupation of a banker's clerk when he carries his eye up and down the columns of pounds shillings and pence, calculates the compound interest, reversionary values, &c. . . . this total absence of ideas exposes even the most eminent analyst to frequent risks of paralogism and physical absurdity."[32]

Some of Robison's views were extreme, but he was certainly not alone in combining a hostility towards algebra with a fear of revolution—political as well as philosophical. Adams, for example, was inspired to revise his book of lectures when he read French "tracts hostile to good order." He slated Aepinus for his "mathematical theory on electricity, [which] has closed the door on all our reserches into the nature and operations of this fluid." He challenged *philosophes* like Condorçet, "who denies the existence of a God, because he cannot perceive him with his corporeal eyes," to explain the invisible mystery of magnetic attraction.[33] Because politics, religion, and natural philosophy were so closely intertwined, they need to be considered together. The polemical injunctions about mathematizing natural philosophy enunciated by men like Adams and Robison were rooted in the same concerns about traditional values, British religion, and a hierarchical social order which drove Burke to condemn French geometrical surveyors or British excisemen. He denounced Richard Price, the pro-revolutionary Nonconformist minister who backed political arithmetic, as a "calculating divine"—a condemnation oxymoronic in its force. "The country of Newton has not exhausted itself," boasted Young, Burke's protégé at the Royal Institution. "The algebra of invention, which Dr Hooke proposed to form into a science . . . has enabled them not only to keep pace with the complicated efforts of their contemporaries in other countries, but in many important instances completely to anticipate them."[34]

Patterns of Experiment

While keys symbolized access to magnetic knowledge, the patterns of iron filings ranged round a loadstone became iconic of the mystery surrounding nature's magnetic power. As illustrated in Figure 4.7 (Fig. 570), natural philosophers frequently displayed these patterns in didactic texts and lectures. But in their own research, experimenters concentrated on accurate mensuration and data collection as they sought answers to two major questions: how does terrestrial magnetism vary, and what are the characteristics of magnetic materials? Their investigations of terrestrial magnetism were tied to maritime improvement and financial benefit; their research into magnetic materials similarly revolved around practical issues, such as making better magnets and determining which substances are magnetic.

Natural philosophers tried to order the apparently capricious behavior of magnets and compasses into neat patterns. Measuring the earth's magnetism was not initially a distinct topic of investigation but formed one aspect of attempts to study the entire globe as a coherent entity. As the century opened, magnetic researchers were unsure what factors affected the perpetual fluctuations they observed, a problem similar to that posed by such meteorological indicators as temperature and barometric height. They tested numerous possibilities, including air pressure and humidity, the time of day, the season, and location on the surface of the terraqueous globe. During a voyage to America, for five months Middleton took daily observations of the temperature, barometric pressure, magnetic variation, wind, and weather. English investigators who spoke Latin could learn about Musschenbroek's three-year meteorological project, in which he kept a daily record of many natural phenomena and tried to relate magnetic dip and variation to the amount of wind and rain as well as to the time of year.[35]

Experimenters also explored the terraqueous globe by modelling it. They continued to assess competing theories of terrestrial magnetism by moving small compass needles over the surface of spherical terrellae, the miniaturized globes recommended by Gilbert for mimicking magnetic nature.[36] Instrument makers sold spheres fashioned from solid natural loadstone or hollow globes with a small internal magnet, either fixed or movable. When used for teaching, these didactic devices corroborated textbook illustrations of magnetic dip with gratifyingly symmetrical patterns, even over the polar regions, where no measurements had been made. By altering the central bar's location and orientation, instructors could draw more detailed comparisons with the patterns displayed on magnetic charts. These three-dimensional representations provided an important foundation for the electromagnetic theories developed by Faraday and his circle during the nineteenth century.[37]

Natural philosophers also introduced new ways of modelling magnetic nature. A few people made elaborate mechanical devices incorporating rotating terrellae, designed to test Halley's theories quantitatively.[38] However, there was greater interest in symbolic representations. For example, to demonstrate that changes in the magnetic variation are governed by the sun, Canton replicated solar heat with boiling water. He placed a magnet on either side of a compass and then observed how the needle moved as the sun rose—by pouring boiling water on the eastern magnet—and then set—by heating the western magnet. Cavallo extended this experimental methodology to imitate environmental fluctuations affecting terrestrial magnetism. He observed the movements of a compass needle surrounded by four earthenware pots as he altered their contents by adding boiling water to a magnet or acid to iron filings.[39]

As in other global studies—seismological, astronomical, and meteorological, for instance—natural philosophers relied on older observations for analyzing long-term changes, and on travellers' reports for spatial differences. They could, however, conduct their own investigations of local patterns, for which they developed specialized instruments. At the Royal Society, a succession of experimenters undertook protracted series of measurements, paying growing attention to the replicability and accuracy of their results.[40] They instituted precautions to reduce error, such as repeating experiments in different places to guard against readings' being affected by nearby iron. Natural philosophers and instrument makers cooperated to produce apparatus which would yield more precise measurements and which could be profitably marketed.

At first, land-based natural philosophers used equipment for measuring variation and dip similar to that which navigators were taking on oceanic voyages. By the early nineteenth century, specialized instruments had been developed to serve the varying requirements of distinct groups of users. This differentiation was analogous to the contemporary evolution of telescopes into small portable ones for use at sea and giant stationary astronomical ones, or to the types of bicycle in the nineteenth century. Examining such alterations in material objects yields valuable insight into shifts in social relationships. Technological changes cannot simply be judged as progress, since people's conflicting interpretations of artifacts mold their eventual forms. The compass sensitivity condemned by navigators was precisely the characteristic which appealed to natural philosophers seeking accuracy: people held different criteria for defining a "better" magnetic instrument.[41]

The modern concept of a scientific instrument was constructed only during the nineteenth century. Before then, as exemplified by the trade card reproduced as Figure 2.7, instrument sellers expanded their traditional role as purveyors of mathematical instruments by adding philosophical and optical equipment to their stock. They retained the tripartite distinction amongst mathematicians' tools for solving practical problems, philosophical apparatus for investigating and controlling nature, and optical devices for enhancing natural vision.[42] Investigating alterations in how they marketed magnetic apparatus during the eighteenth century reflects the multiple transformations taking place as natural philosophers constructed the new science of magnetism.

Traditionally, instrument makers categorized compasses and loadstones as mathematical instruments, ones sold mainly to navigators. Even the new artificial magnets were initially listed as mathematical instruments.[43] But towards the end of the century, the structure of London's instrument trade changed as demand rose steeply. While small waterside shops specialized in such nautical goods as compasses, general retailers with extensive catalogs catered for a variety of purchasers. They started to reallocate their magnetic products: navigators requiring steering or azimuth compasses would find them classed as mathematical instruments, while natural philosophers and the polite consumers of rational entertainment could buy magnetic kits and variation compasses listed as philosophical instruments. During the nineteenth century, still clearer boundaries were constructed between three distinct groups of instruments used by the separate communities of surveyors, navigators, and magnetic scientists.[44]

The instruments for measuring variation and dip changed in different ways during the seventeenth and eighteenth centuries. Dip needles essentially retained their original appearance but came to be used almost exclusively by natural philosophers. In contrast, by the end of the eighteenth century both navigators and natural philosophers were measuring variation, but with specialized compasses which looked completely different from each other.

Magnetic dip was first investigated at the end of the sixteenth century by Robert Norman, an experienced seaman and nautical instrument maker. Figure 5.2 shows Gilbert's illustration of Norman's dip needle, crucial for this natural philosopher's experimental justification of his magnetic cosmology.[45] During the eighteenth century, people recorded and wrote about dip less frequently than about variation. Maritime observers found dip hard to measure accurately, because they had to align sensitive needles with a magnetic meridian: on land the meridional direction could be marked, but at sea the ship was constantly moving. Halley set an influential precedent by not charting dip on his oceanic voyages, perhaps because his theory could not account for its terrestrial pattern. He sympathized with Derham, who complained that his

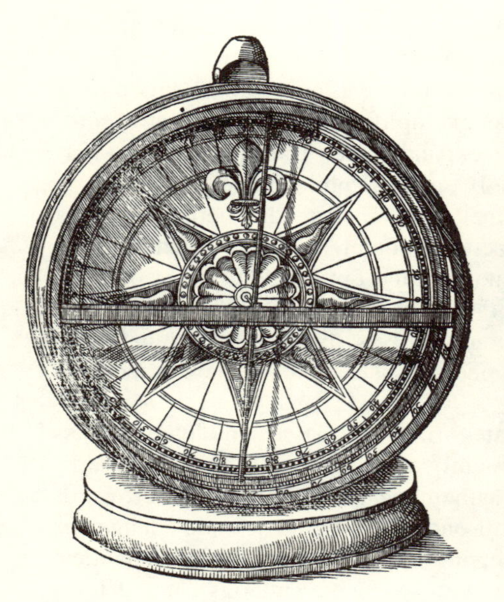

Figure 5.2 (*left*). William Gilbert's Illustration of a Dip Circle. William Gilbert used instruments like the dip circle—invented by the mathematical practitioner Robert Norman—to advertise the power of a new natural philosophy based on observation and measurement. Gilbert instructed readers how to suspend an iron bar, to be magnetized with a loadstone, on a circular planed board at least six finger lengths in diameter. Gilbert, *De Magnete*, p. 277. Science Museum, London.

Figure 5.3 (*right*). Dip Circle of Edward Nairne and Thomas Blunt. One of London's most prestigious international instrument makers, Edward Nairne was a Fellow of the Royal Society, where he reported on his magnetic experiments. His apprentice Thomas Blunt later became his business partner. In this model of around 1775, the steel needle is 12 inches long (30 cm), encased in a brass surround fitted with two spirit levels. The silver scale is divided to 20′; the small arcs near the expected reading are for preventing errors of parallax. Science Museum, London.

own experiments were hampered by the lack of a good instrument.[46] Whiston, on the other hand, committed himself to measuring dip in the hope of winning some longitude money. Pointing out that making the needle longer was one way of increasing accuracy, he travelled round England and France with needles several feet long and compiled maps showing lines of equal dip. He did persuade four mariners to take his cumbersome apparatus overseas, but they had little incentive to devote the attention needed for obtaining readings.[47]

But on land, natural philosophers compiled records of how the angle of dip varied with different times of day and also from year to year. Interested in measuring small local fluctuations, they modified Norman's design to yield more precise measurements. Eliminating the maritime decorative rose, they converted his wooden dip circle into a glazed instrument of natural philosophy crafted from fine metals (Figure 5.3). Equipped with various devices for increasing accuracy—makers paid particular attention to reducing friction at the needle's suspension—the scales were finely divided to read fractions of a degree. The Royal Society commissioned explorers to collect international measurements: the Navy Board paid for Cook's astronomers to take dip needles on his 1772 voyage. But these delicate instruments were unsuitable for use at sea. Stationed in Hudson's Bay, Thomas Hutchins regretted with masterly understatement that his experiments "could not be done with that exactness I could wish." Struggling for hours to align his dip needles, he explained: "These observations were made on a large piece of ice, to which the three ships were grappled. . . . a breeze springing up gave the ice a circular motion, which made it impossible to keep the instrument exactly in the magnetic meridian." As London instrument sellers marketed sensitive dip circles all over the world, they became essential apparatus for the well-equipped natural philosopher but continued to be rejected by navigators.[48]

As discussed in Chapter 3, navigators were also skeptical about the azimuth compasses introduced by land-based natural philosophers like Knight and Smeaton. Towards the end of the century, an increasing number of specialized compass inventors, many of them with nautical experience, concentrated on designing azimuth compasses more suited for navigators. For example, Figure 5.4 shows the compass designed by Ralph Walker, which was adopted by the Royal Navy and widely recommended in navigation manuals. One of his major innovations was the sighting device promoted as making the compass easy to use—even during hazy weather—and eliminating complex calculations; he thus claimed to answer maritime criticisms of earlier models.[49]

Natural philosophers demanded different facilities from their variation compasses. They already knew the local variation fairly accurately and were not troubled by problems of robustness and stability. They wanted compasses which would provide extremely precise measurements of small changes of variation. In the 1720s, the clockmaker George Graham built delicate instruments for his project of recording diurnal changes in terrestrial magnetism over several months. He narrowed the angle through which his variation needles could move and encased them in a protective rectangular box. Knight introduced further modifications, including an integral microscope to read the scale more accurately (Figure 5.5). By mid century, Knight's model was being marketed to natural philosophers at Adams' shop. Climbing the Alps, Henri de Saussure was forced to rebuild the granite pedestal he described as an altar for his Knight variation compass, since the ice was constantly moving. In London, untroubled by such problems, Canton used this type of instrument when he measured the magnetic variation several times a day for almost two years. From nearly four thousand readings, he concluded that, apart from irregular fluctuations linked with the aurora borealis, the variation altered systematically with the time of day and the time of year.[50]

Instrument makers and natural philosophers continued to contribute their various skills to the construction of magnetic apparatus, making it increas-

Figure 5.4. Ralph Walker's Variation Compass for Navigators. Walker returned from Jamaica in about 1792 to present his compass design and magnetic theories to the Longitude Board. The Board awarded him £200 for the compass, which was made by George Adams. Although the elaborate sighting mechanism is the most obvious feature, the decorated rose also makes it apparent that this is an instrument for navigators, not natural philosophers. This interwar replica is now at the Greenwich Maritime Museum, London. Courtesy of Antony Fanning.

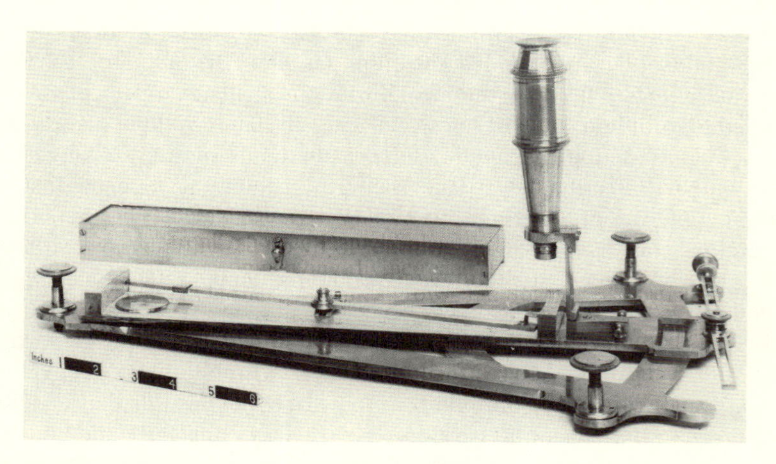

Figure 5.5. Gowin Knight's Variation Compass for Natural Philosophers. George Adams made and sold this variation compass, invented by Knight, around the middle of the century. Unlike in navigation compasses, the needle moves through only a small angle. It includes two vernier scales, a screw for making fine adjustments, and a magnifying viewer. The ten-inch bar magnet has a brass counterbalance slide to compensate for the effects of dip and is suspended on a steel pivot with a chalcedony cup. Science Museum, London.

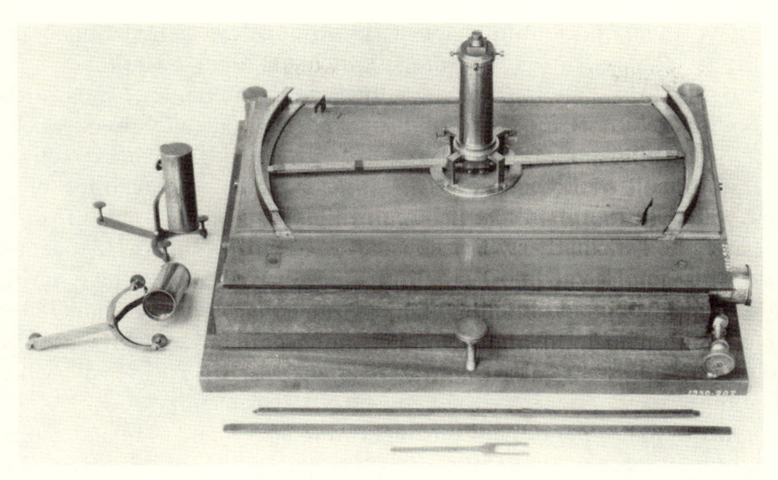

Figure 5.6. Variation Compass Belonging to Henry Cavendish. By the end of the century, international observers were using sensitive magnetic apparatus to record patterns of terrestrial magnetism throughout the globe. An experimental station—completely free of iron—had been built in Sumatra, equipped with the latest literature and instruments and with special wooden candlesticks for nocturnal measurements. This unsigned instrument has a twelve-inch bar magnet on a unifilar suspension. The scales, divided to 10′, are read by independent microscopes. Science Museum, London.

ingly sensitive for conducting international mensuration projects. Henry Cavendish, renowned as a brilliant yet reclusive aristocrat, supervised magnetic measurements during a ten-year meteorological investigation based at the Royal Society. Although he did not publish much of his work, he invented techniques for analyzing mathematically the errors associated with magnetic observations. Collaborating with the acclaimed instrument maker Edward Nairne, he further modified Knight's variation compass to yield measurements of an even higher degree of accuracy. For instance, he used a telescope for aligning the instrument with the meridian and designed a needle which could be inverted for taking the average of two sets of measurements.[51] Cavendish owned the later model illustrated in Figure 5.6, so strikingly different in appearance and performance from the contemporary variation compass for mariners in Figure 5.3.

Natural philosophers advertised their sophisticated instruments as internationally marketable commodities attesting to the fruitful cooperation of commerce and natural philosophy in promoting British interests. As they consolidated and refined older magnetic knowledge, they searched for ways of making their terrestrial measurements still more accurate. Their instruments became suitable only for skilled specialists able to take the elaborate precautions required to prevent their reacting to vibrations or air currents. The Reverend Abraham Bennet suspended his needle by a spider's thread, checking his readings by twisting it eighteen thousand times with a flax-spinning spindle.[52]

Whereas the content of scientific magnetic knowledge was being publicly disseminated, its production was becoming restricted to a narrow élite.

Experimenters also developed precise measuring techniques for assessing a substance's magnetic properties. This research was framed by practical considerations, mainly assessing a mineral's value and designing more reliable instruments which could be marketed. At the end of the seventeenth century, loadstone was used by miners for distinguishing between qualities of iron ore and by natural philosophers for examining mineral specimens. Mechanical philosophers explored the effects of physical operations, such as hammering, heating, and twisting iron.[53] Practitioners gradually replaced these qualitative tests by more precise criteria. Natural philosophers and craftsmen needed to find the most appropriate steel for magnetic needles and to ascertain which substances were impervious to magnetic effects, since only these could be used for making instrument casings. Miners were concerned to relate an ore's magnetic properties to the amount of iron it contained, since this affected its monetary value.

One outstanding Enlightenment cliché is the taxonomic urge, a rationalizing fever supposedly overwhelming Europeans engaged in a gamut of intellectual activities. But far from being a process of academic systematization, classification involved conflict amongst communities. People produced diverse schemes for furthering their own material welfare as well as for substantiating epistemological positions.[54] Mineralogical classification was bound up with miners' attempts to improve yields and with chemists' bids to boost their low status. In England, practitioners did not immediately adopt the taxonomies developed for pedagogic purposes at Abraham Werner's Freiberg school, but assessed rival schemes for their usefulness. Their debates were shaped by alternative models of mineralogical practice: people who embraced Enlightenment ideals of an open community reliant on simple apparatus opposed the creation of a group of experts trained in sophisticated experimental techniques.[55] English investigators of magnetic materials were not interested in taxonomic schemes for their own sake but were seeking to answer practical questions: how to identify the best loadstone and iron ores, and which substances are magnetic. Classifying magnetic materials entailed commercial decisions.

People living in the early eighteenth century did not draw the modern distinction between animate and nonanimate matter. They adhered to older Neoplatonic and Aristotelian interpretations of the natural world as a fixed continuous hierarchy linking human beings downwards through simple animals and plants to stones; solidified by lapidifying fluids, minerals grew like plants. Definitions could have important financial consequences. For example, a disgruntled mid-century lead miner recounted how forty years earlier a court witness had defended clerical taxation of miners' earnings because "the Ore grew (not in the Sense as we mean, when we speak of the Generation of Mettles) but after the manner of other Vegitables, as we see Annually on the

Surface of the Earth . . . and this was sufficient Reason enough to make it Titheable, as indeed it now continues in some Places."[56]

Early commentators were less concerned with the nature of loadstones than with their appearance and value, describing them—along with many other found objects—as stones or "marvellous Fossils." John Woodward, investigator of natural and artificial antiquities, declared that gentlemen should take over fossil classification from "Miners and meer Mechanics" so that they could benefit from England's subterranean wealth. Claiming that Continental research was stultified by state appropriation of discoveries, he directed profiteers to the "large Masses of good Loadstone found on Dartmoor."[57] Loadstone was reported to grow from petrified wood, while evidence from church crosses demonstrated its regeneration from iron. As Knight explained, "Iron Mines are pregnant with Loadstones not yet come to Maturity, and may in Time become Mines of good Loadstones." Natural theologians argued that this regrowth of minerals, one of England's most lucrative natural resources, demonstrated divine design: "Minerals, Metals and Stones . . . have Power of *growing* in their respective Beds: That as the Beds are robbed and emptied by Miners, so after a while they Recruit again." To enable native loadstone to compete with stronger foreign varieties, it needed to be nourished with iron filings as though it were a vegetable crop.[58]

When mineral experts introduced new taxonomies in the second half of the eighteenth century, they paid scant attention to loadstone, which was no longer valued either for cabinets or for navigation. If included at all, loadstone was rather ambiguously listed under iron. Although steel producers referred to an ore's magnetic strength, natural philosophers did not include magnetic behavior amongst the set of external characteristics recommended by Wernerian specialists for describing specimens.[59] But around the turn of the century, natural philosophers constructing the new science of geology promoted studies based on chemical methods of analysis. They started to quote oxygen content as a quantitative criterion for distinguishing types of magnetic ore.[60] This enabled mineralogists to cite an ore's magnetic strength as a numerical guide to its richness, and so they adopted magnetic behavior as a standard descriptive measure.[61]

Natural philosophers also tried to use accurate measurement for deciding whether other materials besides iron and steel were magnetic. Contrary to their public rhetoric, natural philosophers found that increasing precision did not guarantee answers to their questions. Brass, for instance, exhibited magnetic behavior, but inconsistently. Secluded in his Norfolk home, Arderon experimented for months: "I have tried to communicate the Magnetick Vertue to hamered Brass, & I find, I can, enough to prevent a good Needle from traversing well, & as the Compass is of such great use, I think no Compass Box should be made of this metal." Acting as his gentlemanly patron, Baker encouraged Arderon to continue his investigations: resolving this issue was not only important for navigation and philosophical instruments but could also affect the price of metals on the stock market.[62] Cavallo felt the problem was of such interest to his contemporaries that he discussed it in three of his thirteen annual Bakerian lectures at the Royal Society.[63] He gave detailed instructions for making brass, explaining that he had "described this operation in so particular a manner for the sake of those workmen, who want brass free from magnetism for the construction of certain philosophical instruments." But although experimenters used increasingly sensitive equipment and carefully monitored the composition of various alloys, no consensus had been reached by the end of the century. Cavallo found that, with his sophisticated measuring techniques, he was unable to replicate the results obtained by a locksmith making alloys by rule of thumb.[64]

Other substances also defied quantitative analysis, for instance the black magnetic sands which promised a lucrative alternative to iron ore. As instruments became more sensitive, tests indicated that gold and platinum might be magnetic. In 1763, Baker wondered if "perhaps no powdered Mineral is intirely destitute of Particity—will adhere to the Loadstone, tho' speaking comparatively and at large one would say, that such Mineral is not at all Magnetical."[65] At the end of the century, this dilemma was compounded by reports that Continental researchers had found other metals to be magnetic and claimed they could make nonrusting compass needles from cobalt and nickel. Citizen Coulomb claimed to have demonstrated experimentally that "all Bodies, whatever may be their Nature, are obedient to the Action of Magnetism, and that this Action is sufficiently powerful to admit of being measured."[66] But some English investigators were skeptical. They argued that since magnets were such sensitive detectors of iron, an experimenter could not distinguish whether another material—nickel, for example, or iron pyrites—was inherently magnetic or contained minute particles of iron. They asserted that magnets, the traditional diagnostic tool, still provided a more reliable test for traces of iron than did chemical analysis.[67] Natural philosophers held conflicting opinions on whether experience or expertise should guide their conduct. But although they privately disagreed about the epistemological limitations of mensuration, they maintained their public image of experts skillfully controlling precise instruments.

Outlining Magnetic Power

By examining communities of practitioners and their texts and the audiences for whom they were designed, historians can examine how the consolidation of scientific knowledge relates to power struggles in society. Instead of colluding with rhetorics of scientific progress, we can retrieve the muted voices of those who opposed the replacement of traditional skills. Through interpreting change as the outcome of conflict, we can represent the past not solely as the necessary prologue to modern practices but as an active scene including diverse participants. Natural philosophers sought to convince people of the validity and usefulness of their approach to nature. By persuading those who were not themselves participating in the production of magnetic facts of the legitimacy of their experimental program, they established themselves as the authoritative custodians of magnetic knowledge.[68] In particular, natural philosophers negotiated with naval practitioners for control of magnetic expertise. They increasingly asserted their control over navigational affairs and simultaneously incorporated maritime practices into their own texts and instruments.

In a mutually profitable promotional campaign, instrument makers, natural philosophers, and booksellers cooperated to endorse and advertise each others' wares. Natural philosophers used a variety of media to present themselves as magnetic experts. In the second half of the century, encyclopedias, books, and lecture courses on natural philosophy routinely came to include sections on magnetic experiments. Articles appearing in the *Philosophical Transactions* were summarized in a wide range of journals of varying degrees of erudition. These also carried promotional features, lightly disguised as editorial material, on magnetic innovations—methods of making artificial magnets, new compass designs, proposals for solving the longitude problem. Naval educators associated with the Royal Society wrote navigation manuals endorsing the techniques and instruments of the Fellows. To complement these self-vaunting tactics, natural philosophers criticized magnetic texts written by maritime authors and compared them unfavorably with those written by their colleagues.

Although natural philosophers were producing an increasing amount of printed information about magnetic experiments, little of it was substantially new. The savants who enthused about Knight's ability to manipulate magnetic poles with his new bars could have learnt by glancing at a popular encyclopedia that Boyle had performed the same feat with a piece of loadstone.[69] Virtually every text purveying magnetic knowledge described Halley's seventeenth-century theory of terrestrial magnetism, which was repeatedly tested, questioned, and refined. Even those who disagreed with his model agreed that no satisfactory replacement had been found.[70] Over two-thirds of Cavallo's didactic book described laws and experiments which were in essence no different from those of far earlier experts such as Boyle and Musschenbroek; in his own research, Cavallo thought it worthwhile to repeat experiments described by Kircher almost 150 years earlier. In his short section on new experiments, he focused on topics important for practical applications, such as suspensions for compass needles, modelling terrestrial magnetism, and using magnets to test for iron.

Authors made their magnetic information appear authoritative by laying it out as a set of laws, presented as a paradigmatic exemplar of established factual knowledge. For example, Whiston's 1721 codification was reprinted in its entirety in encyclopedic texts right up to the end of the century.[71] From time to time, writers modified these sets of magnetic laws by incorporating quantitative or more detailed observations. A clear demonstration of this process of fleshing out an existing corpus of experimentally determined knowledge is provided by comparing the first edition of Chambers' 1728 *Cyclopædia* with Abraham Rees' 1778 updated edition. The 1728 entry included a numbered list of thirty-six magnetic phenomena. In the revised version, exactly the same list was retained, although about a third of the items were expanded.[72] For instance, far more precise instructions were given for locating the poles in a bar magnet, accounts were included of Musschenbroek's attempts to find a quantitative law of magnetic force, and instructions for arming magnets were supplemented with technical details about steel and artificial magnets. By the end of the century, the authors of many didactic texts had marshalled magnetic information into numbered series, supported by printed tables of precise data collected from around the globe. This format displayed to polite audiences that natural philosophers had organized magnetic phenomena into a coherent body of knowledge which could be experimentally demonstrated.

Natural philosophers brought maritime expertise into their own domain and relabelled it as scientific knowledge. Successive editions of encyclopedias not only reflect such hegemonic processes but also reveal how the encyclopedias themselves contributed to the formation of new disciplinary boundaries. In the 1728 edition of Chambers' *Cyclopædia*, the article on navigation dealt almost entirely with its historical and commercial importance. But by 1778, a long section had been included on the activities of people linked with the Royal Society. Similarly, although the 1778 entry on variation was linked with "geography and navigation," it included the work of natural philosophers like Canton and Leonhard Euler. Between 1728 and 1778, the entry "compass" was greatly altered and expanded to incorporate Knight's modifications. Land surveying provides a useful comparison. Although instrument sellers had developed new surveying equipment during this period, expertise still lay in the hands of the army and local practitioners, many of whom preferred their traditional methods. Unlike the maritime parallel, the descriptions of surveying compasses in encyclopedias remained unchanged.[73]

At the beginning of the century, terrestrial magnetism was covered mainly in books on navigation, while loadstones belonged in mechanics.[74] Encyclopedia entries on magnetic topics became progressively longer and more closely related to the practices of natural philosophers. Discussions of magnetic natural philosophy encompassed information formerly viewed as belonging to maritime discourses. In the 1704 edition of his innovatory *Lexicon Technicum*, the Whig entrepreneur John Harris wrote eleven lines about magnets, referring mainly to Boyle. Six years later, now an ardent recruit to the Newtonian cause, he produced a heavily revised second edition. Expanding the entry to a page, he discussed terrestrial magnetism at length; in addition, he placed magnetic effects in a separate entry on attraction. By the 1736 edition, there were two pages on magnetic attraction, while a long and completely different article dealt specifically with magnets, much of it identical to the entry in the first edition of Chambers' *Cyclopædia*.[75]

By the second half of the century, most general books on natural philosophy were allocating a separate—although short—section to magnetic effects, both mineral and terrestrial. The first edition of the *Encyclopædia Britannica* found space for only nine lines on magnet, but a decade later there were over ten pages, with two major and separate entries, "magnet" and "magnetism." These were probably written by James Tytler, a Scottish ex-Glasite and religious writer, more famous nowadays for his brief career as a balloonist. Tytler may also have contributed the couple of pages on variation, a standard account focused on Halley and easily purloinable from a variety of sources.[76] Halfway through the production of the third edition, Robison took over. He had gained personal experience of measuring terrestrial magnetism, particularly while observing an aurora borealis in the St. Lawrence River. His six-page entry on variation placed a new emphasis on theoretical explanation and reinforced the role of natural philosophers by emphasizing that local experiments could yield valuable information about global phenomena.[77] In the two-volume supplement published four years later, he completely rewrote the entry on magnetism. Over forty pages long, this article was an original critical exposition, including his own experiments, reviews of Continental studies, and an unprecedented geometrical technique of analysis. He frequently referred the reader to his own entry on variation in the full encyclopedia. Like several others from this supplement, Robison's piece on magnetism was later reprinted as part of a university textbook.[78]

For about the next forty years, articles dealing with various aspects of terrestrial magnetism appeared in different places in encyclopedias because they were related to different communities of practitioners. For example, accounts of interest to navigators which described patterns of magnetic variation and how to measure it were treated separately from discussions for natural philosophers of Halley's theory of the earth's internal structure. But in the *Encyclopædia Britannica* of 1842 and the *Encyclopædia Metropolitana* of 1845, variation was no longer seen to merit an individual navigational entry: the topic was subsumed within the long scientific treatments of magnetism. John Herschel suppressed the maritime origins of terrestrial magnetism when he called it a distinct science with its own founding father, Halley—a natural philosopher.[79]

As well as appropriating maritime magnetic knowledge, natural philosophers reformed naval practices. Navigators had already started to lose command of the helm by the end of the seventeenth century, when Halley overrode the objections of his mutinous crew by insisting on captaining the *Paramour* himself and Petty decreed that a mathematical education was required for naval officers. Naval promotion came to depend on being able to use the techniques and instruments of natural philosophers: when the Portsmouth naval academy was enlarged, a Senior Wrangler was appointed a professor.[80] Naval educators associated with the Royal Society edited popular navigation manuals and produced new ones of their own. They advertised their colleagues' inventions—Knight's compasses, for instance, and Canton's method of making artificial magnets—and spelt out their aim of introducing more mathematical knowledge into nautical texts, claiming that this would enable their students to cope with unforeseen events.[81]

But maritime practitioners used their own texts to protest about these innovations by natural philosophers. As discussed in Chapter 3, they resented natural philosophers for intruding on their territory with instruments unsuitable for "a moveable observatory such as a ship." Nautical men regarding themselves as experts challenged the wisdom of displacing traditional skills by methods based on mensuration and trigonometrical manipulation. James Fergusson, an instrument maker from Stepney, argued that printed tables were treacherously unreliable: "[T]he execrable and dangerous errors they swarm with . . . have doubtless been the real occasion of the loss of many a ship, which its officers have continually attributed to other causes." In his practical manual for officers, the Liverpool dock-master, William Hutchinson, complained that the qualifying examinations demanded by the Board of Longitude eliminated many of the most accomplished navigators who had acquired their expertise at sea. Experienced mariners questioned the appropriateness of mathematical methods in an environment where unpredictable fluctuations in tides, winds, and magnetic variation could render carefully calculated values wholly inaccurate. Such techniques could be positively dangerous, explained Hutchinson, since the "errors, which every man is subject to in nice observations and long calculations among a multitude of figures . . . may be productive of effects contrary to the design; especially by putting people off their guard, and giving too much confidence to the vain and positive part of men, who cannot bear to be thought wrong in their reckoning."[82]

At a time when other rebellious groups were seceding from the monopolizing Banksian learned empire at the Royal Society, Fergusson led a rearguard action waged by mariners and instrument makers. The rival British Naval Society would collect and analyze observations to benefit the nautical practitioner. Naval experts produced books specifically designed for their seagoing colleagues, full of practical tips, worked examples, and mnemonic rules for negotiating the calculations. Hutchinson protested that "MASKAYNES [Maskelyne's] Nautical Almanack" was useless for skilled masters competent to sail round the world but unable "to understand the characters, signs, and terms, &c. . . . which terms, &c. are in an unknown tongue to one who is a mere english reader."[83] Ralph Walker contemptuously resisted scholarly intrusion: "As this treatise is not intended for the learned, but for those of my own profession, it is therefore divested of the tinsel and technical terms of the professional philosopher, that it may be the more easily understood."[84] Navigators continued to divide compass cards into points rather than degrees and to measure distances in nautical rather than land miles. Naval authors sometimes avoided trigonometrical abstraction by using labelled pictures of ships instead of geometrical diagrams. Many teachers recommended as an introductory text the rhymed annotated verses of Falconer's popular epic *The Shipwreck*, written by a sailor with firsthand experience.[85]

Natural philosophers adopted various strategies for asserting their superiority. One tactic was to discredit nautical beliefs as ignorant superstition. They advertised the power of quantified experiments for disproving the maritime tradition that garlic or onion juice affects magnetic power.[86] Canton scathingly dismissed the suggestion that tallow could be magnetic, although sailors had found by experience that wax dripping from the binnacle candles affected their compass readings. He sarcastically accused the Exeter philomath William Chapple of moving his needle by breathing on it, implying that only élite metropolitan natural philosophers could conduct experiments correctly. However, maritime men continued to publish warnings about this potential hazard when reading instruments at night.[87] Another stratagem paralleled that of the naval authors' own concern with group identity and style. Gentlemanly reviewers warned mathematical practitioners teaching navigation and surveying not to stray into the preserve of natural philosophers. Smarting from Walker's accusations, they criticized his "lame mode of arguing and conjecturing" but judged this not too important "in a work written primarily for those whose minds are very little cultivated, and which conveys information of very general importance."[88]

The construction of magnetism, like other new disciplines, was entrenched in the shifting relationships amongst heterogeneous communities and their audiences. Displaying intellectual mastery of the terraqueous globe substantiated bids for power in the smaller world of English society. By the beginning of the nineteenth century, natural philosophers had consolidated and systematized earlier magnetic information, much of it contributed by people with a maritime background, declaring it to be authoritative scientific knowledge. They had adapted navigational instruments to suit their own objectives of precise mensuration and—despite protests by nautical practitioners—had molded navigational texts and techniques to conform to their own ideology. The image of finding the key to unlock nature's secrets remained a potent one. Young contrasted the way in which English researchers would gently "touch the secret spring, by which the door of truth will be unbarred," with the violence of French revolutionary mathematicians, who would "exert all the powers of machinery in order to force it open by direct violence."[89] Exhibiting appropriate restraint, English natural philosophers quietly reformed magnetic practices, establishing themselves as the new leaders of a public science based on analyzing numerical data recorded with instruments of precision.

CHAPTER SIX

God's Mysterious Creation: The Divine Attraction of Natural Knowledge

In 1795, Coleridge gave a lecture in a Bristol coffeehouse on the historical evidence for the life of Christ. He remarked that the Stoics discoursed on the "properties of God, with the same wisdom, with which we might suppose a Mole after turning up a few Inches of Soil might describe [the] central fires, or magnetic Nucleus of this Planet."[1] For discussing a theological problem, Coleridge selected a magnetic image, framing it in a confident affirmation of progress since the age of Greek philosophy. Such pronouncements display how intimately religious, historical, and magnetic investigations of the terraqueous globe were intertwined. Just as Coleridge's magnetic nucleus affected the behavior of iron, loadstone, and compass needles, so too God was central to debates about magnetic phenomena. Establishing laws describing the constantly changing terrestrial magnetic patterns promised not only to guide travellers around the earth's surface but also to provide a clock for exploring backwards in time. People interpreted divine magnetic nature in different ways to support their divergent attitudes towards fundamental issues concerning invention and progress. They viewed this mysterious power as God's gift to navigators but also as a challenge to human comprehension. By constructing conflicting versions of when compasses had been invented, writers substantiated biblical chronology, boasted about the advances made by natural philosophers, or celebrated Britain's imperial expansion.

Numerous commentators agreed that loadstone, with its unique properties, was a divine donation particularly benefiting Britain, a commercial nation at the heart of an international trading empire. At the end of the seventeenth century, like other natural theologians stretching their ingenuity to furnish adroit justifications of deserts and mountains, John Ray extolled loadstone: "[O]f how infinite advantage it hath been to these two or three last ages, the great improvement of navigation and advancement of trade and commerce by rendering the remotest countries easily accessible, the noble discovery of a vast continent or new world, besides a multitude of unknown kingdoms and islands, the resolving experimentally those ancient problems of the spherical roundness of the earth." Over a hundred years later, as he puffed his new longitude compass, Ralph Walker apologized for a short digression: "Nothing shews the Supreme Architect in a more exalted point of view. . . . By it he enables us to behold his works, and our fellow creatures, in all the different corners of the world . . . to colonize and carry on commerce for our benefit and happiness, stirring up our minds to activity and industry; above all, expanding our ideas, and giving us a just sense of his greatness and government of this world."[2]

Authors repeatedly used nature's magnetic powers as evidence for an omnipotent God, one who was infinitely wise and benevolent. These frequently reiterated comments were not restricted to the practitioners and teachers of natural philosophy. For example, the eminent juror Matthew Hale devoted half a book to an Aristotelian exegesis of the divine implications of loadstone, arguing that only God could have imbued this outwardly insignificant stone with the multiple powers of attraction and direction. In his famous moralizing dialogue *Theron and Aspasio*, the Calvinist James Hervey reinforced commercialized privilege under the guise of charity: God achieves "the most *important* ends by the most *inconsiderable* means. . . . Through this channel, are imported to our island the choice productions, and the peculiar treasures, of every nation under heaven," thus—he continued—simultaneously enriching the houses of the wealthy and giving employment to the poor.[3] There was no clear demarcation between the concerns of theology and natural philosophy, although writers advocated different routes for learning about God and His divine creation. Natural philosophers imbued their theories of magnetic ethereal fluids with religious significance and justified exploiting the earth's magnetic supplies for commercial expansion as a project which was providentially underwritten. With its navigational benefits, loadstone substantiated the translation of God's provision of natural materials into a holy commandment to utilize them for human advantage.[4]

Whereas we live in a world shaped by electromagnetic fields, many educated people of the eighteenth century perceived a universe permeated with a divine magnetic power. Like God Himself, magnetic effects were mysterious. In common with other didactic authors, Margaret Bryan chose terms reminiscent of occult explanations to remind her pupils of God's beneficence:

> The magnet's potent spell attracts the ore,
> Whose strong affinity obeys its power;
> Possess'd, diffus'd, its laws impress'd exacts;
> The needle points where'er its power directs . . .
> To God Omnipotent! whose gracious word,
> Created all!—and saw that all was good![5]

Ontologies varied: writers described terrestrial globes bathed in ethereal magnetic fluids, penetrated by minute circulating effluvia, or radiating forces through space. To understand magnetic texts, being aware of these differences in theoretical approach is often less important than appreciating the cultural implications of the magnetic vocabulary. Magnetic terms were laden with inherited connotations which enriched people's discussions but can be difficult to retrieve.

Although it formed the frontispiece of a seventeenth-century German text, the image reproduced as Figure 6.1 illustrates components of magnetic symbolism referred to by many later English authors. Kircher continued to be

Figure 6.1. Athanasius Kircher's Magnetic Kingdom of Nature. Kircher's image portrays magnetic symbolism referred to by later writers. The magnetic chains—or Plato's rings—dangling from God's hand sympathetically unite Kircher's harmonic universe. The Latin motto means "The world is bound with secret knots." Kircher, *Magneticum Naturæ Regnum*, frontispiece. By permission of the British Library (shelfmark 33.b.14).

acknowledged as a major authority on practical magnetic affairs, and many people were familiar with his hermetic universe, in which the animate and the inanimate were bound together by magnetic chains controlled by the hand of God.[6] Writers juxtaposed accounts of magnetic phenomena with sympathetic explanations of the long-distance synchronicity of maturing grapes and fermenting wine or of the instinctive antipathies of certain animals or people. In his long and celebrated physico-theological poem, *A View of Death*, John Reynolds placed this couplet immediately after his discussion of "Magnetick virtues":

> Why flow'ring vines oblige, from distant soil,
> Their blood, in *Britain*, to ferment and boil

His detailed notes about magnetic experiments were next to his explanation that "*effluvia*, flying hither, as far as from *Spain*, or from the *Canaries*" caused "[t]his odd phaenomenon of wine's working and fermenting in *England*, while the vines are in flower beyond the sea." Citing Newton, John Trenchard devised rational explanations of religious enthusiasm: "There is a certain Sympathy and Antypathy in Nature. . . . The Loadstone draws Iron to it, Gold Quicksilver; The Sensitive Plant shrinks from your Touch; Some sorts of Vegitables, though set at a distance, attract one another . . . ; a Ratlesnake fixing his Eyes upon a Squeril, will make him run into his Mouth. . . . Some of the Quakers have arrived to a great proficiency in this natural Magnetism, or Magick." Although more common near the beginning of the century, such associations continued to appear in encyclopedias.[7]

The words "magnetic" and "sympathetic" were used interchangeably in some contexts, echoing earlier Neoplatonic beliefs in an animistic, synchronized universe. Writing in the guise of a medical quack called Duncan Campbel, Defoe described how loadstone draws "Iron to its self by the Power of Sympathy, or the natural Disposition it has to Embrace that particular Metal." At the end of the century, authors still reiterated verbally Kircher's magnetic imagery of sympathetic plants and animals. Discussing Campbel's magnetic therapies, a journal article commented that, just as palm trees grow towards each other, so "[t]he steel and loadstone are known and plain truths of sympathy." Such ideas reached wide popular audiences through the medical astrological texts of Ebenezer Sibly, a Masonic physician. Like Kircher, he gave magnetic accounts of the natural sympathies and antipathies between creatures like cockerels and lions, and of the control of the sun over plants.[8]

As the century opened, Newtonian gravity had not entirely displaced Gilbertian magnetism as the cosmic binding power. People could still buy books which proffered magnetic repulsion as an explanation for the retrograde motion of the planets or invoked magnetic attraction to explain how the moon affected the tides. In the syllabus of Worster's London lectures, recommended university reading for much of the century, the brief reference to loadstone immediately followed the discussion of tides.[9] "Magnetic" was loosely equivalent in meaning to "attractive." Extolling such mysteries as evidence of divine wisdom, Grew described how "a kind of *Magnetisme*" persuaded plants to grow upwards and made materials like amber and jet attract straw when rubbed. Chemical texts often cited the magnetic attraction between metals, describing gold, for example, as the magnet of quicksilver; Newton followed earlier alchemists by referring to the star regulus of antimony as a magnet. Alchemical and Paracelsian cosmologies informed references to the "secret Magnetism" drawing particles together at the Creation or the magnetic attraction of the cold ethereal spirit drawing mercury up a barometer tube.[10] Medically, so-called magnets were valued for their pulling action: therapists prescribed arsenical magnets (made from antimony, sulphur, and arsenic) for drawing poisonous substances from venereal ulcers, or soothed injuries with magnetic plasters—only some of which included powdered loadstone.[11]

Encyclopedists debating the etymology of the word "magnet" drew on classical sources to place its origins in different places called Magnesia. Often relating Pliny's tale of a shepherd whose hobnailed boots fixed him to the ground, they listed several Latin names for loadstone, most commonly *lapis Heracleus*, referring variously to its Herculean strength or to a city near Lydian Magnesia.[12] Linguistic confusion led to magnetic properties' being attributed to other substances. For instance, writers conflated manganese, magnesia, and *ferrum magnes* (loadstone). The white stone *magnes albus* was commonly called loadstone because it stuck to the lips, so drawing flesh as a loadstone drew iron; authorities explained that manganese was used to decolor glass because it drew out the greenish tint.[13] Less obviously, because medieval transcribers had compounded different Latin words resembling the English word "adamant," people associated diamonds with loadstone: Johnson's *Dictionary* gave three meanings for "adamant"—a hard stone, diamond, and loadstone.[14]

Older beliefs survived most strongly in medical practices. Paracelsian weapon salves were alleged to cure a wound magnetically or sympathetically by being applied to the instrument causing it; promoted by Kenelm Digby in the mid seventeenth century, sympathetic powders, supposedly healing at a distance, were still advertised a hundred years later. Reynolds introduced his magnetic verses by referring to weapon salves:

> That teaches bleeding steel to wound by stealth,
> Or greeting send, and sympathetic health

His annotation explained, "They that study the laws of sympathy, pretend, by a different management of the bloody knife . . . either to vex, or cure the wound that was made thereby."[15] Therapists valued the drawing action of loadstones and artificial magnets for a variety of afflictions, particularly toothache. Even in the nineteenth century, London instrument makers advertised natural loadstone for preventing gout and rheumatism, tailoring specimens to suit individual requirements.[16]

Several links bonded magnets, sexual attraction, and childbirth. Adam Walker's divining circles (Figure 4.9), magnetically indicating the sincerity of declared love, illustrate the endurance of older imagery. In his *Lexicon Technicum*, Harris retained the Gilbertian vocabulary of coition to describe how male and female poles move towards each other. Formulating new views of conception, William Harvey had attributed to semen a powerful magnetic ability enabling it to act at a distance; reversing his gendered imagery, eighteenth-century midwifery texts pictured the womb attracting male seed like magnets attracting iron filings.[17] Magnetic imagery featured in debates about the sympathetic effects of the maternal imagination on the unborn fetus: "And as well surely may we deny some Effects of the *Magnet* . . . to which (being ignorant of their Causes) we have given the Names of *Sympathy* and *Antipathy* . . . as these wonderful and surprising ones of the Mother's *Fancy* over the *Fœtus*."[18] Women strapped loadstones to their thighs to aid delivery or guard against miscarriage; writers conflated loadstone and eagle stone (*aetites*), also used to protect pregnancies because it is hollow with a smaller stone rattling round inside.[19] Traditionally, suspicious spouses used loadstone for detecting adultery. Newton owned a magnetic signet ring, mounted with a powerful chip of loadstone instead of a diamond. Remembering loadstone's adamantine affiliations, this may have symbolized the custom described by Bacon of exchanging rings to guarantee fidelity.[20] Gentlemen could read Claudian's descriptions of Roman love potions and engraved charms; passing references to "Orphean magnetism" reveal that a wider audience was familiar with the version of the myth relating how Orpheus' bride was drawn to him like iron to a loadstone.[21] As will become apparent during these last two chapters, recognizing this multiplicity of connotations is essential for considering how people used magnetic imagery. Even though élite authors scoffed at many of these beliefs, they drew on them in constructing their texts. Magnetic effects remained an unsolved mystery smacking dangerously of the occult. Writers took advantage of this uncertainty to deploy magnets and compasses polemically in debates about epistemology, religion, and human history.

Natural philosophers explored the earth's magnetic past to gain information about fluctuating patterns of terrestrial magnetism. Once a law had been deduced, some investigators applied it retrospectively as a clock to synchronize biblical and mythological accounts with other forms of historical evidence. English people did not view history as an academic subject but constantly examined previous cultures to fashion their ideals. Particularly in the first half of the century, antiquaries systematically collected data about the natural world and earlier human societies in a single enterprise of historical study. As they perused magnetic records, they concerned themselves with topics modern scholars perceive as distinct: a single study might embrace, for instance, scriptural analysis, the discovery of the compass, and Greek mythology. Until Knight's work of 1748, the only English book of the eighteenth century devoted to magnetic theory remained Whiston's vindication of biblical chronology. With his *Decline and Fall of the Roman Empire*, Gibbon produced the first work which effectively combined the traditional genre of narrative history with this antiquarian interest in documented evidence. By the end of the century, people held changed attitudes towards historical transformation and the function of exploring the past. Mutually demarcating their specialized areas, practitioners constructed disciplines with distinctive labels, such as philology, magnetism, and theology.[22]

As part of his attack on Newton's supporters, Hutchinson rhetorically inquired: "Who would have thought or imagined, that the use of a small Stone contrived for the benefit of Mankind, should have been produced as an

Evidence, to destroy the validity of Revelation?"[23] Initially opaque to modern readers, his question indicates how investigating references to magnets can help illuminate larger issues: people arguing about the discovery of the compass were not indulging in pedantic diversions but were vitally engaged in debates of major significance. Retranslating Hebrew texts to support his own biblical interpretations, Hutchinson stressed the intrinsic divinity of the earth's magnetic powers. He opposed polemicists of progress who hailed the recent invention of the compass as demonstrating the value of experimental philosophy; instead, Hutchinson insisted that God had long ago favored His chosen people by revealing loadstone's directional properties. As scholars delved into ancient texts, they constructed conflicting versions of the compass's origins to substantiate diverse interpretations of the earth's history and of human achievement.

This chapter examines how people exploited the flexibility of magnetic evidence to make it serve various ends. Exploring Coleridge's magnetic nucleus, the first section analyzes the role of terrestrial magnetic theories in religious controversies. The second section retrieves magnetic symbolism to demonstrate how writers cited the mysteriousness of magnetic behavior to impart moral lessons and to attack the activities of natural philosophers. Finally, comparing narratives of compass discovery illustrates how authors incorporated them into contrasting interpretations of biblical chronology and historical transformation. Although protagonists often relied on the same sources of information, they reached conflicting conclusions because they had different objectives in mind. Like the Bible itself, the text of God's magnetic world could be interpreted in numerous contradictory ways, ensuring its similar longevity as a powerful cultural manifesto.

The Magnetic Terraqueous Globe

Addressing a Royal Society audience steeped in classical culture, Halley (Figure 6.2) quoted Latin descriptions of the underworld to corroborate his new magnetic model of the earth. "I am sure," he claimed, "the Poets *Virgil* and *Claudian* have gone before me in this Thought, inlightning their *Elysian Fields* with Sun and Stars proper to those infernal, or rather internal, Regions." His infernal pun exemplifies the religious ramifications of terrestrial magnetism. Writers related the earth's structure to biblical narratives and to the location of hell; they sought to reconcile the irregular patterns of terrestrial magnetism with their conviction of a perfect creation. From Worcester Cathedral, the pious electrical philosopher Lovett reprimanded skeptics who alleged that the separation of the earth's magnetic and geographical poles demonstrated divine imperfection. On the contrary, he insisted, it provided evidence of God's foresight in providing a way for navigators to measure longitude.[24] While charts portraying contemporary variation guided explorers around the globe, historical records of its alteration provided a magnetic clock for those who wished to investigate the past. People studied the earth's magnetic structure to confirm scriptural chronologies, validate theological arguments, and construct new theories about terrestrial history.

As Coleridge's casual reference indicates, educated people regarded Halley's nuclear model—by then a century old—as essentially correct. Over a period of ten years, he put forward three versions, the second of which proved to be by far the most important. In his first presentation at the Royal Society, in 1683, he suggested that the earth is itself one large global magnet with four poles, two placed asymmetrically near each of its geographic poles. Nine years later, he discussed two shortcomings of this hypothesis. For one thing, no natural magnet had more than two opposite poles; secondly, other magnets all had fixed poles—how, he wondered, could terrestrial poles apparently shift their location inside solid loadstone? Retaining his four-pole principle, he proposed instead that the earth consists of an outer magnetic shell enclosing an inner nucleus, separated from each other by a fluid buffer. He attributed the observed changes in variation to the slow relative rotation of this shell and nucleus, each with two fixed poles.[25] As earlier chapters have discussed, most commentators lent qualified approval to this theory, which received wide public coverage. Although observers collected data which were hard to accommodate, in the absence of more satisfactory competitors, nineteenth-century magnetic experts retained a four-pole explanation.

While Halley was justifying this second attempt to use natural philosophy for describing magnetic changes in God's created world, Thomas Burnet was embroiled in controversies surrounding his own naturalistic explanations of scriptural events. Eventually forced to renounce his hope of becoming Archbishop of Canterbury, Burnet continued to forge a correspondence between historical accounts derived from the Bible and from nature. In Burnet's prelapsarian paradise, the earth's poles were equally inclined to the sun. Once ravaged by the flood—punishment for a sinful world—its axis tilted. In a fleeting reference, Burnet maintained that the current magnetic imbalance characterized the chaotic terraqueous globe before the millennial reign of Christ.[26]

When Burnet retired from his post at court following a decade of turbulence surrounding his activities, Halley faced similar threats to his career. Concerned to confute the charges of atheism surrounding his failure to win the professorship of astronomy at Oxford, he sought to provide convincing evidence that the universe is not eternal. He concluded his discussion of terrestrial magnetism with a third model, the more elaborate scheme illustrated in Figure 6.3. Although Cotton Mather ensured its survival in American literature, in England neither Halley, his colleagues, nor his successors discussed it in detail again. However, Halley evidently attached great significance to these multiple internal layers: it was this illustration of his life's achievements which, aged eighty, he chose for the portrait reproduced as Figure 6.2. In this version, Halley suggested that the earth contains not just a single nucleus but three concentric shells proportional in diameter to the planets Venus, Mars,

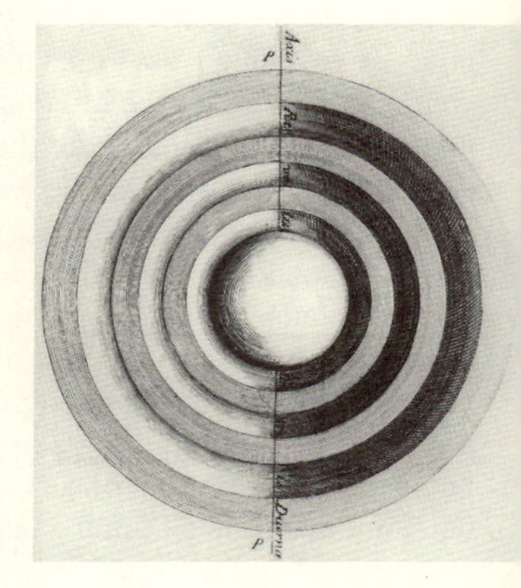

Figure 6.2 (*left*). Portrait of Edmond Halley. Michael Dahl painted this portrait in 1736, when Halley was eighty years old. He is holding a model of the earth's internal magnetic structure which he had described to the Royal Society in 1692. Royal Society, London.

Figure 6.3 (*right*). Edmond Halley's Model of the Earth's Magnetic Structure. The diagram shows three internal rotating magnetic spheres, sized in proportion to Venus, Mars, and Mercury. Estimating Earth's diameter at 8000 English miles, Halley allowed 500 miles for the thickness of each shell, with the central globe of Mercury as 2000 miles across. He proposed that God had designed this structure, perpetually illuminated by special light, to support an unknown form of subterranean life. In 1716, when a spectacular aurora borealis stunned the British nation, he suggested it might be caused by luminescent vapors escaping through the terrestrial cortex. *Philosophical Transactions* 16 (1692), facing p. 555. Whipple Library, Cambridge.

and Mercury. Displaying his diagram, he explained that the shaded portions represented the strong "Magnetical Matter" constituting hollow rotating spheres.[27]

Halley cited work by Newton to support his claims, but his major arguments were religious ones. He relied on divine design to forestall likely criticisms: "[N]o Man can doubt but the Wisdom of the Creator has provided for the Macrocosm by many more ways than I can either imagine or express." But he also confronted the problem of God's intentions more rigorously. "To those that shall enquire of what use these included Globes can be," he answered, "solid matter . . . is so disposed by the Almighty Wisdom as to yield as great a Surface for the use of living Creatures as can consist with the conveniency and security of the whole." Just as city dwellers built multistory houses, he argued, so God provided additional living space by using the interior of the globe. Articulating an opinion which later became a commonplace amongst Newton's promoters, Halley maintained that God had also made the planets habitable. For this subterranean life to survive, he suggested the existence of unknown sources of light.[28] Halley also made this magnetic theory support the orthodox scriptural view that the universe has a finite age. He pointed out that one consequence of his proposed internal structure of the earth would be to reduce its specific gravity relative to that of the moon. In an argument involving ethereal resistance to motion, this was a vital first step in claiming that the planets were slowing down. A few months later, accusing ancient observers of falsifying their calculations of longitude to conceal this retardation, he concluded "that the eternity of the World was hence to be demonstrated impossible."[29]

During the eighteenth century, some natural philosophers used Halley's magnetic model of the earth's structure for relating biblical and terrestrial history. For instance, shortly after Burnet's downfall, Whiston had published his own reconciliation of biblical and natural philosophical accounts of creation. According to his 1696 *New Theory of the Earth*, the earth was forged from cometary chaos and subsequently hit by two comets, responsible for initiating diurnal motion and for the flood. After turning to magnetic phenomena in 1718 as a potential source of longitude money, Whiston paid increasing attention to Halley's theories. Claiming that Halley's model confirmed his own ideas, he now suggested that the first of these comets might have been made of loadstone. Although sufficient to cause diurnal motion, its impulse had not yet reached the earth's center, hence explaining the slower rate of nuclear rotation. Whiston used his experiments with a dip needle to introduce pages of mathematical calculations—based on Halley's scheme—triumphantly proving that the earth is 6200 years old. Backed by this gratifying corroboration of contemporary biblical scholarship, he confidently revised Halley's quantitative conclusions about the rate of rotation of the inner sphere.[30]

Like Halley himself, many natural philosophers continued to regard dating and explaining the biblical flood as important issues. Alexander Catcott of Bristol, for example, an ecclesiastical author whose theological convictions sustained his geological field work, carefully copied out everything Hutchinson had to say about magnetic phenomena. As he defended revealed truth, Catcott provided diluvian origins for Halley's magnetic model. He explained that the central nucleus was made up of debris washed down by the Noachian flood, "surrounded on all sides by the water of the Abyss. To which, or to a similar kind of nucleus, moveable in a fluid medium Dr *Halley* ascribes the *Cause* of the *Variation of the magnetic needle*."[31] In contrast, a Scottish episcopalian writer, Patrick Cockburn, portrayed Moses as a historical narrator rather than a divinely inspired scribe. He maintained that on the third day of creation, liquid poured into a single subterranean cavity. Concerned to refute any argument Whiston had used for substantiating his cometary theories, Cockburn scorned Halley's hypothesis: "I confess it seems unaccountable to me, how a magnetick globe (of what circumference you please) in the *center* of the earth should . . . affect the needle at the *surface* of it . . . or indeed why it should be there at all? This is an Hypothesis which can never be proved." Towards the end of the century, as Churchman promoted his magnetic charts in England, this American surveyor used the alterations in magnetic variation to validate biblical chronology. Drawing on historical accounts of extraordinary tides and floods, he showed how they corresponded to his cyclical theory of the earth's rotating magnetic poles. He thus provided a magnetic explanation for the Noachian deluge.[32]

In theological debates during the eighteenth century, people rarely wrote directly about the location and nature of hell: Halley's own allusion leant on Latin mythology rather than on scriptural texts. But they did incorporate questions about eternal damnation in their discussions of divine mercy and individuals' relationships to society. Turning to Halley's model, Whiston used ancient texts to conjecture that Hades is located in the illuminated space between Halley's central nucleus and the earth's outer shell: "[N]othing seems so agreeable both to Nature and Revelation, as this Hypothesis; which supposes such a Receptacle for Invisible Beings beneath." Whereas Whiston was amongst those who insisted that comets were penal sites of torture for guilty souls, by the end of the century writers viewed them as habitable domains for the blessed. In his magnetic treatise, Knight sardonically explained that his theories implied the sun and stars were "no longer frightful Globes of Fire, but inhabitable Worlds: Those Philosophers who thought them too hot for the Habitation of Salamanders, and those sublimer Genii, who thought them to be Hells, will now perhaps be in Pain, lest the Inhabitants should freeze with Cold." But he immediately recovered himself: "[L]et us be under no Concern; all-wise Providence has taken care of all Things for the Good of his Creatures."[33]

Like Knight, natural philosophers continued to base their own designs of the earth's magnetic structure on the blueprint drawn up by God. Because of its providential significance, protagonists in religious debates used models of terrestrial magnetism to support their arguments. Concealed beneath the surface of the terraqueous globe, the source of the earth's magnetic strength remained a divine challenge. When Mary Shelley wrote that Captain Walton hoped to "discover the wondrous power which attracts the needle," she romantically articulated the fascination of deciphering this natural secret with theological, philosophical, and navigational implications.[34]

The Mystery of Magnetic Nature

In the middle of the century, privileged Oxford students could attend a course of lectures on poetry given by the High Church Tory propagandist Joseph Trapp at the Schools of Natural Philosophy. By translating his Latin text into English, Trapp influenced a wider audience. "What can be more suitable to the Dignity of a Poem," he asked, "than to celebrate the Works of the great Creator? What more agreeable to the Variety of one, than to describe . . . the attractive Force of the magnet . . . and innumerable other Wonders, in the unbounded Storehouse of Nature." For modern researchers, the widespread genre of religious didactic poetry endorsed by Trapp provides its own storehouse of moralizing natural philosophy. Theological authors incorporated magnetic exemplars because they relied on Derham's successful *Physico-theology*—reflecting Derham's own experimental involvement, this book included a relatively long section on magnetic topics. Whereas poets converted topics like optics and astronomy into twin celebrations of God's beneficence and Newton's magnificence, they used magnetic phenomena to highlight the limitations of rational natural philosophy. Like many versifiers, an anonymous critic of the Royal Society chose rhyming couplets for portraying the mysterious behavior of loadstone and iron as a salutary reminder to extol God's omnipotence:

> From God alone all *active* pow'r proceeds,
> Ev'n *human reason* his assistance needs . . .
> See *Steel* the *magnet* seek with amorous *chace*,
> With other *metals* shun the like *embrace*.[35]

In both poetry and prose, propagandists used the inexplicability of magnetic behavior to elaborate the impenetrability of providential design and the fallibility of human enterprise. In a diversity of critiques, theological exegetes preached about the limitations of human reason for fathoming God's purpose, while resourceful satirists played on loadstone's unique properties to mock natural philosophy.

By far the most frequent comments on loadstone concerned its mysteriousness and its usefulness. Operating in a manner "confessed to be still unknown by all true and unbyassed Philosophers," loadstone should impress the ignorant philosopher: "[W]ould he not then when he saw the merchandize and Product of other Countries, which are only attainable by the help of this Stone, pronounce it to be one of the most useful things in the World, and own himself, with the utmost Gratitude, obliged to receive it as a most valuable Present from a generous Benefactor."[36] Such sentiments became commonplace homilies in disparate texts, including didactic works of natural philosophy, religious poetry, and navigation manuals.

Christian apologists adapted this rhetoric for delivering various rectifying messages. Countering a perceived threat of atheism, they reasoned analogically: like magnetic effects, God was inexplicable but indubitably existed. As arguments raged about miracles, several writers noted that uneducated people in the past might have regarded magnetic phenomena as miraculous. Polemicists of progress pictured natural philosophy successfully conquering superstitious belief in magical powers. But in an argument appropriated by Coleridge, the physician and ethical philosopher David Hartley used the same observation to defend the reality of biblical miracles. Like Bishop Butler, he maintained that just as witnesses of magnetic activity were disbelieved, so the narrators of miracles might well have been describing authentic effects.[37] Rhetoricians gained further moralizing mileage from the dull appearance of natural loadstone: condemning human pride, they emphasized the hidden virtues of this "mean, contemptible, and otherwise worthless *fossil*," which was more valuable "than all the Precious Stones and Gemms which . . . make such a goodly Shew in the World with their Lustre and Brightness."[38]

Several pious philosophers cited the mysterious nature of magnetic phenomena in order to underline their insistence that human reason is insufficient for comprehending divine creation. One important example of this approach was the Tory opposition to Newtonian natural philosophy in Cambridge at the end of the seventeenth century. Challenged by the apparent absence of divine planning in the Glorious Revolution, and threatened by the increasing Whig success in college elections, High Church Tories stressed the importance of unquestioningly accepting God's decisions. John Edwards, for example, maintained that "here particularly we are gravell'd with the Attraction of the *Load-Stone*, and if we speak freely and ingenously, we must own that we know not how to render an Account of it; which without doubt was thus design'd by Providence, that we might look up to the Original Founder of all Beings, and acknowledg his Superintendency and more immediate Agency in this and some other strange Events which we meet with in the World."[39]

From attacking natural philosophy in general, it was a short step to criticizing Newton, its symbolic figurehead. Influenced by Law, the Behmenist pietist, the Oxford scholar Thomas Hobson wrote a long annotated poem demonstrating the inability of "presumptuous *Reason*" to understand God's world. Who, he queried, can

> Even tell, why smallest atoms prone descend,
> Or sympathetic magnets forceful draw
> The trembling steel?—The known *effects* are plain:
> The deep, mysterious *cause* a *Newton* shuns.

Other poets levelled more specific criticisms at Newtonian philosophers. In his *Creation*, the controversial physician Richard Blackmore consciously styled himself after Lucretius, the Roman poetic philosopher whose Epicurean ideas were familiar to all educated gentlemen. Aiming to reach a more popular readership, Blackmore introduced his atomic rebuttal of Newtonian attraction with pages of magnetic couplets providing evidence of divine design:

> The mass, 'tis said, from its wide bosom pours
> Torrents of atoms, and eternal showers
> Of fine magnetic darts, of matter made
> So subtle, marble they with ease pervade . . .
> What can insult unequal reason more

Than this magnetic, this mysterious power? . . .
Can this be done without a guide divine?
Should we to this hypothesis incline?
Say, does not here conspicuous wisdom shine?
Who can enough magnetic force admire?
Does it not counsel and design require . . .[40]

Unlike Blackmore, Christopher Smart won a prize for his far subtler theological poetry, although his contemporaries rejected his densely woven imagery—composed while he was confined in an asylum—so appreciated by modern critics. He urged Christians to celebrate God's power:

Survey the magnet's sympathetic love,
That wooes the yielding needle; . . .
baffled here
By his omnipotence, Philosophy
Slowly her thoughts inadequate revolves.

A member of Johnson's circle, this intensely religious Tory poet frequently articulated his aversion to Newton's philosophy. Who, he asked rhetorically as he mocked philosophical aspirations, is the pilot chosen by Philomela, the migrating nightingale?

Her science is the science of her God.
Not the magnetic index* to the North
E'er ascertains her course, nor buoy, nor beacon.
She heav'n-taught voyager, that sails in air,
Courts nor coy West nor East, but instant knows
What Newton, or not sought, or sought in vain.[41]

When he visited the Royal Society in 1700, the London wit Ned Ward was amused by the Royal Society's magnet, but he soon became "Glutted with the Sight of those *Rusty Reliques*, and *Philosophical Toys* [at] *Maggot-mongers Hall*." Like many satirists, he mocked the Fellows as unrealistic virtuosi. Ward devoted only a couple of sentences to that particular "Bauble in this Store-*House*," but other writers used magnetic phenomena for more sustained attacks on natural philosophers and their colleagues.[42] For instance, Thomas Brown—often compared with Ward—contributed several scathing pamphlets to the frequent condemnations in Stuart London of medical practitioners, particularly the College of Physicians. Like his fellow critics, he jeered at the obscurity of their jargon, the ineffectiveness of their treatment, and their institutional protectionism. In common with writers across the social spectrum, Brown couched his satires in scatological and erotic terms alien to modern sensibilities. Appreciating this Enlightenment sexual culture is important for retrieving concealed textual meanings and for understanding people's attitudes towards bodies and medical treatments. Figure 6.4 shows Brown's illustration of an imagined operation to cure a boy who had accidentally

Figure 6.4. "A Description of the Coledge of Physicians." Thomas Brown first published this satire on the College of Physicians in 1697, in his *Physic Lies a-Bleeding*. This caricature of an imaginary operation with a magnet also mocks John Woodward. T. Brown, *Works*, facing p. 99. By permission of the British Library (shelfmark 12271.a.15).

swallowed a knife. The physicians contemplated a chemical potion but decided "the other Remedy was more Philosophical, and therefore better approv'd; and that was, to apply a *Loadstone* to his *Arse*, and so draw it out by a *Magnetick* Attraction; but which of the two was put in Practice I know not, for I did not stay to see the noble Experiment, tho' my particular Friend Dr W——d [John Woodward] was the first that proposed that Remedy, and he is no Quack, I assure you." This medical use of a magnet was not so unequivocally preposterous as it seems today. Brown may have read debates—reported in all seriousness—about Dutch operators who applied a powdered loadstone plaster to attract a ten-inch knife towards the surface before removing it surgically from a boy's stomach. Derham, hardly one to joke about such matters, cited the anal excretion of swallowed knives as displaying divine perfection in the design of bodily passages. As his text makes clear, Brown was caricaturing not only the College of Physicians but also Woodward, currently widely lampooned as a pedantic virtuoso, a philosophical fossil gatherer. Brown may well have selected this magnetic motif to reinforce accusations of homosexuality being levelled against Woodward.[43]

Johnson also aimed at several targets in a *Rambler* article demurely introduced by a Claudian stanza praising the hidden wonders of outwardly unattractive loadstone. This letter, purportedly from a reclusive failed projector called Hermeticus, concerned specific contemporary controversies about the entrepreneurial pretensions of magnetic researchers. It also commented more generally on the objectives of experimental philosophers and their use of language, as well as wider social issues, particularly moral standards and gender roles. In his international *Journal Britannique*, Matthew Maty singled out for praise this "fiction sur l'aiman conjugal [tale of a marital magnet]," in which a Hermetic philosopher sells magnets "destinés au plus dangereux des usages [intended for the most dangerous of uses]."[44]

As discussed in earlier chapters, Johnson frequently questioned the activities of natural philosophers. At the time, he was engaged in soliciting Admiralty support for the magnetic longitude scheme devised by his protégé, Williams. In this essay, although he did not name them, Johnson focused on the magnetic researchers Knight and Canton, then marketing their artificial magnets. He mocked their promotional exercises to enhance his gibes about unrealistic philosophical schemes, personal ambition, and the Royal Society's recent linguistic prescriptions. Johnson also satirized lax social mores. Supposedly quoting from a rabbinic sage, Hermeticus reminded his readers that husbands traditionally used loadstones for detecting adulterous wives. He despairingly concluded that men would not be interested in buying his magnets: "[F]ew ever sought for virtue in marriage, and therefore few will try whether they have found it." So he decided to turn to women, ingeniously calumniating a gamut of offending hypocrites—male and female—by considering how "the three classes of maids, wives, and widows" might be interested in detecting marital infidelity. Widows, he quipped, "will be confederated in my favour by their curiosity, if not by their virtue; for it may be observed, that women who have outlived their husbands, always think themselves entitled to superintend the conduct of young wives." Parodying Knight, he offered a selection of magnets made from his secret recipe: "I have some of a bulk proper to be hung at the bed's head as scare-crows, and some so small they may easily be concealed. . . . Some I can produce so sluggish and inert, that they will not act before the third failure." But he sarcastically reported that all his customers "started with terror from those which operate upon the thoughts."[45]

Figure 6.5 shows Captain Lemuel Gulliver, amazed to see a flying circular island four and a half miles across tacking through the sky above the kingdom of Balnibarbi and the Academy at Lagado. This floating palace of Laputa, with its lower adamantine plate, was steered by a central "load-stone of a

Figure 6.5. Laputa, the Flying Magnetic Island. The upper picture, taken from a 1727 French translation, shows Gulliver waving his cap when he first saw Laputa, the flying magnetic island with a shiny adamantine base. He described how galleries around its sloping slides were connected by stairs to the palace at the top. Swift included the map in the first English edition of 1726. In the text, parodying the style recommended by natural philosophers, he explained how *c* represents the attractive end, and *d* the repelling end, of the loadstone travelling above the dominions of Balnibarbi. Whipple Library, Cambridge.

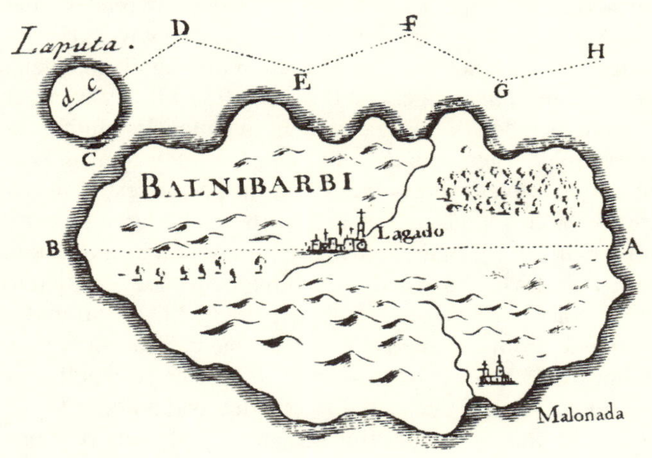

prodigious size, in shape resembling a weaver's shuttle . . . sustained by a very strong axle of adamant . . . poized so exactly, that the weakest hand can turn it." Open to multiple interpretations, Swift's account of his hero's Laputan experiences displays a complex use of magnetic marvels as a satirical resource. Most obviously, Swift was parodying the Royal Society: the Lagado project to make heaters by storing sunbeams in cucumbers has become a cliché of virtuoso absurdity. But he also structured this rich allegory as a probing exposition of the relationships amongst natural philosophy, language, and power.

Sharing Johnson's Tory High Church reservations, Swift composed this scheme for magnetic propulsion as a mocking allusion to the technical ventures being proposed by projectors for commercial advantage. Although he faithfully reproduced the proportions of the large terrella at the Royal Society, the physical details of the Lagado Academy more closely resembled those of Exchange Alley than of Gresham College. Swift simultaneously attacked aerial speculation—symbolized by the flying island—and the increasing control of natural philosophers. The core of Halley's magnetic earth paralleled the island's magnetic heart but also its internal storehouse of instruments for regulating the world. Concerned about the threat to writing which he perceived in the Fellows' pronouncements, Swift mimicked the nonnarrative nature of their discourse and mocked attempts to devise a universal language. Expressing the opposition to mathematical abstraction voiced by Lockean critics like Berkeley, he reported that the Laputan philosophers ate food cut into geometric shapes and lived in lopsided houses. This entire episode can be interpreted as an acute political reading of Francis Bacon's influential *New Atlantis*, the book which had described the ideal community of natural philosophers central to Royal Society ideology. Laputan confidence in the triumph of human ingenuity, which had enabled the magnetic island to control the far larger territory below, would be shattered as irretrievably as the adamantine (or magnetic) fortress when it crashed to the ground. The very name Laputa means "whore" in Italian. Swift implicitly opposed Jerusalem, the whore of Babylon, the city which brought destruction upon itself, to the Baconian utopia of New Jerusalem. He used religious metaphor to embellish his apocalyptic predictions of the consequences incurred by turning away from God to practice natural philosophy.[46]

These examples demonstrate that the incomprehensibility of magnetic nature was not an isolated problem confined to the Royal Society meeting rooms. Writers with various political and religious agendas deployed unsolved magnetic mysteries to stress the inadequacy of human reason for deciphering providential intentions. Critics emphasized this continuing ignorance to mock natural philosophers' epistemological strategies, entrepreneurial ambitions, and moral probity.

Compass Histories

Georgian people frequently appealed to classical authorities for their ideas as well as their ideals. Like Halley and Johnson, authors routinely introduced magnetic commentaries by discussing Greek and Roman experts, particularly Lucretius and Plato. Performers of natural philosophy displayed Plato's rings, a chain of iron loops suspended from a magnet (see Figure 2.8). Addison related Lucretius' account of the sympathetic correspondence between two friends who, hundreds of miles apart, communicated by means of magnetic dials allegedly capable of affecting each other over long distances. He commented on the style rather than the content of this narrative, because—like other Whig advocates of an expanding public sphere—Addison was concerned to remodel classical republican virtues for redefining citizenship in a commercialized nation.[47]

From Greek texts, it was clear that the attractive relationship between loadstone and iron had been known for many hundreds of years. However, people engaged in extensive debates about the discovery of the directional properties of loadstone. The most widely disseminated view stemmed from the humanist establishment of printing, the compass, and gunpowder as an exemplary trilogy demonstrating Renaissance innovation. As Baconian polemicists proclaimed the benefits of harvesting nature, they continued to extol this triplet of inventions as manifest proof of human progress. But competing narrators related conflicting versions of the origins of printing and also of magnetic compasses.[48] Although protagonists were apparently concerned to determine whether compasses had been used in the distant past or whether they had been introduced relatively recently, more was at stake than a chronology of magnetic discovery. People disagreed not only about when the compass appeared but also how. Was it due to human ingenuity or to divine dispensation? They were not interested in establishing the age of the compass merely as an academic exercise: they perceived it as valuable evidence in discussions about the triangular relationship amongst progress, natural philosophy, and God.

Ideologists of human progress cited the compass as a supreme example of modern achievement; the Renaissance rhetoric meshed perfectly with the self-promotional campaigns of natural philosophers celebrating the fruitful union of invention and commerce. Samuel Madden, an Anglican priest who helped to found the Dublin Society of Arts, advertised the possibility of progress when writing his futuristic *Memoirs of the Twentieth Century*. Defending himself against criticisms of imagining impossible advances, he stressed "how few years are past . . . since the compass and the needle trac'd out the mariner's unerring road on the ocean, and war join'd fire to the sword, or muskets banish'd bows and arrows; since the invention of printing gave new lights and aids to the arts."[49]

As Swift satirized so viciously, in the earlier years of the century opponents were still embroiled in the debates about history and progress commonly labelled the Battle of the Books, or the Ancients versus the Moderns. Although they agreed that the compass was a recent invention, they incorporated this innovation into opposing arguments. Defending classical learning, William Temple—Swift's employer—conceded that navigation was much improved but castigated modern projectors for allowing such successes to lead them into fanciful speculations, like travelling to the moon or telepathy. On the other hand, William Wotton—Temple's critical rival—stressed the benefits of experimental philosophy:

> There would have been a stop put to the Progress of Learning long ago, if immediate Usefulness had been the sole Motive of Men's Enquiries. Whatsoever our Great Creator has thought fit to give a Special Being to, is, if accessible, certainly worth our searching after. . . . He that first pick'd up a Magnet, and perceived that it would draw Iron, might then perhaps be laugh'd at, for preserving a Child's Play-thing; and yet the Observation of that noble Quality, was necessarily previous to the succeeding Observations of its pointing towards the North, which have proved so unspeakably useful in Civil Life.[50]

Compasses were embedded in mutually reinforcing rhetorics of British nationalism, commercial supremacy, and progressive natural philosophy. It was in these contexts that compass discovery stories featured most prominently. In his *Dictionary*, Johnson quoted Locke's assertion: "He that first discovered the use of the compass did more for the supplying and increase of useful commodities, than those who built workhouses." Locke's popular expositors sustained this message of progress through human ingenuity. A document attributed to Locke insisted that loadstone's directive powers had been discovered around 1300, probably by the Italian Flavio Gioia; citing several authorities, it firmly rejected arguments that compasses were used earlier. Defoe, aiming to inspire budding inventors, wrote a polemical history displaying the commercial advantages gained from earlier discoveries. Celebrating the Renaissance as an epoch of accomplishment, he dwelt at length on the benefits brought by the compass. Denying Chinese priority, he placed compass invention in the fifteenth century: "[I]n consequence of that great Discovery of all, I mean *the Compass*, Navigation being as it were let loose, and the Seaman's Hands unty'd, which were fetter'd and manacl'd before by their Ignorance, not daring to venture far from the Shores; I say in consequence of this great Discovery, all the *European* Nations went to work, spreading the Seas with Ships, and searching every part of the Ocean for new Worlds."[51]

Like Defoe, writers used compass invention to emphasize the burgeoning of European culture. Lecturers like Priestley taught that Portuguese improvements to navigation had directly contributed to European commercial, military, and cultural superiority. In Scotland, conjectural historians selected dates for compass discovery which substantiated their accounts of human development progressing through successive stages culminating in modern commercialized societies. Adam Smith, for example, was amongst those who wished to push back compass invention to support contentions that major changes were initiated by the rising middle classes rather than Renaissance courtiers. William Playfair—younger brother of John Playfair, famous writer on the earth's natural history—promoted Smith's accounts of human history. In his book "designed to show how the prosperity of British Empire may be prolonged," he compiled an elaborate colored chart displaying chronologically the movements of wealth and power during the past three thousand years. Placing the compass in the fourteenth century, he accorded it a key role in effecting the transition of culture from the climatically favorable Mediterranean countries to northern Europe, where the more egalitarian social structure enabled commerce to flourish. Another Scottish economist, Adam Anderson, included a similarly chauvinistic account of compasses in his celebration of mercantile improvement, written for the Society for the Encouragement of Arts "to compleat our Superiority over the rest of the commercial world." Barry's painting for the Society (Frontispiece), one of a series illustrating progressive stages in human history, endorsed visually the claims that the discovery of the compass was central to British supremacy.[52]

One obvious coup would be to establish national ownership of the compass, and British patriots competed with those from other maritime European countries to claim priority. British chauvinists, knowing about the evidence pointing to an Italian or Portuguese origin, relied heavily on etymological analyses. Lacking any named inventor, they stressed the importance of early magnetic experimenters—notably Roger Bacon, the thirteenth-century scholarly monk—and the unique improvements contributed by the English to early designs. Writers confidently reinforced national pride: "[T]he mariners compass was invented, but after such a manner that it is hard to decide by whom, or in what nation. Yet this is certain, the *English* may justly claim the honour of having brought it to perfection. . . . from the *English* word box, the *Italians* have made their Bossola."[53]

Near the beginning of the century, Chinese eulogists suggested that the compass had been brought to Europe around the time of Marco Polo's voyage. But as the Jesuit-inspired cult of the Chinese sage lost prestige, satirists used Chinese customs for judging English practices. Reinforcing sinophobic sentiments, patriotic writers reiterated Defoe's glib rejection of Chinese priority, although interested experts like Lord Macartney, ambassador to China, published detailed assessments of the conflicting evidence. Towards the end of the century, as travellers like Sir William Jones fostered the incorporation of Indian and Arab literature into polite learning, Oriental experts debated the Asian spread of magnetic navigation.[54] In their search for a heroic discoverer, writers quoted classical sources, travellers' documents, and foreign poems in diverse journals and books. Although there was no universal consensus, by the end of the century most authorities maintained that the compass had been invented by Flavio Gioia, nowadays viewed as a composite hero constructed during earlier investigations into the compass' origins.[55]

Although these advocates of experimental philosophy for commercial expansion disagreed about details, they all celebrated the power of human ingenuity for effecting improvement. But not everyone concurred in this rhetoric of progress through invention. Cheyne, for instance, condemned global trading enterprises as disseminators of contaminating luxury, articulating the diagnoses of decline purveyed by defenders of civic virtue. The neo-Harringtonian Andrew Fletcher urged the need for a standing army to protect a virtuous republic against vice. Attributing Italian decadence to printing, he warned that there was "a preceding Invention, brought into common use about that time [which] produced more new and extraordinary Effects than any had ever done before; which probably may have Consequences yet unforeseen, and a farther Influence upon the Manners of men, as long as the World lasts; I mean, the Invention of the Needle." As new luxuries were imported from Asia and America, "all Ages, and all Countries concurred to sink Europe into an Abyss of Pleasures."[56]

Other critics praised the benefits brought by compasses but insisted that God had determined the timetable for inventions such as gunpowder, printing, and the compass. Often emphasizing the fallibility of postlapsarian human reason, they declared that just as God intentionally prevented people from deciphering the mysteries of magnetic nature, so too He had decided when to release the secret of magnetic navigation. As Derham explained, God "hath been pleas'd to lock up these Things from Man's Understanding and Invention, for some Reasons best known to himself, or because they might be of ill consequence, and dangerous amongst Men." Some writers used these arguments to substantiate their conviction that the English were divinely favored. Perplexed by God's strange choice of the Portuguese to make dramatic navigational advances, they maintained His Renaissance timing had helped the British become rich and powerful. They could thus simultaneously exonerate the Greeks from failing to discover magnetic navigation while thanking God for reserving "so infinitely benevolent and kind a gift" until "industry, manufactures, and trade, drooping with all the languor impressed from war, devastation, and mock-heroism, were just expiring, to permit the discovery of this valuable property in the magnetic needle."[57]

But some pious proponents of revealed truth disagreed that God had deliberately delayed His dispensation of this magnetic information. They maintained that biblical texts corroborated other historical testimony of compasses' having been in use for many hundreds of years. Faced with deist assertions that Christianity was a relatively new religion, guardians of religious orthodoxy felt it was crucially important to establish that the Hebrew race was the oldest and the best. They sought to corroborate the Bible's validity by comparing it with other ancient narratives. People engaged in diverse theological and antiquarian inquiries shared the goal of establishing reliable chronologies. As they made evidence drawn from the natural world tally with that derived from mythological and biblical accounts, they incorporated objects such as fossils, compasses, and Roman coins—nowadays studied separately—within a common enterprise of studying former ages. Investigations of the earth's magnetic past blended imperceptibly with biblical exegeses demonstrating scriptural truth and with antiquarian reconstructions of Britain's pre-Roman origins.[58]

To confirm the antiquity of compasses, they drew eclectically on earlier researches which had been executed for different ends. The Oxford scholar Thomas Hyde, for instance, had been interested in portraying the Arabs as preservers of Greek culture; as supportive evidence, he concluded that ancient races had used compasses for navigation. Choosing to ignore contradictory opinions, analysts cited Hyde's expertise to back their own arguments.[59]

Referring to Hyde's authority, Hutchinson insisted it was patently obvious that the Jews had had magnetic compasses: "What Objections have been made against the Veracity of the Scripture, upon a Supposition that the *Jews* had not the knowledge of the Use of the *Loadstone* . . . in Navigation; . . . that they could not have sailed to the Land of [Hebrew characters] *Opher* (the Dust Coast) without it, everyone knows." Like other promoters of revealed wisdom, he trawled the Bible for texts like "Turn us, and we shall be turned" which could be given a magnetic twist. Duncan Forbes, Lord President of the Scottish Sessions, summarized Hutchinson's magnetic aims in a book which won many converts to his cause: "As an instance of the Perfection of the natural knowledge that is to be met with in the Scriptures, and of the Absurdity of those who charge the Writers of them with Ignorance; he avers, that the Loadstone and its Effects are frequently, at least six times, directly spoken of; That the Reason and Cause of the mysterious Phenomena of Magnetism are clearly to be gather'd from the Reveal'd Philosophy."[60]

Investigators of England's past leant on Hyde to corroborate assertions that English civilization was of privileged provenance, imported by superior races who had been spared the punishment of God's diluvial wrath. How, they wondered, could early travellers have reached England? As he delved into Greek mythology in order to consolidate his research into the Druids, William Stukeley provided the answer: "I suppose it all ends in the mysterious invelopement of the knowledge of the magnetic compass." When the chaplain William Cooke defended Christianity against "Free-thinkers ever boasting the Superiority of their own Reason, and full of their Pretensions to Philosophy and polite Learning," he incorporated Hutchinson's magnetic scholarship into his arguments that Christianity, the oldest true religion, reached England with ancient traders, such as the Phoenicians.[61]

Stukeley tramped round Stonehenge and Avebury with his measuring equipment, but he also scoured a range of literary and historical texts for references to compasses and consulted magnetic experts like Halley. He judged that the best evidence for dating these ancient temples stemmed from their alignment at a slight angle to the sun's direction at the summer solstice. The Druids, he argued, must have been relying on compasses which gave different readings from present ones, because of the intervening alterations in the earth's magnetic variation. Basing his calculations on Halley's figure of seven hundred years for a complete magnetic terrestrial cycle, Stukeley concluded that Stonehenge had been constructed in 460 B.C. and Avebury in 1860 B.C. As he triumphantly pointed out, these dates chimed neatly with the chronologies of scholars like Archbishop Ussher and Newton, and confirmed his own assertions about Druid history based on other types of evidence.[62]

Various people were interested in this sort of information. To supplement his reports to Baker on artificial magnets, Arderon carefully copied out several pages of Stukeley's magnetic investigations of Stonehenge. This interest in terrestrial history overlapped with Euhemerist attempts to adjust traditional chronology by historicizing ancient myths. Newton, controversially truncating Greek history, sought to establish the greater antiquity of the Hebrew race by redating the Argonauts' expedition. Paralleling his insistence that they navigated with an astronomical measuring device, other writers identified several mythological objects with the compass, including Apollo's cup, Jason's fleece, Jupiter's ram, and Pythagoras' dart. At the close of the century, scriptural interpreters pondered whether the Flood was an effective divine punishment if antediluvian navigators had been blessed with compasses; they shared sources with Oriental specialists haggling over which ancient merchants owed their success to magnetic navigation across oceans and deserts.[63]

Natural philosophers advertising their contribution to British mercantile expansion sought to block the propagation of views opposing their rhetoric of progress through human inventiveness. When Desaguliers endorsed a natural theology text "drawn by an Honest Plain Dutch Philosopher," he congratulated the translator on omitting whole sections arguing from revealed truth. Rejecting notions of inspiration by the Holy Ghost, Desaguliers cited "the Instance of the Load-Stone, the principal Property of which was utterly unknown during so many ages; but being now discovered, has furnish'd us with Means to find out a New World, from which great Treasure is transported." In addition to using the tactic of censorship to separate providential intervention from invention, élite reviewers mocked writers who claimed that divine revelation lay behind magnetic discoveries: "[A]ccording to him [Lovett], Providence appears to have laid a train, ever since the days of Burrows and Gunter, for the completion of it by him—nay, he gives some pretty plain hints as if there was a *revelation* in *this* case too; by which we do not think he has taken the most effectual way to recommend it to notice in this very unbelieving age."[64] Entrepreneurial philosophers emphasized that narratives of discovery should be read from God's book of nature rather than from the Bible. Case Billingsley, a longitudinarian projector, denied divine intervention: "[T]ho' this might at First be esteem'd a Miracle, because the Use of it was not before known to us; yet I doubt not but that *Surprising Stone* had the same Virtue *in itself*, ever since the World began; and there may be some rational Account given of it." Knight furnished just such a rational explanation. He claimed that his circulating fluids explained why terrestrial magnetism might be increasing "perpetually, from Year to Year: Probably it has been improving ever since the Creation; which may be one Reason why the Use of the Compass was not discovered sooner."[65] Progressive natural philosophers continued to tie compass invention to human rather than providential agency.

So discussions of magnetic phenomena were by no means confined to natural philosophers or maritime practitioners. Georgian commentators turned to the past for lessons on moral as well as natural philosophy. Magnets and compasses featured in debates concerning the relationships amongst accounts of terrestrial and human history based on material, scriptural, and mythological evidence. Writers stressed the intractable nature of magnets' behavior in order to display the limitations of human reason for deciphering God's creation, or to attack specific approaches in natural philosophy. Polemicists tapped the discovery of the compass as a versatile resource for substantiating diverse claims—not only about the commercial benefits of inventive natural philosophy but also the providential dispensation of knowledge, progressive histories of civilization, and scriptural validity.

Constructing a science of magnetism entailed resolving such conflicts between factions holding opposing views. Like modelling other facets of a refined public sphere, defining scientific roles did not simply involve ensuring the predominance of a single set of unchanging values. Some didactic authors transmuted pious rhetorics into moralizing lessons teaching the benefits of studying natural philosophy. God has created these magnetic mysteries, preached Bryan to her pupils, "to exercise our faith, and excite our attention to the other duties of religion." In a text combining ideas drawn from the Hutchinsonian as well as the Newtonian *Principia*, Adams delivered multiple magnetic messages. He emphasized the commercial advantages recently brought by compasses, rebuked French materialists for denying the divine cause of magnetic attraction, and interpreted the earth as a natural book of divine instruction embodying a harmonic descent from the spiritual to the physical world. Arguing from design, he insisted that "there are other uses of the magnetic effluvia, besides those we discern. Here again . . . we have a further confirmation of the littleness of human knowledge, and see how much pains God has taken, so to speak, *to hide pride from man*."[66]

Towards the end of the century, new disciplinary boundaries were mapped out as specialists mutually shaped their domains of expertise. By defining themselves against each other, they molded the content of restricted areas of knowledge, prescribed acceptable codes of practice, and identified recognized practitioners. As commercializing British society ceased to resemble a classical republic, commentators modified concepts of citizenship. David Hume had already articulated the reconciliation of increasing luxury, an expanding public sphere, and inventive improvement: "Thus *industry, knowledge*, and *humanity*, are linked together by an indissoluble chain, and are found, from experience as well as reason, to be peculiar to the more polished, and, what are commonly denominated, the more luxurious ages."[67] Accommodating altered views of the past and its lessons, magnetic experts reformulated earlier differences while forging a collective identity as élite dispensers of knowledge for national progress.

CHAPTER SEVEN

A Powerful Language: Images of Nature and the Nature of Science

As Johnson embarked on his ambitious project to compile a definitive version of English usage, he intended "to embalm his language, and secure it from corruption and decay." When he eventually published his famous *Dictionary* in 1755, he acknowledged the impossibility of this task. Magnetic vocabulary illustrates how the meanings of words are constantly shifting. To exemplify the words "magnetical" and "magnetick," Johnson selected a quotation from one of the period's favorite authors, John Milton, describing how the constellations obey the sun's power:

> Turn swift their various motions, or are turned
> By his magnetic beam

For Milton, immersed in Gilbertian cosmology, these lines signified something different from their implications for Johnson's post-Newtonian readers. Whereas Milton perceived a universe bonded magnetically, Georgian poets—living in a gravitational world—referred to the sun's magnetic beams metaphorically. Johnson explained that "[a]s by the cultivation of various sciences, a language is amplified, it will be more furnished with words deflected from their original sense; the geometrician will talk of a courtier's zenith . . . and the physician of sanguine expectations and phlegmatick delays." Promoting natural philosophy to a wider audience irreversibly altered, but also enriched, the English language. Like Milton's solar power, discarded theories of magnetic action endured tropologically.

Johnson's linguistic plans also resonated with political objectives. Aware that meanings do not simply diffuse outwards from élite discourses, Johnson used emotive phraseology to portray linguistic interchanges as corrupting processes of deterioration. He described how the "tropes of poetry will make hourly encroachments, and the metaphorical will become the current sense; pronunciation will be varied by levity or ignorance, and the pen must at length comply with the tongue; illiterate writers will at one time or other, by public infatuation rise into renown, who, not knowing the original import of words, will use them with colloquial licentiousness, confound distinction, and forget propriety." Johnson associated ignorance of classical languages with the breakdown of class divisions: while the written language declined from its traditional refinement by incorporating spoken vernacular innovations, so too moral standards would disintegrate as established hierarchies crumbled. Language became an important weapon in struggles for cultural authority. In particular, as they strove to establish their credibility as experts, natural philosophers developed linguistic criteria for defining the boundaries of a public magnetic science.[1]

Sharing Johnson's appreciation of the power of language, many eighteenth-century writers hotly debated its origins, its relationship to human thought, and its function in a changing society. The major authority to whom they referred was John Locke, who, in a forceful denunciation of figurative speech, pointed to the misuse of words as a prime source of error. They elaborated Locke's insistence that languages are conventional, with words arbitrarily attached to ideas and meanings determined by agreement. At the early Royal Society, natural philosophers sought to tie together words, objects, and their mental representations to yield a one-to-one correspondence between the signifier and the signified. They preached that understanding the terraqueous globe entailed reading the divine text of God's natural book. Introducing his magnetic treatise, Knight reiterated their polemical message that this was a straightforward operation: "The Language of Nature, though more copious, is far more simple than ours; and a few Characters once decyphered will help us to discover the rest." Rejecting classical techniques of oratorical persuasion, the Fellows recommended a plain style of writing, aiming to achieve unambiguous communication using terms derived from experience.[2]

When Swift satirized this program of language reform in *Gulliver's Travels*, he articulated widespread reservations about Lockean concepts of language and human experience. Like Johnson, critics of a broadly High Church Tory persuasion warned that deteriorations in literary style must inevitably reflect declining religious and political attitudes. Opposing the rhetoric of the Royal Society, they maintained that metaphor and analogy play key constitutive roles in understanding God's world. Berkeley, for example, is now predominantly studied for those aspects of his thought seen as relevant to the modern discipline of philosophy. But his religious beliefs underpinned his philosophical denunciation of Lockean abstraction. He suffused works like *Siris*, *Alciphron*, and his writings on vision with moral and theological injunctions. For him, color and light form part of a powerful natural language through which God communicates knowledge about His creation.[3]

Diverse clerics, most famously Bishop Butler, shared a similar emphasis on divine analogy. As they reinforced orthodox reliance on scriptural revelation against the perceived threat of deism, they preached that God made the physical, moral, and spiritual worlds analogous to each other and to Himself. They held that explicating prefigurative language and emblematic imagery was vital for understanding the natural world, although the imperfection of human vision made it difficult to discern the hidden relationships. In contrast to Knight, John Wesley taught that "the book of nature . . . consists not of words, but things which picture out the Divine Perfections." The Hutchinsonian natural philosopher Jones of Nayland denied that experimental investigations could unambiguously decode nature's secrets. Articulating a Berkeleyan position, he told his students that "an experiment in nature, like a text in the Bible, is capable of different interpretations, according to the preconceptions of the experimenter."[4] For men like Jones, the human, physical, and spiritual worlds were intertwined in a complex web knotted by scriptural typologies. After the French Revolution, High Churchmen with Tory allegiances delivered sermons resonating with prophetic figurative symbolism. Jones explained how biblical texts forecast current events: the "popular commotions" brought about by "the philosophical politician" were God's millennial warnings of the world's physical destruction as society deviated from a hierarchical structure divinely emblematized by the solar system.[5]

Locke himself had been reacting against earlier essentialist views of language which accorded power to words themselves. During the eighteenth century, some writers retained aspects of this formerly prevalent belief. They attached particular importance to the biblical account that, through Adam, God named all His creatures in a single day. Although their interpretations varied, many pious opponents of Lockean conventionalism were united in attributing divine significance to human language. Pointing to the universality of language before the fall, they stressed the imperfection of lapsed human reason and clouded senses.[6] Controversies about language formed one aspect of disputes concerning the adequacy of revealed theology, in which writers like Law and Hutchinson identified Locke as a major enemy. The Behmenists and the Hutchinsonians emphasized that God had dictated the Bible, the only valid account of reality. They differed about which techniques were most appropriate for exposing the truths concealed in these divine texts, but both groups subscribed to a linguistic theory of dual signification. They maintained that a complete coherence of language and truth operates between domains, so that a signifier can simultaneously denote material and spiritual entities. For example, they envisaged God's holy light pervading the earth not just *like* the sun but *as* the sun: biblical imagery of God as the fountain of light and wisdom underpinned physical theories of light as a fluid.[7]

Being aware of eighteenth-century debates about language is important in considering modern discourses. As natural philosophers constructed disciplinary science, they reiterated Locke's dismissal of rhetoric as an instrument introducing confusion into texts designed to convey knowledge. They sought to establish a distinctive style of writing to set sciences like magnetism apart from other practices. Eschewing figurative imagery, they claimed to use language as a transparent vehicle for unambiguously communicating information about the world. Their polemical messages have continued to frame judgments that scientific texts are fundamentally different from literary ones. But critics now challenge the divorce epitomized by C. P. Snow as the two-culture divide. Scientific writings are as historically embedded as literary ones: contrary to the claims of many philosophers, scientific statements are not universal propositions. Rhetorical techniques frame apparently objective scientific theories and practices, so that texts which seem remote from fiction are similarly suffused with ideological assumptions. Both literature and science are mutually shaped by each other and by the communities which generate them.[8]

Many scholars exploring the relationships between scientific and literary texts have focused on metaphor. Rather than marginalizing it as a rhetorical device for embellishing poetic writings, commentators now view metaphoric thought as an essential cognitive tool for coping with novel situations. Since language constrains us to describe the unknown in terms of the known, we must use existing vocabulary for describing previously unencountered phenomena or for formulating new ideas. This means that, far from being devoid of metaphors, scientific analyses depend upon them. Metaphors are not a straightforward comparison between the familiar and the unfamiliar but juxtapose two terms in an unexpected way so that the meaning of both is permanently affected. As technology and neurobiology have changed, people no longer compare the brain with a telephone switchboard but with a computer; these changing images have modified our vision of computers as well as of brains. Relying on constantly shifting associations, metaphors operate interactively, both transforming and generating scientific ideas.[9]

Newton set a significant precedent in magnetic symbolism. In the 1660s, while he was exploring Cartesian models of the universe, he appropriated Hooke's notion of the "sociability" between opposing poles. He drew on this concept in developing his subsequent work on attraction, itself a persuasive metaphor. Magnetic attraction provides a clear example of the diachronic instability of metaphoric phraseology and of how it participates in scientific as well as literary discourses. During the seventeenth century, people labelled magnetic poles as male and female because—like people—they attracted their opposites. Newton transferred this notion derived from an image of human attraction to one describing how material objects appear to be drawn towards each other. He made gravitational attraction a metaphor so central to science that it has largely lost its figurative aura; now forming part of most people's view of reality, it has itself become a basis for comparison. In contrast to Hooke's imagery, modern writers often rely on current physical theories to provide metaphors for the attraction between people, although there is no simple flow of meaning from one referent to the other. A theatre reviewer recently wrote: "The play enacts an intense ballet of attraction and repulsion between Miss Julie and the family servant. They are magnetised to each other in a lusting/loathing way until the famous offstage copulation makes the force-field between them go haywire." Coined in an era of electromagnetic fields, this quotation shows how literary tropology may be rooted in contemporary scientific models; in addition, it illustrates how meanings reverberate between the terms of metaphors. Although this image is informed by modern physics, its evocative richness also depends on a literary heritage which has interactively shaped magnetic tropology as theories have changed.[10]

Natural philosophers of the eighteenth century contributed towards separating scientific and literary disciplines. Examining people's conflicting views about the nature and function of language provides valuable insight into how distinctions between discourses are entrenched within struggles for legitimation. Magnetic texts offer rich material to analyze in this way because—as discussed in the preceding chapter—magnetic vocabulary carried multiple associations; furthermore, magnetic effects were described with such loaded terms as virtue, power, and attraction. This connotative wealth had a double consequence: although it enabled many people to deploy magnetic metaphors, natural philosophers lacked a precise vocabulary in which to discuss magnetic phenomena. Advocates of Lockean conventional language sought to pin down magnetic terminology and give words a unique meaning; they insisted on the possibility of separating figurative and literal accounts. But those who favored divine linguistic interpretations of the natural world deliberately exploited the metaphorical potential of magnetic terms.

Taking language as its central theme, this chapter explores ways in which writers deployed magnetic imagery and how natural philosophers used linguistic criteria to police the boundaries of magnetic science. By investigating this topic in three complementary aspects, the discussion exposes the power of language to shape disciplinary practices. The first section examines how natural philosophers relied on metaphorical language for debating theories of magnetic action, while the second illustrates how authors intentionally drew on these theories to enrich their texts. These two sections are reciprocally related: their close intertwining itself demonstrates how natural philosophy, poetic writing, and religion were tangled together in a tropological web. The final part of the chapter considers how natural philosophers used linguistic style as a discriminatory tool for monitoring the new discipline of magnetism. As a case study, it focuses on animal magnetism, which flourished briefly in London after Franz Mesmer's downfall in Paris. By revealing ways in which the practice was excluded from legitimate science in England, it exemplifies how projects to redraw maps of knowledge are rooted in bids for cultural authority. Natural philosophers establishing themselves as experts ensured that only practitioners speaking the right language were allowed to share in constructing their version of magnetic knowledge.

Theories of a Magnetic Nature

As natural philosophers speculated about magnetic nature, they disagreed on the role which language played in formulating their ideas. Locke had himself denounced rhetoric in rhetorical terms, and his proponents found it equally hard to avoid figurative language. To advocate a plain style of writing, Boyle used elaborate magnetic imagery:

> And as the Load-stone not onely Draws what the sparkling'st Jewels can not move, but Draws stronglier, where Arm'd with Iron, than Crown'd with Silver: so the Scripture, not onely is Movinger than the Glitteringst Human Styles, but hath often-

times a Potenter Influence on Men in those Passages that seem quite Destitute of Ornaments, than in those where Rhetorick is Conspicuous.

Like Boyle, writers found that magnetic metaphors exerted an irresistible power.[11]

By the beginning of the eighteenth century, natural philosophers had redefined the terminology of occult qualities; "occult" no longer referred to insensible agencies but to unintelligible ones. As their doubts proliferated about magnetic activity, they continued to use terms which derived from occult origins. During the first half of the century, natural philosophers used the word "virtue" as the equivalent of magnetic power. Writers employed the same vocabulary for describing magnetic action as that used in moral and political discourses.[12] The Earl of Shaftesbury, leading exponent of civic humanism, explained that if people felt themselves misgoverned by an unjust "Power," they would become "animated to exert a stronger Virtue." Similarly, Grew instructed readers how to revive a weakened loadstone by gradually increasing the weight of iron it suspended: "[A]s in Morals, the exercise of Virtue, makes it more generous."[13] The emblematic religious poems of Francis Quarles, first published in 1635, were continually reprinted and rewritten throughout the eighteenth century. He played metaphorically on divine power and magnetic virtue:

> But hath the virtu'd steel a pow'r to move?
> Or can the untouch'd needle point aright?
> Or can my wand'ring thoughts forbear to rove,
> Unguided by the virtue of thy Sp'rit?[14]

Paralleling the way in which luxury lost its moral overtones to become a category of economic expenditure, magnetic philosophers came to employ "virtue" as a neutral term indicating theoretical ignorance. By the end of the century, "virtue" had been replaced by dynamic but ill-defined words like "power," "strength," and "force": an editor of Quarles's poems had to explain the meaning of "virtu'd steel."[15]

For explaining magnetic phenomena, natural philosophers used figurative terms like "fire," "fluid," "positive" and "negative," and "incitement." Nevertheless, their opinions varied about the correspondence between this vocabulary and physical reality. When constructing theories, natural philosophers inevitably relied on metaphorical thought: they were interpreting effects which had no visible causes, a power unfelt by the human body. They converted this enigmatic mystery into a useful analogical weapon in epistemological battles. Attacking Newtonian attraction as an occult quality, the Hutchinsonian Julius Bate declared that "that there are *Powers* inherent in Matter acting without Means, are two too large Steps for an honest Man in his Senses to take." Every "Man of Sense and Candour" must agree that "Experiments [are] the only *Method* of giving People *sensible* Proofs in Philosophy." In his

Dictionary, Johnson quoted Locke's assertion that magnetic attraction is essential to the nature of loadstone and iron. But Bate's plays on the word "sense" point to a major problem for Lockean natural philosophers: the power of a magnet could not be directly observed by the human senses. Andrew Wilson, a pro-Hutchinsonian medical practitioner from Newcastle, ironically commented that it "may as justly be alledged, that it is the essential property of animal substances to live." Jones argued that people could neither see nor feel magnetic powers, "for we are not made of iron. If we speak accurately, we must not say that these powers are in iron or the loadstone, but that the powers themselves are in nature, and the *effects* of them in iron and the loadstone."[16]

To overcome this human limitation, many natural philosophers claimed instead that the patterns of iron filings around magnets (see top center of Figure 4.7) revealed magnetic nature unambiguously. But these allegedly transparent configurations provided a flexible resource. Diverse writers drew on them for validating interpretations which conflicted, polemically taking advantage of metaphorical language to strengthen their arguments. Rather than devise explanations, they preferred to enlist magnetic observations as ammunition in other debates.

Although many writers retained vague descriptive terms for magnetic action, some of them gave words like "virtue" a more tangible meaning. Desaguliers, for instance, explained that an iron bar can be magnetized by hitting it because "it will receive the Virtue, as if there were in the Iron several Threads or Beards fix'd at one End." Knight claimed that "the Manner in which Steel-Filings dispose themselves round a Loadstone expresses very exactly the Direction of the magnetic Virtue, and very well corresponds with the Circulation . . . of the repellent Fluid." Just as magnetic virtue circulates about a magnet, he argued, streams of fluid constantly revolve through and round that "great Loadstone" the earth. Figure 7.1 shows Halley's illustration of how the curved arrangements of iron filings on a piece of paper display the effluvia circulating around a spherical terrella in the same way as around the terraqueous globe. During the eighteenth century, such images of free circulation formed a prevalent metaphor uniting a balanced mobile nature with physical health and cultural prosperity. Harvey's demonstration that the blood circulates through the body provided a natural model for physicians to preach the virtues of fresh air, for Smithian economists to visualize the free market of labor and goods, and for city planners to place a new emphasis on motion.[17] As English natural philosophers contemplated the patterns of iron filings round magnets, the most common interpretation they made was that some sort of circulating fluid swept them into place. They envisaged terrestrial fluids circulating through the atmosphere, magnetizing vertical iron window bars and streaming like air through the pores in steel magnets. Relying on figurative vocabularies, they constructed ethereal accounts which superficially resembled one another but concealed a wide disparity of sources and significances.[18]

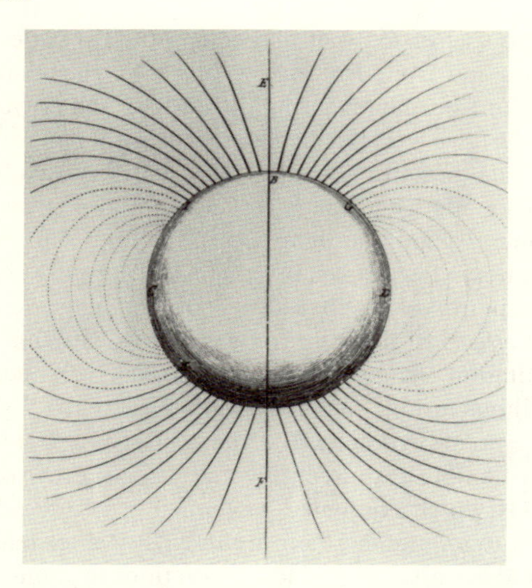

Figure 7.1. Edmond Halley's Circulating Effluvia. Halley's diagram shows a terrella of natural loadstone surrounded by steel filings scattered on a sheet of paper. He explained how, by gently tapping the underside of the paper, the filings would gradually arrange themselves to display the patterns of a circulating effluvial fluid. Similarly, he argued, subtle matter circulated around the magnetic globe of the earth. He used this model to explain the aurora borealis as luminous effluvia escaping from the terrestrial interior. *Philosophical Transactions* 29 (1716), Plate 2 (facing p. 389). Whipple Library, Cambridge.

Concerned to banish occult concepts, Descartes had elaborated a detailed mechanical explanation of magnetic activity. He postulated two types of tiny particles like screws, circulating through parallel channels in loadstone; threaded in opposite directions, these effluvia flowed in contrary streams to push similar poles apart and dissimilar ones together. Chauvinistic English experimenters mocked this Cartesian scheme, but since Newton had bequeathed only scant and inconsistent theoretical pronouncements about magnetic nature, his successors could cite passages supporting their own explanations, whether corpuscular or dynamic. Struggling to explain the spectacular aurora borealis of 1716, Halley maintained that iron filings were ordered round a loadstone by circulating "*Magnetic Effluvia*, whose Atoms freely permeate the Pores of the most solid Bodies." Using analogical phrases like "a kind of Circulation" and "as with an Atmosphere," he tried to sketch an image of two types of atoms "of a contrary tendency" while simultaneously denying the validity of Descartes's model.[19]

But most authors of natural philosophy books, disturbed by the resemblance of such a notion to Cartesian hypotheses, remained silent or admitted ignorance. They preferred to convert magnetic effluvia into anti-Cartesian polemic—"that Romantick Solution of the *French* Philosopher."[20] Yet outside this specialized group, many people during the first half of the century professed the reality of effluvia. English audiences had ready access to Descartes's magnetic model, unlike other aspects of his philosophy. Translated into English, several Cartesian texts of natural philosophy—notably that of the seventeenth-century Jesuit Jacques Rohault—were highly successful. Heavy-handed censors liberally amended sections contradicting their Newtonian perspective, but, lacking guidance from the master, they left the magnetic chapters essentially unscathed. Popular journals adapted Cartesian diagrams and commentaries; experimenting in Bristol, John Browning recounted how he had "been a Dabler in Artificial *Magnets* myself from Rohaults Chapter of Magnetism."[21] The eminent physician Browne Langrish accepted unquestioningly that effluvia existed: "how instantaneously the *Effluvia* of a Loadstone affect the Fileings of Iron laid on a Table; that it will make them all stand upright in a Moment, and if you withdraw the Stone, or intercept the *Effluvia* with an Iron Plate, that these Effects cease as soon." So sure was he of the reality of these invisible particles that he continued analogically: "If we consider that these Things are demonstrably true, it may easily be conceived that the animal spirits are capable of performing *Muscular Motion*." Similarly, Adam Smith used a magnetic analogy when discussing the mental links between apparently unrelated events, likening them to Cartesian effluvia acting as invisible bridges to explain the perplexing motion of iron towards loadstone.[22]

At successive periods of his life, Newton embraced different explanations of attractive phenomena in general. However, the behavior of magnets rarely entered either his published or his unpublished writings as effects to be explained, but appeared instead as analogical examples supporting arguments about other types of force. For the rest of the century, even natural philosophers sharing a declared Newtonian allegiance put forward conflicting versions of magnetic nature.[23]

When it suited his purposes, Newton chose to assume the reality of magnetic effluvia. In successive editions of the *Opticks*, he developed his theory that apparently solid matter is actually porous. In an important part of this argument, he asserted in 1704 that "Bodies are much more rare and porous than is commonly believed. . . . Gold is so rare as very readily and without the least opposition to transmit the magnetick Effluvia." Newton's critics found it advantageous to turn his own arguments against him. Why, they demanded, should materials transparent to magnetic effluvia be opaque to light? Why could magnetic effluvia not pass through iron, and why should they be affected by heat? Since they often posed such questions polemically, they did not necessarily feel obliged to provide lucid alternatives.[24]

But on other occasions Newton opted to take it for granted that magnetic activity is due to forces acting through space. In the 1706 *Opticks*, he based his new proposal of a particulate ether on the eloquent metaphor of attraction. He opened his Query by asking: "Have not the small Particles of Bodies certain Powers, Virtues, or Forces, by which they act at a distance . . . for producing a great part of the Phænomena of Nature? For it's well known, that Bodies act one upon another by the Attractions of Gravity, Magnetism, and Electricity." In an extended argument by analogy, he suggested that other short-range forces, as yet unobserved, might be responsible for attracting tiny ethereal particles to each other. He developed these magnetic comparisons in a new Query of 1717: "As Attraction is stronger in small Magnets than in great ones in proportion to their Bulk, and Gravity is greater in the Surfaces of small Planets than in those of great ones . . . [so t]he exceeding smallness of its [Æther's] Particles may contribute to the greatness of the force by which those Particles may recede from one another."[25]

As is well known, Newton's critics denounced gravitational attraction as an occult quality, a figurative description rather than an explanation—"a Simile or rather a Cover for Ignorance," as North expressed it. North was amongst those who reinforced these arguments by examining magnetic phenomena: "[L]ittle are wee helped by the late word In use, attraction; for that is Idem per Idem. and If one asks why one thing draws an other—It is answered by a certain drawingness it hath." Bate challenged Newton's supporters "to shew us the Chain which fastens . . . Iron to the Loadstone." Conveniently ignoring Newton's assertion that magnetic and gravitational attraction are different, he mockingly asked why mercury and diamonds do not behave like loadstone: "Nor ought the Virtue in it to be more inclined to draw Iron, than to pull a Feather to it."[26] These debates about attraction were theologically and politically laden. Some of the religious writers who attacked the materialist implications of Newtonian gravity also discussed magnetic attraction, since attributing an intrinsic magnetic power to loadstone implied limiting God's role in the universe. For instance, at Cambridge University, Matthew Prior—an anti-Lockean Tory medical lecturer—leant on magnetic simile when he characterized Newtonian attraction as a deterministic system threatening human free will:

> Great Kings to Wars are pointed forth,
> Like loaded Needles to the North.[27]

As Newton's expositors elaborated his ether, they investigated how repulsive forces might help explain phenomena such as the expansibility of gases, evocatively termed their elasticity, or springiness. Natural loadstone provided an ideal analog for hypothesized particles simultaneously exhibiting attractive and repulsive properties. As Desaguliers explored the elasticity of solids, he modelled steel springs from small spheres behaving like tiny grains of magnetic black sand. He drew diagrams illustrating how a spring's properties might be reproduced by altering the shape and polarity of adjacent magnetic particles. Like other authors, Langrish based his discussion of crystallization and chemical reactions on magnetic repulsion. When lemon is poured on salt of wormwood, he explained, "Ebullition" will take place until "all the Particles, both acid and alkaline, have met with each other at their proper Poles of Attraction."[28]

By far the most extensive magnetic expansion of Newton's ether was

Knight's 1748 treatise *Attraction and Repulsion*. His fundamental assumption was that matter is composed of small particles which are either attractive or repulsive. Clustering round one another to build up larger corpuscles of varying size and net force, these combine to produce fluids and solids exhibiting different characteristics. Knight aimed to provide a comprehensive text: his discussions ambitiously ranged over the creation of matter, comets, gravity, heat, and light, although he balked at tackling electricity. Knight devoted the last third of his book to magnetic activity, which he attributed to a fluid of mutually repellent corpuscles flowing through the pores of solid materials. Slipping from visible iron filings to hypothesized streams of virtue, Knight used terms like "the Quantity of relative Magnetism" to lend his account authority. Writing from his own observations of magnetic behavior, he gave detailed accounts of the fluid's flow for several configurations of pairs of magnets, describing how the repellent corpuscles rushed out of pores at the South ends, curved round, and entered again at the North poles. Like other fluid theorists, he confronted a major problem: although South poles would be pushed apart by the exiting streams, it is harder to explain why two North poles should repel each other. Perhaps it is rather mean to select his solution to this conundrum for conveying the flavor of his work, although its opacity does suggest why his contemporaries never quoted him: "[T]here must be a Conflux of repellent Matter towards both Poles to supply their Streams, which must make a double Flux of it towards the intermediate Space; and because both the North Poles are supplied by Streams coming from their contrary South Poles, the Stream coming to the one will be opposite to that coming to the other." This quotation also illustrates how Knight, like other writers, permeated his writing with figurative imagery—even poles were named by analogy with the earth.[29]

As discussed in Chapter 2, Knight's contemporaries largely ignored his theories, although they frequently acclaimed his practical expertise. Modern historians, on the other hand, have focused on his ideas rather than on his inventions. Enrolling him in their own research projects, they have described Knight variously and incompatibly as the author of "a mechanistic . . . protoscientific" theory, a precursor of Roger Boscovich, a precursor of James Hutton, a systematic exponent of Newton's ether, an "unreconstructed Cartesian," and the speculative herald of "the attempt to realize the ideal of the Last Query."[30] Appraising Knight's text in its entirety, rather than retrospectively isolating aspects conforming to one particular interpretation, yields a broader picture. Chapters 2 and 3 explored the roles of Knight's book in his commercial ventures of self-promotion. Studying the magnetic section of *Attraction and Repulsion* underlines how Knight—like many of his colleagues—was engaged in multiple activities: distinctions between science and technology, or between theory and practice, are inappropriate.

Involved in producing artificial magnets, Knight was familiar with techniques of ore refining and steel manufacture. He drew directly on the studies of two men who were similarly engaged in practical operations: the metallurgical research carried out at the Paris Académie by his fellow magnetic expert René Réaumur, concerned to demonstrate the utilitarian potential of natural philosophy; and the phlogistic theories developed by the German chemist Georg Stahl for explaining smelting processes in mines. Basing his arguments on Réaumur's illustrated microscopic investigations of various metals, Knight accounted for the different magnetic properties of iron and steel by distinguishing their internal pores. Later in the century, English chemists studying combustion modified phlogistic models, but Knight incorporated Stahl's original version of phlogiston in his own magnetic explanations. Whereas we perceive phlogiston as an imaginary, even laughable, fluid with conveniently flexible properties, for Knight it was real and visible and made metals look black. "That there is in Steel a very large Quantity of the *Phlogiston*, is beyond Doubt. . . . soft Steel has a dusky appearance; so that we see the *Phlogiston* in its Pores with the naked Eye." Evidently recounting his personal trials, he continued: "When it is very hard, the Appearance is quite otherwise. . . . Where now is the *Phlogiston*? Is it fled away, or is it only retired behind the Scenes? . . . Take the hardened Steel, and lay it upon Charcoal, or any thing that is red-hot. . . . It comes out again into the visible Pores, as the Steel grows hot; and even spreads itself all over the outward Surface." Knight's apparently speculative account was rooted in observations made for practical purposes—his own experiments with iron filings for improving compass needles, Stahl's experiences of processing ores, and Réaumur's investigations of metal production.[31]

During the second half of the century, natural philosophers increasingly turned to ethers when explaining natural phenomena. Couching their descriptions in figurative terms, English authors used few diagrams and gave less precise details than Continental ones. Their magnetic ethers often played a subsidiary role in discussions of other ethereal fluids, such as phlogiston, light, and electricity.[32] Although these writers shared a vocabulary of effluvia, fluids, and circulation, they were engaged in diverse projects, and they propounded very different ethers. John Dalton, for example, envisaged rings of magnetic fluids arching like rainbows high up in the atmosphere. Giving them a metallic quality, he postulated their existence to explain the aurora borealis. The physician William Hillary, on the other hand, hoped that magnetic effects would corroborate his vision—based on the ideas of the Dutch physician Herman Boerhaave—of a ubiquitous fiery ether, a unique fluid of minute elementary particles. Attributing similar prominence to the sun's role, Adam Walker claimed that the magnetic fluid marshalling iron filings into their ovoid patterns demonstrated how nature constantly strove towards equilibrium in a universe ruled by a solar ethereal fire. Robert Harrington, a surgeon from Carlisle, portrayed electrical fire circulating through the great terrestrial magnet but being repelled by elementary fire which united with the earth to create life.[33]

As natural philosophers eclectically combined the views of earlier authorities, they generated an enormous variety of magnetic ethers. Classifying their physical characteristics is less fruitful than considering their theological significance. Loosely modelled on Newton's ideas, ethereal fluids circulating through a magnet often provided an intermediary agent enabling God to move matter without constant intervention; they were "the universal Mover of all gross Bodies, and the immediate Cause, under God, of all natural Actions." Pious philosophers drawing on Behmenist and Hutchinsonian views stressed the importance of revelation and the figurative interpretation of the natural world. Denying that matter can be intrinsically active, they rejected Newtonian attraction at a distance. Frequently citing Berkeley's *Siris*, they described divine ethers operating by contact. They envisaged the universe as a machine, operated by circulating ethereal streams driven by the sun. The surgeon John Freke, for example, emphatically stated that fire is "the Origin and power of Magnetism . . . the only intelligible Cause." Like Hillary, he drew on Boerhaave's fiery vision, but he also explicitly related his fire to the ancients' *anima mundi* [soul of the world]. Circulated by the sun, his Behmenist *flamma vitalis* [vital flame] was God's direct agent in the universe; just as a budding elder stick was primed with life, so a magnetized steel bar was laden with activating fire.[34]

The year after Freke's treatise appeared, Francis Penrose, an Oxfordshire surgeon, produced another magnetic tract. He cited several sources, but he evidently derived many of his ideas from Hutchinson's works. An ardent pulpit defender of the Hutchinsonians, Penrose stressed that effluvial fluids were responsible for magnetic action, loadstone itself being entirely passive. Penrose's books attracted only a limited readership, but his contemporaries gave them substantial reviews. These indicate the importance of appraising the prevalence of Hutchinsonian ideas for appreciating how natural philosophers shaped the characteristics of legitimated magnetic science.[35]

Hutchinson himself only confided magnetic "Hints and Conjectures" to his private papers. Deliberately exaggerating the differences between his views and Newton's, he maintained that magnetic effects were due to the interaction of an effluvial ether with the larger particles of iron and loadstone. Endowed with motion by God, this ether "of infinitely small Corpuscles put into Motion by the Sun" comprised a mechanical agent moving the gross matter visible to human beings. Hutchinson suffused his ether with religious and linguistic significance, conceptually bonding it to the Trinity by Hebraic ties. He believed that the Bible, the unique source of revealed knowledge, had been dictated by God in a powerful natural language. Through linguistic research, he aimed to uncover God's plan by removing human corruptions of the divine Hebrew text. Maintaining that biblical words simultaneously represented material and immaterial entities, he derived physical as well as spiritual accounts of the world from analyzing the Bible. In a passage sprinkled with biblical quotation, he explained that "of what moves or is moved to a Loadstone, whose Parts the *Aleim* have mercifully contrived, and mechanically disposed to be turned by the Air: to be a Guide for Sailors in broad Seas . . . which Disposition . . . gives the Air a Power, or Opportunity of driving other dense Bodies towards it." Thus, in Hutchinson's cosmology, divine power was realized as physical power. Such phrases reflect his attempts to translate Hebraic words into English: he insisted on using *Aleim*, the plural form of "God," because the scriptures describe the Creator as a Trinity.[36]

As he searched for the original divine version of scriptural texts, Hutchinson became convinced that the word usually translated as "pearls" or "rubies" did, in reality, mean "loadstone." The magnetic rephrasing which aroused most controversy was his replacement of "for the price of wisdom is above rubies" by "the attraction of wisdom, is greater than that of loadstones." Although Hutchinson's revisions were unusual, he was participating in a widespread practice of biblical commentatary. Like other proselytizers of revelation, the druidical investigator Cooke welcomed Hutchinson's clarification of this obscure passage. He used it to underline the superiority of moral virtues over worldly benefits: "[T]he Attraction of Wisdom is to the well-disposed greater in Effect, and of more Value, than the Attraction of the Loadstone, tho' it draws the heaviest Bodies, and by It all the Advantages of Navigation and the Wealth of Commerce are obtained."[37] Hutchinson's translations reached a wider audience through the renowned Hebrew lexicon of John Parkhurst. This Cambridge Hebraist, sponsored by the Oxford Hutchinsonian Horne, interspersed his linguistic analyses with illustrations of how contemporary experiments in natural philosophy corroborated biblical texts. Annotated by Parkhurst, Hutchinson's comments on loadstone provided a source for religious exegetes as well as for philological scholars.[38]

Towards the end of the century, ethereal models proliferated. Appropriating ideas from a variety of sources, diverse English authors continued to use figurative language revealing the combined theological and physical functions of their ethers. In addition, widely acclaimed writers of books and encyclopedias endorsed—albeit with reservations—the magnetic fluid formulated mathematically by Aepinus. They presented his fluid of mutually repulsive particles as analogous to Franklin's electrical fluid, explaining the magnetic action of iron by a surplus driven to one end of a bar.[39]

But natural philosophers increasingly challenged these circulating fluids which had once seemed self-evident. Reinterpreting patterns of iron filings, they claimed that each filing was incited to become a small magnet; these small individual magnets then aligned themselves in chains around the larger magnet. Calling for a return to Gilbert's techniques of Baconian observation, Robison introduced a new methodological approach. Using diagrams like the one reproduced by Young in the top left of Figure 4.7, he argued that the curves could be derived by analyzing the filings' mutual interactions mathematically. Trained in the Scottish commonsense school of philosophy, Robison built on the investigations of Johann Lambert, the German mathematical experimenter, to construct geometrical systems determining the orientation of a small needle moved round a pair of magnets. His innovative diagrams of magnetic space influenced the work of Faraday's associates and hence the Victorian development of field theories.[40]

Robison's rejection of invisible fluids formed part of his larger project of basing natural philosophy on general principles derived from observation. His prescriptions included promoting geometry as the best mathematical language for developing new ideas. Robison was also concerned to strip verbal accounts of ambiguity. Paralleling his denigration of French algebra as mindless symbolic manipulation, he maintained that "metaphorical language has affected the doctrines of mechanical philosophy . . . [and] the only way to decide this dispute is to avoid, most scrupulously, all metaphorical language, though at the expence of much circumlocution." He apparently judged that his own magnetic vocabulary of incitement and attraction provided an unequivocal representation of physical reality. As English natural philosophers fashioned magnetic nature following Robison's geometric analyses, they implicitly endorsed his linguistic criteria for fashioning their community's practices.[41]

Powerful Magnetic Metaphors

Figure 7.2 shows an angel welcoming a human soul—personified as female—clutching a compass as she strives towards God. This emblematic image accompanied two of Quarles's verses comparing an oscillating compass needle with a restless Christian: just as the needle eventually settles at true North, so finally the soul "points alone to thee." Quarles's book *Emblems* was repeatedly published throughout the eighteenth century; new pictures were drawn, and at least one editor rewrote the poetry as well. Religious polemicists drew on these devotional poems for composing hymns, instructing their pupils, and selecting tombstone inscriptions. Although modern literary critics do not focus on this book as a source, numerous writers adopted very similar compass metaphors, perpetuating a tradition of magnetic emblematic imagery stemming from as far back as Petrarch's love poems.[42]

John Norris, a Cambridge Platonist and a clergyman renowned for his inspirational verses as well as his criticisms of Locke's *Essay*, very probably had Quarles's symbolism in mind when he wrote his devotional poem *The Aspiration*. During the eighteenth century, appreciative audiences admired Henry Purcell's musical rendition of Norris' lines describing the uncertainty of an equivocating spiritual pilgrim:

> Ev'n here thy strong magnetic charms I feel,
> And pant and tremble like the amorous steel;
> To lower good, and beauties not divine,
> Sometimes my erroneous needle does decline;
> But yet so strong the sympathy,
> It turns again and points to thee.

Those who preferred rather lighter fare might be entertained at Marylebone Gardens by the playwright Dorothea Dubois's musical entertainment about

Figure 7.2. "I am my Beloveds, and his desire is towards me." Francis Quarles's moralizing 1635 text combining emblematic pictures and verses was studied so heavily that no copies survive of the earliest editions. This magnetic emblem, showing an angel guiding a spiritual pilgrim towards God, typifies Quarles's aim to convert readers by focusing on the relationship between Christ and the human soul. One eighteenth-century editor converted the motto into "The Christian's Loadstone." Quarles, *Quarle's Emblems and Hieroglyphicks*, p. 134. By permission of the British Library (shelfmark 130.c.16).

male infidelity called *The Magnet*. Like Johnson, she relied on other sources for her magnetic imagery, referring to the popular practice of using loadstones for detecting adultery:

> Now to ye, married Fair-ones,
> Our Counsel is due;
> Of the Magnet be careful,
> 'Twill keep your spouse true.[43]

Writers used a variety of magnetic figurative language to enhance their pronouncements on diverse topics, including divine power, female characteristics, and the limitations of Newtonian attraction. People held different opinions about the nature of metaphor. Those who adhered to a Lockean notion of language and endorsed a plain style of writing viewed their metaphors as unambiguously comparing distinct entities with stable definitions; they intended their metaphors to operate unidirectionally. This broadsheet writer was intending to rally support, not alter views of magnetic attraction: "*[t]he political Magnet*: . . . *or Loadstone*, which (*it is wished*) might have the *Power* and *Property* to *attract to itself*, (that is to say . . . *in Favour and Support of that System of Government, which now obtains*,) and consequently to *direct* and *govern* the Use and Application *of all the Swords* in the Kingdom." But other writers denied such clear demarcations; they deliberately sought to capture ambiguity. For them, figurative language not merely served to decorate a text but functioned as a vehicle of meaning. Like the government polemicist, Hutchinson also used a military metaphor as he played on moral vocabulary. He likened ignorant people who interpreted iron's motion as magnetic attraction to biblical witnesses who saw "that Virtue went out of the *Essence* that was in *Christ*." He ironically pictured a loadstone with "Power to irradiate, send out Virtue" forcing iron filings to align themselves "as Centinels do with their Pikes or Muskets, when their *Theos*, the King passes." Far from being a simple comparison, his image parodied eulogies of the Newtonian solar system and mocked a natural philosophy which attributed divine power to inanimate objects. In this evocative trope, Hutchinson reflexively used a parallel based on familiar magnetic action to comment on magnetic theories themselves.[44]

Like Norris and Hutchinson, many poetic and pious authors envisaged metaphors as yielding a stereoscopic vision of human experience by juxtaposing two incongruous referents, so that each could be seen in a new way. Constantly shifting in meaning, figurative language functioned interactively, both transforming and generating theories of the natural world. The effectiveness of magnetic imagery depended on blurring or even eliminating any boundary between a physical phenomenon and the concept with which it was being associated. Particularly for Behmenist and Hutchinsonian writers, metaphorical language bonded together the spiritual, biblical, and material worlds. Quoting Christ's words on the cross—"I, if I be lifted up, will draw all men unto me"—Horne, the Hutchinsonian Bishop of Norwich, preached: "The virtue of his death . . . composes a divine magnetical influence . . . which is to act upon the mass of mankind, and draw them upwards from the earth. It acts in a due and appointed order, first upon their souls, and afterwards upon their bodies."[45]

The intense religious writings of Law and Smart—inspired respectively by Behmenist and Hutchinsonian ideologies—provide powerful examples of rich magnetic tropology. Law enhanced his pietist attacks on deist rationality by comparing magnetic attraction and human volition. In a long allegorical passage, he increasingly obfuscated the ties linking a loadstone and a compass needle with those uniting God and a believer's soul. Employing a deceptively straightforward analogy, he started by separating faith and reason: "If the *Needle* touched with the *Loadstone* was an intelligent Being, it could reason and make Definitions of *itself*, of *Attraction*, and of the *Loadstone*; but it would be easy to see, that the *Attraction* in the Needle, or the Virtue of the Loadstone that was left in it, was something in its *whole Nature* really different from this reasoning about it." He then elaborated this portrait of a deluded deist, emphasizing that just as a needle which has lost its magnetic virtue will be unable to turn towards a loadstone, so "if the Soul loses its . . . *Attraction* to God, all its Reasonings and Definitions about God and Goodness are of no Use to carry it to God." Over the next few paragraphs, he tightened the metaphorical bonds until God essentially *became* a divine loadstone controlling the universe: "[A]ll things have *magnetical* Effects and *Instincts* both towards God and one another. This is the *Life*, the *Force*, the *Power*, the *Nature* of everything."[46]

This passage, with its resounding message that "all is *Magnetism*," almost certainly influenced Freke's decision to make magnetic power central to his cosmology. Historians have focused on the controversies aroused by Freke's 1746 tract on electricity, but six years later he sandwiched an edited version between essays on fire and on magnets. At this stage, Freke himself claimed that "[m]y Intent of writing the First Part of this Treatise on Fire [was] chiefly to guide the Mind to this Subject of Magnetism," to which he devoted a substantial essay. This apparently retrospectively stated aim suggests that Freke's ontology, in which "Fire is the Instrument or Cause, under God, not only of Magnetism, but of all the Phænomena in the Universe," was progressively shaped by Law's mysticism.[47]

In his magnificent celebration of nature, *Jubilate Agno*, Smart also portrayed a harmonious universe driven by a fire which was simultaneously physical and spiritual. Poetically articulating Hutchinsonian opposition to Newton's philosophy, he entangled biblical prophetic imagery within his figurative account of a holy creation resisting explanation. As he wrote about longitude schemes, he intertwined the religious and nautical significance of magnetic direction:

> *For the Life of God is in the Loadstone, and there is a magnet, which pointeth due EAST.*
>
> Let Martha rejoice with the Skallop—the Lord revive the exercise and excellence of the Needle . . .
>
> For due East is the way to Paradise, which man knoweth not by reason of his fall . . .
>
> *For the Longitude is (nevertheless) attainable by steering angularly notwithstanding.*

In these lines, the scallop is emblematic of the Christian pilgrim, aiming towards perfection like a needle striving towards the pole, magnetically directed to the spiritual East of the prelapsarian Garden of Eden. Perhaps with Hutchinson's retranslation of "pearl" as "loadstone" in mind, Smart—like Law—used magnetic language to integrate divine power, human affairs, and the material world.[48]

Jubilate Agno remained unpublished during Smart's lifetime, but some of the century's most successful writers figuratively expressed Behmenist interpretations of magnetic power. In his very popular *Universal Beauty*, a long physico-theological poem with a mystical aura, Henry Brooke criticized Newton for preferring to explain the world rather than worship its Creator. Reiterating Milton's solar image, Brooke portrayed a Behmenist divine sun sympathetically unifying an obeisant paradisiacal nature:

> Here, winding to the Sun's magnetic ray,
> The solar plants adore the lord of day,
> With Persian rites idolatrous incline,
> And worship towards his consecrated shrine;
> By south from east to west obsequious turn,
> And mov'd with sympathetic ardours burn.[49]

Similarly, as Richardson associated with Freke and other members of Law's pietist circle, he participated in their revival of Neoplatonic ideas. He wrote: "There is a kind of magnetism in goodness." Unlike the "rope of sand" binding bad men together, "trust, confidence, love, sympathy, twist a cord, by a reciprocation of beneficent offices, which ties good men to good men, and cannot easily be broken."[50] This imagery recalls Law's universal divine magnetic power and also Kircher's harmonious universe, sympathetically bound with magnetic chains (Figure 6.1). Like other Behmenist authors, Richardson was almost certainly familiar with the *Ion*, Plato's Socratic dialogue picturing poets receiving divine inspiration along a chain of magnetic rings linking them to God. Although probably unknown to performers of natural philosophy displaying the trick colloquially known as Plato's rings (see Figure 2.8), the *Ion* was a key text in theories of acting, and one which Shelley thought sufficiently interesting to translate. Echoing Behmenist love, Shelley personified himself as the earth magnetically ruled by Emily and Mary, the sun and moon of his life:

> Twin spheres of light who rule this passive Earth,
> This world of love, this *me*; and birth
> Awaken all its fruits and flowers, and dart
> Magnetic might into its central heart.[51]

Like Shelley, many writers used magnetic imagery for representing asymmetrical power relationships. Rather than perceiving mutual interaction, they envisaged a strong magnet drawing a smaller one towards it, or loadstone controlling a piece of iron. Metaphorically, therefore, magnets symbolized dominating figures, whereas iron and compass needles denoted subservient partners. Compiling his *Dictionary*, Johnson chose a couplet by John Dryden as his first illustration of the word "magnet."

> Two magnets, heav'n and earth, allure to bliss,
> The larger loadstone that, the nearer this

A recent convert to Rome and a defender of James II, Dryden published his famous allegorical poetic fable *The Hind and the Panther* the year before the

Glorious Revolution. Like Milton, he subscribed to Gilbert's view that the earth's magnetic powers stretch through the universe. By depicting the sun as a giant loadstone, he portrayed the small Catholic Church as the victim of Anglican rapacity.[52]

More commonly, magnetic power represented divine omnipotence. As Norris put it, "We must conclude, that *God* is the true great *Magnet* of our Souls; that he . . . has imprest a Motion upon our Intellectual Heart, only to incline it towards himself." Like Norris, the metaphysical poet Thomas Ken often used metaphors based on magnets and compasses. In his 1721 *Preparatives for Death*, he pondered on his destiny:

> Or shall I mount to endless rest,
> By heaven'ly Force impress'd?
> Or shall I by magnetick might,
> Be drawn to endless Light? . . .
> The Magnetism of heav'nly Love,
> Draws some to God above

In an unusual reference to the power of evil, he figuratively described how

> Sin grew the constant Pondus of the Will,
> The Needle turn'd from God, to point at ill.[53]

Writers increasingly applied such magnetic images not to divine attraction but to the attraction between people. Unlike the harshly abstract laws of modern science, natural laws during the eighteenth century mapped the behavior of a benevolent nature as well as a harmonious society. Newtonian promoters portrayed the sun gravitationally controlling the planets like the king ruling his subjects. Francis Hutcheson, Scottish exponent of Shaftesburian virtue, based his social theories on the sympathy bonding together people of sentiment. He proclaimed that "[t]he *universal Benevolence* toward all Men, we may compare to that Principle of *Gravitation*, which perhaps extends to all Bodys in the Universe."[54] But unlike gravity, magnetic power operated specifically; furthermore, it carried traditional sexual associations. As civic humanist models became less prevalent, authors used magnetic tropology not for portraying the moral sympathy binding communities together but for representing the relationships between a particular couple.

Because writers perceived magnetic action as asymmetrical, their symbolizations mirrored gendered hierarchies of power. John Cleland, for instance, used a compass metaphor as he relished Fanny Hill's "rage and tumult of my desires, which all pointed strongly to their pole: man." Shelley's more subtle imagery reflected the fluid cosmology of his former teacher, Adam Walker, who framed his magnetic discussions in Behmenist terms of polar opposites and solar fire. In *Prometheus Unbound*, Shelley portrayed magnetic power blissfully uniting a man to a woman whose lunatic subordination paralleled the moon's devotion to the controlling masculine sun:

by a power
Like the polar Paradise
Magnet-like of lovers' eyes;
I, a most enamoured maiden
Whose weak brain is overladen
With the pleasure of her love,
Maniac-like around thee move[55]

But female novelists like Fanny Burney and Elizabeth Hervey reversed the dynamic direction: it was their heroines who wielded power as "the general loadstone of attention." Hervey's "lovely Emma was the magnet that attracted" rival suitors; as Lord Orville courted Evelina, he declared: "Far be it from me . . . to dispute the *magnetic* power of beauty, which irresistibly draws and attracts whatever has soul and sympathy: and I am happy to acknowledge, that . . . we have *goddesses* to whom we all most willingly bow down. . . . they know the attraction of the magnet that draws me."[56]

Like geometrical compasses, magnetic compasses had long been emblems of constancy. For a maritime people, whose fortunes depended on overcoming stormy weather, confidence in magnetic fidelity was essential. In this couplet, John Gay portrayed a man's unswerving dedication in the face of a feminized stormy nature:

Change, as ye list, ye winds! my heart shall be
The faithful compass that still points to thee.

In his didactic *Astronomical Dialogues*, John Harris wrapped a moral message of constant rectitude in his natural philosophy instruction for flighty female pupils:

So when the Needle hath been once drawn oe'r
The Loadstone's Poles, and felt its wondrous Power,
'Twill e'en in Absence keep its Truth and Worth,
And always point *tow'rds its beloved North*

As Reynolds visualized his soul navigating on a cosmic voyage of self-discovery, he regretted fickle human frailty:

I see why the touch'd needle scents about,
Till it has found the darling quarter out;
And why, unconstant grown, it sometimes takes
New-sprung amours, and its dear north forsakes[57]

Almost a century earlier, Quarles had eloquently described how a compass needle does not settle immediately but

First franticks up and down, from side to side
And restless beats his crystal'd Iv'ry case,
With vain impatience; jets from place to place,

And seeks the bosom of his frozen bride,
At length he slacks his motion, and doth rest
His trembling point at his bright Poles beloved breast.

By personifying instruments which record change—such as barometers, thermometers, and compasses—writers could model human sentiments. Quarles gave a masculine identity to the authoritative needle which guided mariners across the oceans; but, following the classical verdicts of epigrams fondly reiterated by English gentlemen, his female soul typified characterizations of feminine behavior:

Ev'n so my soul, being hurried here and there,
By ev'ry object that presents delight,
Fain would be settled, but she knows not where;
She likes at morning what she loaths at night

Later writers shared this perception of female changeability: Smollett, for instance, sardonically judged that, for a vacillating female admirer, "Humphry is certainly the north-star to which the needle of her affection would have pointed at the long run."[58]

But as men and women fashioned new gender roles appropriate for a culture of sensibility, authors increasingly used compass metaphors to celebrate the sensitivity of female spirits and mold their behavior into a desirable constancy. Thus the female novelist Frances Brooke constructed a heroine whose fashionable emotions were devastatingly attractive to this male admirer: "What a charm, my dear Lucy, is there in sensibility! 'Tis the magnet which attracts all to itself . . . 'tis sensibility alone which can inspire love." But the poet Frances Greville begged to be relieved of the anguish experienced by a sensitive heart:

Nor ease nor peace that heart can know
That, like the needle true,
Turns at the touch of joy or woe,
But, turning, trembles too.[59]

In contrast, male authors concentrated on female commitment. Darwin fantasized about the "chaste Mimosa" growing in his botanic garden, visualizing her as a veiled Moslem bride trembling towards the mosque:

There her soft vows unceasing love record,
Queen of the bright seraglio of her lord . . .
So turns the needle to the pole it loves,
With fine librations quivering, as it moves.

By confining the discarded yet faithful Julia in a convent, Byron could also imagine her concealed in religious vestments, trapped like Quarles's frantic soul in an ivory case. As he captured her writing to Don Juan with "small white fingers [which] trembled as magnetic needles do," he articulated a masculine version of the ideal tamed woman:

> My heart is feminine, nor can forget—
> To all, except your image, madly blind;
> As turns the needle trembling to the pole
> It ne'er can reach, so turns to you, my soul.

Grounding their visions in familiar navigational instruments, Darwin and Byron romantically steered their readers towards a new society; they presented exoticized religious versions of sensitive wives faithfully tending a more domestic sanctuary. As natural philosophers transformed the laws of nature from sympathetic embodiments of enlightened ethical guidelines into ruthless abstract forces, poetic magnetic imagery gained a new power. Writers converted the former theological symbolism of faithful magnetic pilgrims willfully turning themselves towards God into prescriptions for female conduct. By citing the needle's predetermined orientation, they endorsed these role models by appealing to the immutability of women's created nature, inescapably governed by scientific laws.[60]

Exclusive Powers

Drawing maps entails deciding what should fall outside boundaries as well as what should be placed within them. As natural philosophers contributed towards defining a science of magnetism, they policed its borders to determine which practices and practitioners should be allowed entry to this new domain. One way of doing this was through language. Following Johnson, bids to establish a standard English notionally offered equality to all of Britain's people but were also political ploys enabling the polite to protect their own interests. Just as judges punished breakers of criminal laws, so those who spoke the correct language exerted power over transgressors of linguistic rules. Elite authors reinforced their own status by decreeing suitable ways of writing about natural philosophy, so helping to shape new class and gender distinctions.[61]

Natural philosophers aiming to revise the language of their discourse were also seeking to reform its participants. One concern was to reject figurative imagery. As Robison carefully restricted the vagueness of terms like "power" and "force," he lamented that "metaphorical language has affected the doctrines of mechanical philosophy." Darwin had his own versified Linnean system in mind when he declared that "Science is best delivered in Prose, as its mode of reasoning is from stricter analogies than metaphors and similies." By purging their work of metaphor, natural philosophers differentiated their writings from literary genres but also distinguished acceptable practitioners: they expelled pious writers who deliberately invoked metaphorical ties between the spiritual and material worlds, and also people who relied on traditional perceptions of magnetic activity. Consolidating linguistic definitions entailed erecting social boundaries: deciding on the form of magnetic knowledge and how it should be communicated also determined who were the accredited experts.[62]

As they advertised the power of reason, enlightened philosophers promoted their exclusive role as society's guardians by dismissing older beliefs as superstition. Like faith in astrological and alchemical ideas, belief in magnetic therapeutic powers was effaced from most published texts at the end of the seventeenth century, although it survived in practice. Editors deleted entries in popular encyclopedias and wrote denunciatory prefaces for occult books reproduced as collectors' items.[63] Ridiculing popular ideas as ignorant, authors held up magnetic medical therapies and weapon salves as exemplars of the need for rational reform. At the end of the century, Cavallo denigrated "people who believe, that the application of the magnet cures the tooth-ach, eases the pains of parturient women, disperses white swellings, &c; and, on the contrary, that the wounds made with a knife, or other steel instrument, which has previously been rubbed with a magnet, are mortal." It was in this context that he declared: "[A] great deal of confusion in the science of magnetism has been also occasioned by the application of the word *magnetism* to other things, which had nothing to do with it." Cavallo was not merely seeking to purify magnetic vocabulary: he was also interested in eliminating magnetic healers and their patients from legitimate magnetic science.[64]

Earlier chapters discussed various strategies natural philosophers adopted to consolidate their expert exclusivity, including several illustrations of the discriminating power of language. As they fashioned their own identities, maritime and scientific communities mutually rejected each other's linguistic style. Navigators criticized natural philosophical texts for being laden with jargon, complex phrases, and Latin names; reciprocally, intellectual critics characterized navigational books as being written in a manner appropriate for less-educated people. Rationalizing polemicists castigated popularizers for veering too far towards amusing rather than instructing their audiences and outlawed performers like Graham and Katterfelto as quacks peddling deception with the inflated language of puffery. Tentatively inspected by fashionable metropolitan circles, provincial educators were derided for their colloquial style: although "Mr Walker the lecturer" was invited to the same polite dinner party as Fanny Burney, she judged him "vulgar in conversation."[65]

Through their own popularizing texts, élite natural philosophers shaped the asymmetrical relationship of scientific education. In didactic dialogues, authoritative tutors used simplified language for transmitting condensed versions of their expertise to less-knowledgeable audiences. When Harris' aristocratic pupil referred to Milton's magnetic sun, she quoted one of the adaptations compiled for women and other readers deemed unable to cope with the original. These authors of natural philosophy fashioned women as recipients of information which they expressed differently from when they communicated with each other. Thus they reinforced other barriers precluding female participation in the creation of knowledge.[66]

From the mid 1780s, practitioners of animal magnetism flourished briefly in England for about a decade. Their contemporary dismissal as fraudulent quacks has been perpetuated by critics finding it useful to label them pejoratively as pseudo-scientists. However, understanding how practices like astrology or phrenology move from recognition to dismissal entails examining them contextually. Modern historians have amply demonstrated the value of studying marginalized groups. Important examples for this period include French mesmerists and English underground radicals, whose reassessment has cast new light on diverse issues, including political activism, popular culture, the demarcation of scientific disciplines, and the genderization of medicine.[67]

Although animal magnetism originated in Europe, its rejection in England was related to local conflicts; the skeptical reception of a therapy which had already been discredited differed greatly from the initial enthusiasm of pre-revolutionary Paris. Examining the impact of the English animal magnetizers provides an interesting case study for exploring the linguistic interactions between excluded practices and élite discourses. Polemicists engaged in a variety of debates enrolled animal magnetism as an indiscriminate term of abuse. They ensured that animal magnetism remained outside the accepted canon of magnetic theories, despite the features it shared with contemporary ideas and practices. This very process of public ostracization contributed to indelible alterations in the connotations of magnetic vocabulary.

Franz Mesmer, still celebrated in some accounts as the pseudo-scientific forefather of psychoanalysis, arrived in Paris in 1778. Like other recognized physicians at the time, he claimed to have cured various ailments by strapping magnets to patients' bodies. Gradually moving away from this literal application of magnets to a more metaphorical vision of a universal magnetic fluid, he developed novel therapies known in France as animal magnetism, mesmerism, or somnambulism. The central feature of Mesmer's clinic was the baquet, an oaken tub filled with magnets, iron filings, flasks of treated water, and aromatic herbs. His treatment had various aspects, depicted in Figure 7.3. Gathered together in his fashionable salon reverberating with harmonious celestial music, wealthy patients clustered round the baquet, transferring healing magnetic powers to their bodies with ropes and iron bars. Mesmer also paid individuals—predominantly women—more special attention. Carefully aligning them to receive the earth's magnetic powers, he passed his hands round them to redirect the flow of magnetic fluid through their bodies. Contemporary accounts report these women experiencing healing crises resembling fits or trances.[68]

With the backing of affluent supporters, versions of mesmeric ideas became enormously popular throughout France. But in 1784, a government-sponsored committee headed by Benjamin Franklin vehemently denounced Mesmer's treatments, and he fled from Paris. For the first time, a few animal magnetizers

Figure 7.3. Franz Mesmer's Baquet. Superintended by Mesmer, standing on the right, patients cluster round his magnetic tub. Supporting himself on crutches, one gentleman wraps a rope round his painful ankle, while another assists a lady in the throes of a crisis. The original French title was "Le Bacquet de Mr Mesmer ou Représentation fidelle des Opérations du Magnétisme Animal." The inscription below this allegedly "faithful Representation" reads like an advertisement. In summary: Mesmer, Doctor of Medicine from the University of Vienna, has invented animal magnetism for curing numerous ailments, including dropsy, paralysis, gout, scurvy, blindness, and deafness. Mesmer can direct a fluid agent with his finger or an iron rod; patients also use his baquet, tying ropes around themselves or placing protruding iron rods near the diseased parts of their bodies. In addition, women in particular benefit from convulsions, while over a hundred of Mesmer's prestigious disciples accelerate the therapy by rubbing afflicted areas. Welcomed by cheerful music, with a tub for the poor on alternate days, sick people from every level of society mingle together, as rich gentlemen mesmerize the diasdvantaged. Undated French engraving by H. Thiriat. Wellcome Institute Library, London.

set up practices in London, rivalling other entrepreneurs eager to convert this discredited therapy into a financial source. William Godwin, for instance, earned some money by immediately—but anonymously—translating the damning Parisian report into English. His friend Elizabeth Inchbald was even more secretive about her translation of a French farce, which was repeatedly staged under her own name.[69]

English Enlightenment medicine was a competitive business, and practitioners of animal magnetism adopted different strategies for attracting patients in this pluralist market.[70] Some of them tried to recreate the charged magnetic atmosphere of Mesmer's salon. The Countess of Minto learnt how her husband had been to visit "Dr. Bell, one of the magnetising quacks, and the first whom I shall have seen." Based in Golden Square—a quarter of declining respectability—John Bell flamboyantly advertised his eight-foot baquet and other magnetic apparatus, emphasizing his Parisian training and his mesmeric allegiance. Although Bell practiced for only a few years, Sibly ensured that his name later reached a wider audience. "Animal magnetism," explained Sibly, "is a sympathy which exists between the magnet and the insensible perspiration of the human body, whereby an æther, or universal effluvia, is made to pass and repass through the pores of the cuticle." Like Bell, other struggling practitioners devised strategies for promoting diverse modifications of magnetic therapies and lecture courses. John Holloway, for instance, was a millenarian Methodist from Hoxton who stressed the spiritual benefits of animal magnetism; he enrolled his brother, a well-known engraver, to help him tap the provincial market.[71]

But by far the most famous animal magnetizer was John de Mainauduc, who repeatedly dissociated himself from the discredited Mesmer, his baquet, and all other magnetic equipment. He elaborated a magnetic therapy in which health was metaphorically bonded with a harmonious nature. Installed in prestigious Bloomsbury Square, de Mainauduc targeted a wealthy clientèle and reputedly earned a fortune from his lectures and magnetic treatments. Amongst his fashionable patients were leisured aristocrats, wealthy Quakers responding to his philanthropic claims, rival surgeons, and a small artistic group fascinated by occult topics. Famous visitors to de Mainauduc's fine residence included the Prince of Wales, General Rainsford (Banks' cousin and governor of Gibraltar), the portrait painter Richard Cosway, and the theatrical artist Philip de Loutherbourg. Unlike other practitioners marginalized as quacks, de Mainauduc was a trained surgeon and midwife, educated at major London hospitals, and a member of the Corporation of Surgeons. Apart from his choice of speciality, his career closely resembled that of the eminent William Hunter, also a self-made medical man of business. Like William Buchan, innovatory author of the hugely successful *Domestic Medicine*, de Mainauduc appealed to self-consciously improving polite audiences by focusing on the preservation of health before discussing the treatment of disorders.[72]

Figure 7.4 illustrates de Mainauduc's basic technique. He placed himself opposite his patient—usually a woman—and, staring intently into her eyes, moved his hands around her body without actually touching her. Lady Catherine Wright insisted to a skeptical physician that some patients were "so wrought on as to fall into a deep Sleep, some are thrown into strong convulsions, in short they suffer differently according to their various indispositions; Little Children cannot be impostors." Betsy Sheridan, the playwright's sister, confided to her journal that the "Duchess of Devonshire was thrown into Hysterics, Lady Salisbury put to sleep the same morning—And the Prince of Wales so near fainting that he turned quite pale and was forced to be supported."[73]

Seeking to validate their practices by establishing historical antecedents, animal magnetizers revived older associations of magnetic vocabulary with sex and the healing touch. De Mainauduc deliberately exploited traditional connotations of magnetic activity in his advertising. Yet despite this strategic appeal to the past, he concentrated on the same issues which preoccupied his

Figure 7.4. "*ANIMAL MAGNETISM*—The Operator putting his Patient into a Crisis." Although this practitioner is unnamed, his technique illustrates de Mainauduc's description of the animal magnetizer's power over his patients, who were predominantly women. This picture shows the powerful gaze therapeutically bonding the couple together and the magnetic effluvia being redirected by the magnetizer's hands. From Sibly, *Key to Physic*, p. 260. Wellcome Institute Library, London.

establishment contemporaries. They shared models bonding the universe, society, and the animal economy, which they pictured with similar imagery. Apart from the relatively brief sections on animal magnetism, the published version of de Mainauduc's theories resembles those formulated by his sanctioned contemporaries.

Like many people of this period, de Mainauduc was interested in analyzing the distinctions between the living, the inanimate, and the dead. Eminent physicians who rejected materialism used magnetic analogies for portraying the relationship between mind and matter. John Hunter declared that "magnetism provides us with the best illustration we can give" of the vital principle which he claimed was superadded to ordinary matter in living beings. Just as the internal structure of an iron bar is the same before and after magnetization, he argued, so inert matter acquires new properties, "a soul to put and continue it in motion." Darwin compared hidden magnetic forces drawing two pieces of iron towards each other with his "spirit of animation" bringing particles of the body together to cause muscle contraction. Far less radical than men like Priestley and Thomas Beddoes, de Mainauduc also taught that life depends on an immaterial principle, an invisible power guiding bodily motion. He described this spirit's commands as "Man's Volition . . . without which the Body must remain silent and passive." The effectiveness of animal magnetism lay in mobilizing this power of volition—the same term which Darwin used to describe how his spirit of animation changed the sensorium to produce muscular motion.[74]

De Mainauduc lectured about a comprehensive scheme of nature in which inanimate matter and living beings were linked through a constant circulation of atoms. He shared other physicians' therapeutic approach of assisting nature to restore a state of equilibrium. According to him, the magnetic therapist exerted his volition to divert the diseased atoms emanating from a sick person's body; by removing obstructions, the patient's normal circulation of healthy atoms would be resumed. Figure 7.4 shows the small particles travelling between the magnetizer's hands and his patient's body. Like John Hunter, de Mainauduc instructed his students that at death the body's atoms are recycled into nature. As with Hunter and other medical experts, de Mainauduc's universe of circulating particles mirrored a harmonious commercialized society, whose health relied—like the body's—on the free movement of self-motivated entities contributing towards the larger well-being.[75]

As they explored the interaction of the mind and the body, many physicians adopted Robert Whytt's doctrine of sympathetic communication along the nerves. Modelling their social theories on Whytt's nervous links between the body's organs, Scottish philosophers developed sympathetic theories of the bonds uniting men of feeling. Reflecting the close associations between magnetic and sympathetic action, John Hunter included animal magnetism in his long discussion of sympathy and referred sensations. Although boasting of his own ability to resist the magnetic influence, he did not deny that other people perceived its effects. Ruled by a king who was apparently becoming insane, English people were fascinated by the relationship between bodily and mental aspects of health. They favored physical explanations of illness, fearing descent into madness if reason were to be overrun by a diseased imagination. De Mainauduc's account of his power over his patients was a comforting one, conventionally mechanical yet devout. Just as sound was conveyed by atmospheric nerves, he explained, sensations were conducted along human nerves, chains of atoms linked together into a single system. With God's acquiescence, the mind's volition could direct this atomic activity.[76]

Medical practitioners of the late eighteenth century sought visual possession over bodies by making the invisible visible. Figure 7.4 illustrates the importance animal magnetizers attached to dominating their patients by eye-to-eye contact. Some adherents literally turned their gaze inwards, diagnosing their own internal complaints as well as those of others. This medicinal stare was not restricted to men like de Mainauduc. George III's unorthodox physician, Francis Willis, demonstrated to Edmund Burke how he controlled his royal patient by fixing him with his eye. Submitting to Willis' gaze, Burke averted his head and acknowledged what a contemporary called Willis' "basiliskan authority." Recalling Neoplatonic beliefs in universal magnetic sympathies—visualized in Figure 6.1—this reference to a basilisk conjured up traditional associations between snakes terrifying their prey into submission and loadstones acting at a distance upon iron. Snakes, venomous creatures which paradoxically symbolized healing, fascinated writers of this period, who speculated about their optical powers of enchantment: the *Encyclopædia Britannica* asserted that the serpent's power of fascination was analogous to magnetic attraction.[77]

Unlike in Paris, there was no official condemnation of animal magnetism. Cavallo, the Royal Society's magnetic expert, was in an awkward position. Already worried that his Catholicism was adversely affecting his career, Cavallo was a frequent visitor to the soirées of émigrés and artists held by the Italian-born "Goddess of Pall-Mall . . . alias the Magnetic Muse" Maria Cosway. The Cosways, well known as disciples of de Mainauduc, lived in Schomberg House, former home of Graham's Temple of Health, with its magnetic Celestial Bed. Although scathing in private correspondence, Cavallo declined to comment in public. Most establishment physicians were similarly reticent. Prestigious practitioners were susceptible to two major accusations levelled at the animal magnetizers: charging exorbitant fees and keeping their methods secret. The king tacitly endorsed empiric remedies through the patent system, and he himself was being successfully treated by a doctor whose expertise was questioned. Furthermore, animal magnetism was reminiscent of the regally sanctioned practice of touching for the king's evil. As with other contested therapies, the frequent allegations of quackery came from outside rather than inside the medical profession.[78]

Numerous skeptics found it advantageous to criticize animal magnetism, which became a favored term of abuse for denigrating a great variety of groups, including physicians, natural philosophers, religious adherents of every allegiance, political factions, French people, and women. In other words, the practice served as a versatile vehicle of mockery for diverse critics. They did not necessarily deny the reality of the immediate effects being achieved by the animal magnetizers; rather they expressed concern about their explanations and the propriety of their promotional and therapeutic techniques.

As members of fashionable society were drawn into de Mainauduc's magnetic salon, satirists converted his therapeutic sessions into entertaining parody, packed with sexual and scatological innuendo. Hostile pamphlet writers boosted their sales by describing in salacious detail the "magnetic conferences" between thinly disguised characters dubbed Major MacNeedle or Lady

Figure 7.5. "Animal Magnetism . . . or Count Fig in a Trance." Within a few years of its arrival in London, animal magnetism had become a familiar term of mockery. An "Inchanted Wire" electrically stunned Count Fig, a grocer called Peter Wheeler sporting a canine physiognomy and addicted to "boyish Pranks." The cartoonist celebrated his downfall by choosing a phrase—animal magnetism—conjuring up duplicity, credulity, and exploitative profiteering. Published 2 July 1789 by W. Price of Tower Hill; 7¾ by 13¼ inches. British Museum.

Bumbustle. The prevalence of these critiques reinforced the magnetizers' own expansion of the metaphorical references of magnetic vocabulary. Animal magnetism entered common parlance as a term synonymous with farcical stupidity, guaranteed to raise a laugh. Soon after Inchbald's pirated skit *Animal Magnetism* opened at the Theatre Royal, the cartoon reproduced as Figure 7.5 appeared. One of a series, it illustrated the fate of Peter Wheeler the grocer, otherwise known as Count Fig, victim of a revengeful prankster's decision to electrify a bellpull. Its title indicates how fully animal magnetism had entered everyday language, carrying multiple associations of trances, gullibility, and chicanery.[79]

Insisting that animal magnetism operated through an explicable sequence of physical events, writers promoting Enlightenment rationality fashioned their élite roles as protectors of less-privileged classes. Interests varied. Godwin, for instance, stressed the importance of studying errors in order to distinguish "what different instruments were necessary to deceive mankind in an ignorant and an enlightened age." On this issue, the radical philosopher was aligned with the Tory Hannah More; writing to Horace Walpole, she included de Mainauduc along with slavery in her "disgraceful catalogue" of "prejudice, ignorance, and superstition." As Swedenborgians, Methodists, and Quakers each accused the others of behaving like animal magnetizers, Walpole himself ironically welcomed therapeutic and religious diversity: "[T]hey inveigle proselytes from one another. I used to be afraid of the hosts of Methodists, but Mother Church is safe if there is plenty of heresiarchs, and physicians pretend to a vocation too."[80]

The great majority of treatments entailed male practitioners' exerting their powers over female patients. As men and women explored this asymmetrical relationship, they molded gender relations. Following the lead of French critics, English authors reinforced images of female inferiority by attributing their perceived susceptibility to physical causes—weak nerves and that conveniently influential organ, the uterus. Endorsing denigrations of pernicious French femininity, chauvinists combined assertions of both male and national supremacy by stressing that it was Parisian women who had initiated this therapeutic fashion. For moralizing Mary Wollstonecraft, animal magnetism was one of the follies—along with sentimental novels and elaborate clothes—to which women fell prey because of their ignorance. How can you believe, she condescendingly demanded, "that these magnetisers, who, by hocus pocus tricks, pretend to work a miracle, are delegated by God?" Warning women to avoid such blasphemies, she prescribed the virtuous route to health: "[I]t is easier . . . to be magnetised, than to restrain our appetites or govern our passions."[81]

Both political parties used animal magnetism to mock the opposition. During the regency crisis, the king's perceived insanity provided a powerful metaphor for a disintegrating society. Seeking to discredit the Whig triumvirate, the *World* ran a scathing series immediately after Burke had been defeated in a key vote on the constitution. The paper interleaved hospital data on insanity, Shakespearean snippets, and pastiches of Sheridan's plays with the prophetic visions of Ann Prescott, de Mainauduc's assistant, while in an animal magnetic trance. "I see," she allegedly forecast, "that derangement will yield to remedies, and that all will be well. . . . coming to himself, [the king] will say, that if my measures had been altered, they would have made me mad indeed!"[82]

Figure 7.6 illustrates how, a few months later, the Whigs employed the same satirical weapon of animal magnetism to caricature William Pitt. With his gouty leg swathed in the bandages of excise, Pitt begged relief from the fumes of discontent directed against his recent bill to tax tobacco. The magnetic doctor was de Mainauduc's pupil de Loutherbourg, notorious for advertising free treatment at his Hammersmith home until besieged by hundreds of hopeful incurables. Coleridge may have had this cartoon in mind when he

Figure 7.6. “Billy’s Gouty Visit, or a Peep at Hammersmith.” William Pitt limps into the Hammersmith home of Philip de Loutherbourg, the theatrical artist mocked for attracting incurable patients by dispensing animal magnetic therapy free of charge. The triangle at top right links de Loutherbourg with two other men marginalized as quacks, Graham and Katterfelto (see Figure 2.9). The scroll across the lower right advertises “Cures by a Touch” for maidens suffering from “a nine months Dropsy,” deaf judges, blind statesmen, lethargic bishops, etc. Published by W. Dent, 20 July 1789; 6⅜ by 13¼ inches. British Museum.

denounced Pitt as "the great political Animal Magnetist, who has most fully worked on the diseased fancy of Englishmen; and by idle shew, and alarming bustle, and many a mysterious trick has thrown the nation into a feverish slumber, and is now bringing it to a crisis which may convulse mortally!"[83]

Because animal magnetism had been imported from France, it gained new life as a token in ideological debates after the French Revolution, when de Mainauduc had already been practicing for several years. The initial élite panic that revolutionary ideas might cross the Channel to disrupt British society was followed by propaganda designed for popular audiences. Linking animal magnetism with negative images of France, polemicists represented the therapy as foreign and dangerously subversive, a "diabolical practice" originating "in that *antichristian Empire of Atheism*, France [which] leads directly to the motto of Atheism, that '*Every man is his own God.*'" Conspiracy theorists like Robison, convinced that civic upheaval was being fomented by a clandestine international network, indiscriminately bracketed magnetizers, Rosicrucians, alchemists, and exorcists as mystical fanatics. Defenders of the status quo spread rumors that John Wilkes had become one of de Mainauduc's disciples; they portrayed mesmerism as a dangerous sect allied with the French Revolution and the visionary Richard Brothers. In his fictionalized political commentary *Letters from England*, Robert Southey linked his long critique of de Mainauduc's animal magnetism to the revolutionary potential of popular leaders.[84]

During the 1790s, people of conflicting political views invested experimental phenomena with different interpretations. Conservatives concerned about the revolutionary implications of natural philosophy ridiculed innovatory ideas, particularly those which originated abroad. The authors of pamphlets attacking men like Priestley and Beddoes intertwined denunciations of their political affiliations and religious beliefs with their research into nature. Animal magnetism, pneumatic chemistry, and galvanism became linked together as part of the smear campaign against materialist philosophers. For instance, "Sceptic" directed his anonymous fantasy of "The Birth of Wonders!" at Humphry Davy. His fable opened with the arrival of the French Revolution, followed by his little sister "*Mesmeria* . . . [whose] eye fascinated and charmed to the spot . . . [and whose] mysterious power of inchantment, subverted the laws of nature." The next baby wonder leapt into a frog Luigi Galvani was eating for dinner, while the brothers Antiphlogiston and little Caloric fooled the world. In the cartoon reproduced as Figure 7.7, Mesmer's aristocratic salon (Figure 7.3) has been transformed into a lower-class scene of sexual licentiousness. The bewigged gentleman on the left, assisting a buxom woman to pull on her exaggerated iron bar, might be de Mainauduc. The men directing operations are wearing Nonconformist hats, recalling Priestley, the Unitarian minister from Birmingham. Their piglike faces identify their political allegiance: following Burke's reference to "the swinish multitude," radicals deliberately adopted a porcine vocabulary for describing their activities.[85]

Figure 7.7. Caricature of Magnetic Treatment. This cartoon of "a room where higher folk / Were toiling in th'impostor's yoke" illustrated a long poetic epistle on animal magnetism sent from metropolitan Squire Quoz to his country uncle. By characterizing animal magnetizers as asinine quacks, the poem satirized the College of Physicians. The portraits are of Philip de Loutherbourg, Dr. Yeldell—a practitioner who used magnetic apparatus in Moorfields—and Ma——almost certainly de Mainauduc. *Attic Miscellany* (1791), facing p. 121. Wellcome Institute Library, London.

As critics forged multiple rhetorical links between animal magnetism and other practices, they indelibly altered the figurative associations of magnetic terminology. While the practitioners themselves deliberately revived older magnetic connotations, antagonistic reviewers ensured that animal magnetism became a familiar term, albeit one which people used in a variety of ways. Even the prefix "animal" became redundant: de Loutherbourg's clinic only needed the label "Magnetism" (lower right in Figure 7.6), while in court circles Colonel Manners "whispered Miss P—— that . . . he intended to introduce magnetizing." Magnetic vocabulary took on new meanings during the eighteenth century not solely because of transformations in natural philosophy but also because of the constant metaphoric interplay amongst discourses. Just as diverse natural philosophers drew on figurative imagery for deciphering nature's magnetic mysteries, so too authors of literary and religious texts devised metaphors based on contemporary theories of the natural world, life, and medicine. The processes of excluding animal magnetism from legitimated magnetic science contributed to its permanent effects on magnetic meanings.

Earlier Neoplatonic connotations were revived by advocates and adversaries alike, transmuted by their individual experiences of animal magnetism. In his popular medical text, as Sibly endorsed Bell's version of Mesmer's baquet he interwove magnetic effluvia with contemporary French vital fluids; he also imbricated them with the occult sympathies bonding plants and animals portrayed in Figure 6.1. Coleridge, at school in London during de Mainauduc's peak, frequently used images referring to the fascinating power of glittering eyes. A wedding guest submitted to the ancient mariner like a patient to a magnetizer: "For that which comes out of thine *eye* doth make / My body and soul to be *still*." Maria Cosway complained to her lover Thomas Jefferson that she felt herself magnetically susceptible to London's depressing influence when he failed to write to her. Godwin, the purveyor of Enlightenment reason, also experienced the pangs of magnetic separation. In private, he assured his wife that animal magnetism drew them towards each other even when they were apart; in public, he described the "magnetical sympathy" powerfully bonding Caleb Williams to the patron who pursued him.[86]

Fleeting references suggest that animal magnetizers may have been practicing in England between this late-Enlightenment period and the nineteenth-century revival initiated by Coleridge after his return from Germany. More significantly, animal magnetism survived in print, both as the continuing subject of hostility by commentators like Southey and in the metaphorical imagery of Romantic authors. Fanny Burney, for instance, no longer simply compared a woman's ability to draw her man with a loadstone's power over iron. Instead, attraction became interaction, "the magnetizing powers of reciprocated exertions." Placing a heroine before a Gothic building, she described "the magnetic affinity, in a mind natively pious, of religious solemnity with sorrow." When Shelley defended the role of poetry in an industrializing society, he modulated the magnetic rings of divine inspiration which he had translated from Plato's *Ion* into a universal mesmerizing agent: "The sacred links of that chain have never been entirely disjoined, which descending through the minds of many men, is attached to those great minds, whence as from a magnet the invisible effluence is sent forth; which at once connects, animates and sustains the life of all." As writers and practitioners used animal magnetism in new ways for various ends, magnetic metaphors continued to shift as they resonated between scientific and literary texts.[87]

Although facilely dismissed as quackery, this brief episode of animal magnetism illustrates the linguistic interactions amongst practices marginalized as pseudo-science, ones which gain entry into the accepted canon, and those apportioned to distinct literary discourses. Diverse critics collectively outlawed animal magnetism, even though it shared many characteristics of contemporary activities which have been retrospectively legitimated as precursors of modern science. De Mainauduc's lectures bore disturbing resemblances to those of accredited physicians and natural philosophers. Animal magnetizers threatened the privileged status of élite practitioners by dangerously straddling fragile distinctions: the genius fired by divine insight and the madman under the sway of his passions; experimental philosophers controlling the forces of nature and theatrical performers duping their audiences; the clinical gaze of legitimated therapists and the mesmerizing stare of powerful charlatans. As natural philosophers struggled to establish a prestigious identity during this period of social upheaval, they located animal magnetism beyond the boundaries of their new scientific disciplines.

CONCLUSION

As British people forged a commercial nation, moral philosophers described the virtues appropriate for this polite society characterized by an expanding public sphere. Many agreed with the arguments of Alexander Gerard, a Scottish associationist who attributed a key role to the power of abstract reasoning. For portraying the genius of imagination, Gerard turned to the physical world, forcefully articulating how magnetic interests extended sympathetically throughout eighteenth-century life:

> As the magnet selects from a quantity of matter the ferruginous particles, which happen to be scattered through it, without making an impression on other substances; so imagination, by a similar sympathy, equally inexplicable, draws out from the whole compass of nature such ideas as we have occasion for, without attending to any others; and yet presents them with as great propriety, as if all possible conceptions had been explicitly exposed to our view, and subjected to our choice.[1]

Building on a framework of current historiographical interests, this book has constructed an image of eighteenth-century England emerging from its past rather than pointing forwards to a future that was still unknown. Such an approach promises a double reward: painting a historical landscape recognizable by the people inhabiting it; and improving our understanding of how our own values have been established. Directed by a concern with the modern preeminence of scientific activities, this study has deliberately focused on a period before the creation of disciplinary science. Natural philosophers actively participated in the massive transformations of English life during the eighteenth century. Their bids for public accreditation meshed with struggles for redefinition by other groups in mutual processes of self-fashioning. Controversies surrounding natural philosophy resonated with religious, economic, moral, gender, and political implications. Exploring these conflicts thus benefits from, but also enriches, the wider historical literature analyzing the century's shifting terrains of power. By exploring the negotiations underlying the construction of scientific disciplines, we gain a deeper appreciation of today's cultural topography.

Before magnetic phenomena became parcelled off into scientific specialties, natural philosophers based their investigations on everyday experiences. They experimented with household objects like pokers and keys and benefited from the practical skills developed at sea by mariners. Revered as vital for navigation and commerce, the magnet's hidden powers conjured up occult virtues and symbolized divine omnipotence. Familiar items for a maritime people, compasses guided the expansion of an imperial nation as well as the spiritual voyages of fallible human souls. Visionary prophets of progress charted out an advantageous course for a greater Britain conducted by this acclaimed invention, while individual entrepreneurs profited by displaying their control over nature's magnetic secrets. Pious believers in divine intervention preached lessons of magnetic virtue from biblical interpretations; but, reverberating with traditional connotations of healing and sex, magnetic imagery also pervaded gendered commentaries about the mysteries of life.

Many of these themes have been illustrated by deciphering the polemical messages conveyed by James Barry's celebration of England's mercantile supremacy, *Commerce, or the Triumph of the Thames*. By including a mechanical artifact in his allegorical vision, Barry extended the scope of civic humanist historical art to embrace the virtuous actions of participants in an industrializing society. This painting forms an appropriate frontispiece because—like this book, but with very different aims—Barry invoked the past to display how compasses united the rhetorics of commercial expansion, national glory, and invention. As natural philosophers reiterated Bacon's insistence on utility, they emphasized how their innovations enabled British fleets to dominate the terraqueous globe. Supplanting navigators as the legitimated guardians of magnetic expertise, they levered themselves into a powerful position by sculpting their identity as indispensable contributors to the nation's prosperity. Just as artistic and literary creations were converted into cultural commodities, so too the products of natural philosophy were marketed as essential purchases for polite consumers. Commercialization and the development of a public science were interdependent. Echoing philosophical and medical models of circulating nature, Burke complacently ratified the hardships suffered under English society's "great wheel of circulation" by declaring that "the laws of commerce . . . are the laws of nature, and consequently the laws of God."[2]

Barry also portrayed how progress should be achieved: he envisaged the future face of science. He gave his unnamed female Nereids the secondary role of carrying industrial products, depicting these scantily clad maidens as "more sportive than industrious, and others still more wanton than sportive." The weightier responsibility of transporting Father Thames' shell was assumed by male heroes, individually identified as famous English navigators. Contemporary image makers constructed similar eulogizing scenes that signalled class and gender differentiations. As they defined scientific disciplines, natural philosophers wrote retrospective histories aggrandizing men of genius for their momentous discoveries. Countless other experimenters and their faceless assistants fell into oblivion. Commentators—male and female—generated texts reinforcing ideological judgments rendering women unsuited both for experimental research and for understanding abstract laws. As élite landscape painters fortified their superior vantage points, they blurred the figures of distant field laborers into anonymity. Writers and artists commemorated the inventors and factory owners producing profitable new goods, yet effaced the workforce responsible for their manufacture.[3]

Before the end of the seventeenth century, educated gentlemen had canonized one of their own as imperial England's magnetic hero:

> *Gilbert* shall live, till *Load-stones* cease to draw,
> Or *British* Fleets the boundless Ocean awe.

Commentators have adapted this iconic figurehead to serve particular ends. By pronouncing that "this work of Dr Gilbert's . . . contains almost every thing that we know about magnetism," Robison underpinned his own insistence on Baconian observation. Jones of Nayland reversed the argument: he contrasted Gilbert's cosmology unfavorably with the Baconian experimental methodology he himself claimed to practice. Perpetuating Gilbert's own suppression of maritime instrument makers, historians of science—notably Thomas Kuhn—have celebrated Gilbert for making Baconian observation the foundation of natural philosophy and hence of mathematized science. By considering magnetic concerns beyond the restricting Whiggish perspective of "prescientific" activity, this book has provided a very different interpretation.[4]

During the eighteenth century, natural philosophers—concerned to reinforce their own status by stressing individual prowess—erected new magnetic champions. One of these was "the truly celebrated Dr Gowen Knight, FRS," then constantly lauded for his magnetic skills, although now excluded from the hall of fame.[5] Well into the nineteenth century, writers praised Knight for applying the techniques of experimental philosophy to navigation, thus facilitating commercial and imperial expansion. Knight exemplified a new class of upwardly mobile entrepreneurs who carved out novel routes to success in many fields. Paralleling the tactics of contemporary innovators establishing careers—such as artists, lecturers, and publishers—he carefully maintained a gentlemanly status by distinguishing himself from popularizers and instrument makers. Knight used Adams' shop as a retail outlet and scarcely mentioned the workers assembling the compasses certified with his name. Natural philosophers increasingly marginalized skilled craftsmen, casting them as technicians no longer welcomed as Fellows of the Royal Society.

The other magnetic hero of the eighteenth and nineteenth centuries was Halley, portrayed as the founding father of geomagnetism. Largely oblivious of mariners' critiques, polemicists have hymned him as the scientific voyager who condensed his observations into useful theory. As Halley's successors constructed a public science of magnetism, they continued to reformulate maritime expertise. Enrolling international travellers as unpaid assistants, they transformed the tools of mathematical practitioners into precision instruments for ordering nature. Replacing traditional navigational methods by quantified techniques, they regulated magnetic information into systematized laws advertised as scientific knowledge.

But it was, of course, Newton who became the quasi-mythological icon of English science. Elevating him to a position of invincible authority, propagandists interpreted his writings to endorse diverse views in multiple debates. Newton himself had secured some tenets of his natural philosophy from being empirically disproved. Jones of Nayland sneered that "N. himself . . . puts his Philosophy into such a dress, that it might be hustled about only among those men who would be ready to deify him: & it answered his purpose." By declaring his work to be certain truth, his successors transformed it into a flexible tactical resource for both explaining and explaining away experimental results. They successfully established a Newtonian ideology which, still enduring, conceals the variations embedded within it. By continually reformulating Newtonian doctrine to suit their own ends, scientists and historians have obscured opposition to Newton's philosophy.[6] From the end of the seventeenth century, natural philosophers solicited approval by labelling their theories with the master's name. They increasingly silenced the voices of numerous critics, many of them Tory High Church Anglicans. As these dissidents pursued their own religious, political, and philosophical agendas, they found it increasingly advantageous to adapt themselves to a Newtonian mold.

A closet Hutchinsonian, Jones belonged to the most aggressive sect challenging the Newtonian orthodoxy. Hutchinsonians valued confrontation for releasing the inherent power of natural language; dissenters of a more mystical or Behmenist orientation believed in the force of prayer, so their objections are less evident. The Hutchinsonians clandestinely plotted strategies for undermining Newtonian domination but publicly circumvented hostility by denying their allegiance when publishing texts or preaching sermons. Jones argued at length against Newton's philosophy of attraction, dramatizing inconsistencies by compiling his opponents' contradictory explanations into a fictionalized interchange of incompatible assertions. He boasted that "Gentleman of the Newtonian side . . . begin to be alarmed about me at Cambridge, & are putting people on their guard." The Tory *British Critic* was the only journal which printed Hutchinsonian views. In other outlets, commentators supplemented philosophical reasoning with rhetoric in order to denounce opinions which undermined the Newtonian edifice. Antagonistic reviewers summarily dismissed books like Jones' which dared to attack the Newtonian system. "The principles of Sir Isaac Newton, form the alphabet of nature's language," thundered one reviewer; any inquiry not building on "those elementary characters . . . is absolutely *wrong*." Hutchinsonian concepts were largely eliminated from the public face of science, although they continued to inform the attitudes of some scientific writers.[7]

Whether claiming Hutchinsonian or Newtonian allegiance, allies fashioned their collective identity by caricaturing opponents as a negative other. Natural philosophers transformed their status by vaunting their own achievements but also adopted various tactics for demeaning practitioners whom they sought to outlaw from the domain of accredited science. So while Newtonian exponents excluded Hutchinsonians as Hebraic enthusiasts, they also distinguished scientific magnetism polemically by comparing it with outdated maritime tradition, superstitious folklore, medical quackery, artisanal mechanics, foolish entertainment, enthusiastic biblical pedantry, or literary imagination. They were not operating in isolation: other groups were similarly claiming authority by delineating their domains from threatening neighbors. Thus naval inventors distanced themselves from landlubberly natural philosophers, preachers extolled a return to faith through prayer, and academic scholars defined such new areas as philology, anthropology, and theology. Towards the end of the century, specialists asserting sovereignty over other fields shaped scientific disciplines while modelling their own identities. Romantic visionaries intended their portrayals of scientific activities to emphasize their distinctiveness from literary and artistic pursuits. As Blake portrayed Newton capturing nature with his mathematical dividers and Wordsworth immortalized "Newton with his prism and silent face," they consolidated visions of scientific genius dispassionately elucidating ruthless natural laws.[8]

While Newton acquired almost superhuman status, writers started to construct new models of the active scientific hero. Natural philosophers of the eighteenth century had validated magnetic research by demonstrating its public benefits. Through successfully justifying their activities, they gave their scientific successors the security to prise apart science and immediate commercial reward. Praising volunteers who braved the rigors of polar travel, the naval administrator John Barrow castigated cynics who clung to financial criteria for judging an expedition's utility. He declared that England, a prosperous, enlightened nation, should encourage the pursuit of scientific knowledge for its own sake. Arguing that "the benefits of science are not to be calculated," he asked: "Who could have imagined that the polarity of the magnet, which lay hid for ages after its attractive virtue was known, would lead to the discovery of a new world?" Heroic explorers would never travel in vain, for they were "the means of extending the sphere of human knowledge; . . . 'knowledge is power'; and we may safely commit to the stream of time the beneficial results of its irresistible influence."[9]

Barrow was writing at a time when natural philosophers were disciplining magnetic nature to construct a new scientific specialty; they also contributed towards disciplining the nature of society. Naval reformers introduced scientific techniques of navigation, but they also replaced paternalistic structures by more impersonal models of organization. Popularizers of natural philosophy constructed receptive audiences of women and the less privileged, characterizing them as inherently incapable of innovative investigation. Whereas devotional poets had magnetically drawn the human soul towards God, literary authors summoned up magnetic laws of nature for reinforcing a woman's role as domestic custodian, ineluctably bound to a single man. Abstract reasoning became celebrated as a masculine capability: scientific laws simultaneously determined the nature of women and ensured that women would be unable to understand feminine nature.[10]

Although rhetorically differentiated, scientific and literary discourses enriched each other. Fanny Burney observed that people often found "the fascination of intelligence" less attractive than "the magnetism of mystery and wonder." Nature's magnetic powers remained an alluringly mysterious secret. Lecturers and instrument makers included magnetic deceptions amongst their rational entertainments, yet scientizing writers sought to purge magnetic vocabulary of magical connotations. In his *Botanic Garden*, the versified counterpart of Barry's allegorical hymn to progress, Darwin carefully distinguished poetic and scientific styles. He stated that his aim was "to enlist Imagination under the banners of Science." Like "magnetism," the word "imagination" was ambiguous. One reason for the intense hostility towards the animal magnetizers was that imagination slipped between physical reality, legal intention, artistic creativity, and mental confusion. Misquoting Darwin, Maria Edgeworth enlisted science under the banners of imagination, which she gave a medical embodiment. By protectively recommending female exclusion from exciting exploration, she endorsed the mirror image of a male scientific hero: "Chemistry is a science well suited to the talents and situation of women; it is not a science of parade; . . . it demands no bodily strength; it can be pursued in retirement . . . there is no danger of inflaming the imagination."[11]

Introducing *Frankenstein* to the nation, Percy Shelley rejected as "imagination" Darwin's notion that life could be restored to dead flesh. Echoing Gerard, he valued Mary's creative use of her "imagination for the delineating of human passions." Subtitling her romance "The Modern Prometheus," Mary Shelley—like Barry and Darwin—rooted her image of contemporary science in the classical past familiar to their audiences. She depicted three types of scientific investigator: Captain Walton, the heroic explorer romantically seeking out the mysteries of the Arctic region; Victor Frankenstein, the alchemical antihero punished for presumptuously harnessing electricity to usurp nature's creative functions; and Frankenstein's creation, the excluded Adamic traveller detachedly observing human nature. By nesting their narratives, Shelley conducted her readers from science's outer appearance to its inner heart.

Magnetic imagery relates to her fiction at several levels. Walton explained to his sister, patiently waiting at home, that he was searching for the secret magnetic polarity concealed beneath the polar ice. Frankenstein, on the other hand, refrained from telling Elizabeth he was penetrating nature's innermost treasures. Darwin, one of the Shelleys' intellectual guides, had compared magnetic powers with the vital principles animating ordinary matter; for Mary, as with the secrets of life itself, nature guarded her magnetic mysteries from masculine scientific inquiry. The creature was invisibly bonded to the creator who pursued him; he resembled both the wandering soul magnetically attracted towards God and Caleb Williams, the persecuted man tied by "magnetical sympathy" to his patron in the novel by Mary Shelley's father, William Godwin.[12]

Many commentators have interpreted *Frankenstein* as a prescient critique of modern science. But writing before the word "scientist" had even been invented, Shelley was exploring the models of scientific activity being constructed during her own lifetime. Natural philosophers and physicians had not yet consolidated any unambiguous status of authority: their identities were still unstable. As they created the public face of science, men like Barrow and

Walton advertised the search for truth that would benefit the nation. With his confident boast that "you cannot contest the inestimable benefit I shall confer on all mankind to the last generation," Walton epitomized Barrow's disinterested idealist willing to sacrifice his life for the sake of powerful knowledge. Through Frankenstein himself, Shelley articulated reservations similar to those expressed by Tory High Church critics of Newtonian projection during the eighteenth century. She reformulated their warnings about investigating the secrets of God as a moral tale describing the consequences of probing deep into the mysteries of nature. Reflecting too late on the outcome of his own inquiries, Frankenstein rebuked Walton for seeking the recipe of life. "'Are you mad, my friend?' said he; 'or whither does your senseless curiosity lead you?'" The creature has come to symbolize monstrosity, yet this disillusioned spectator of society reflected human depravity. Lamenting, "No sympathy may I ever find . . . now, that virtue has become to me a shadow," he disappeared from Walton's cabin into the "darkness and distance."[13]

Like this offspring of Shelley and Frankenstein's creative imagination, magnetic sympathies were excluded from scientific discourse but figuratively remained within it. Modern science is not the monstrous progeny of an instantaneous act of creation but has been collectively fashioned by people engaged in diverse activities in the past. Investigating that past enriches our appreciation of the present.

APPENDIX

MAGNETIC LONGITUDE SCHEMES

The Board of Longitude was established in 1714 to administer government awards, on a sliding scale up to a maximum of £20,000, for the first person to find a method of ascertaining the longitude at sea to an accuracy of 30 miles. In 1774, the Board's scope was extended to cover more general navigational achievements; it was dissolved in 1820. The first meeting recorded in the confirmed minutes was held in 1737 to discuss John Harrison's new chronometer. The early records of the Board are incomplete, and they are sporadic before about 1780. This appendix summarizes claims made betweeen 1714 and 1800 for discovery of a magnetic method of determining longitude. The list is arranged alphabetically, giving major primary and secondary references for each person. It excludes some of the very brief suggestions recorded in RGO 14/42.

BARATIER, JOHN

Wrote to the Royal Society from Berlin to claim a reward for his method based on magnetic dip.

RSJB, vol. 16, pp. 181–2 (26 January 1737).

CHURCHMAN, JOHN

A Philadelphia surveyor who campaigned extensively in America and Britain to sell his new charts of dip and variation, which were designed with gores so that they could be glued to a globe.

RGO 14/42, cut 5. Churchman, *An Explanation of the Magnetic Atlas* and *The Magnetic Atlas.* Ravenhill, "Churchman's Contours?"

EBERHART, CHRISTOPHER

Sent in to the Royal Society a method based on dip: William Whiston said Eberhart stimulated his own researches.

RSJB, vol. 11, p. 261 (30 October 1718).

FRENCH, JOHN

Mathematics teacher whose method entailed turning a compass needle by lighting a fire underneath it.

E.G.R. Taylor, *Mathematical Practitioners of Hanoverian England*, pp. 118–19. Stewart, *Rise of Public Science*, pp. 200–1.

GRAHAM, W.

A man from Newcastle who invented a sliderule to calculate longitude from the variation.

RGO 14/42, cut 4.

Hauxley, Edward

A London navigation teacher and textbook writer who devised a longitude method based on magnetic variation and spherical trigonometry.

Gentleman's Magazine 20 (1750), 78–9, 150, 201–2, 271. E.G.R. Taylor, *Mathematical Practitioners of Hanoverian England*, pp. 179–80.

Jackson, Joseph

A London instrument maker who sent several vaguely worded appeals to the Longitude Board.

E.G.R. Taylor, *Mathematical Practitioners of Hanoverian England*, p. 181. RGO 14/42, cut 1.

Knight, Gowin, FRS

The Board of Longitude paid him £300.0.0 in 1752 for his new variation compass.

Knight, "Description of a Mariner's Compass." See Chapters 2 and 3 of this book.

Lovett, Richard

Lay clerk at Worcester Cathedral who wrote mainly about ethereal theories of electricity and its medical applications.

Lovett, *Philosophical Essays*, pp. 420–9, 445–520. J. Chambers, *Biographical Illustrations of Worcestershire*, pp. 363–4.

Nugent, P. R.

Former surveyor-general of the Island of Cape Breton who proposed a geometric method of ascertaining longitude, based on the asumption that the dip is essentially constant.

Nugent, "A New Theory."

Roberts, James

A lieutenant who invented a compass to measure dip and variation.

RGO 14/42, cut 11.

Tulloch, John

A schoolmaster who invented variation compasses and magnetic perpetual-motion machines to find the longitude.

RGO 14/42, cut 8.

Vico, Charles de

Wrote to the Royal Society from Lisbon about his magnetic needle for finding the longitude.

RSJB, vol. 17, p. 251 (27 November 1740).

Walker, Ralph

A Scottish settler from Jamaica who received £200.0.0 from the Board for his variation compass marketed by George Adams, and who compiled theoretical tables for finding longitude.

RGO 14/4, fols. 17r–19r. Also fols. 24, 31v, 35r, 36r, 41r, 42r, 43r. RGO 14/17, fols. 5–6. Walker, *Memorial of Ralph Walker* and *Treatise on Magnetism.* May, "Longitude by Variation."

Whiston, William

Mathematical lecturer who raised money by subscription to develop a longitude method based on magnetic dip.

RGO 14/5, fols. 6–7. Whiston, *Longitude and Latitude.* Stewart, *Rise of Public Science*, pp. 185–208; Farrell, *Life and Work of William Whiston*, pp. 116–83.

Williams, Zachariah

A Welsh physician, sponsored by Samuel Johnson, who asserted that he had derived the law governing magnetic variation and could use it to find the longitude.

RSJB, vol. 13, p. 323 (7 April 1729); vol. 17, pp. 99 and 428 (8 May 1740 and 17 June 1742). Z. Williams, *Mariners Compass Compleated* and *Attempt to Ascertain the Longitude at Sea. Gentleman's Magazine* 57 (1787), 757–9, 1041–2, 1157–9. E.G.R. Taylor, "Reward for the Longitude"; A. J. Kuhn, "Dr Johnson, Zachariah Williams."

NOTES

INTRODUCTION

1. James Barry, *Series of Pictures*, pp. 59–60. Pressly, *James Barry*, pp. 86–122; Barrell, *Political Theory of Painting*, pp. 163–221 (especially pp. 192–9).

2. Solkin, *Painting for Money*, pp. 157–213 (see pp. 191–2 for the source of Barry's picture, Francis Hayman's 1765 *The Triumph of Britannia*). Spadafora, *Idea of Progress*; Sekora, *Luxury*; Jack, *Corruption and Progress*.

3. Johnson, *The Rambler*, vol. 3, pp. 271–6 (11 February 1752).

4. The word "magnetism" is ambiguous, since it can refer to the physical phenomenon as well as to its scientific study. In this book, the word has been used only to refer to the science. However, the phrase "terrestrial magnetism" has been retained, since there is less room for confusion, and phrases like "terrestrial magnetic power" would be obtrusively cumbersome.

5. C. Smith and Wise, *Energy and Empire*, pp. 151–2. R. Laudan, "Redefinitions of a Discipline." Schaffer, "Scientific Discoveries." Cantor, *Optics after Newton*, pp. 3–11. Whewell, *History of the Inductive Sciences*, vol. 3, pp. 35–8.

6. A typical example is Meyer, *History of Electricity and Magnetism*; this history of theoretical magnetism jumps directly from William Gilbert to Charles Coulomb with one paragraph on artificial magnets. More sympathetic accounts include W. S. Harris, *Rudimentary Magnetism*, Noad, *Manual of Electricity*, pp. 523–93, and Mottelay, *Bibliographical History*. The worst and most recent is Verschuur, *Hidden Attraction*; the best is Home, "Introduction."

7. Lindberg and Westman, *Reappraisals of the Scientific Revolution* (see p. 1 for the relevant quotation from Butterfield's book). G. S. Rousseau and Porter, *Ferment of Knowledge*.

8. Jardine, "Scientific Revolution"; Adrian Wilson, "Integrated Historiography"; Wrightson, "English Social History." Heilbron, *Electricity in the Seventeenth and Eighteenth Centuries*; Cantor, *Optics after Newton*; Porter, *The Making of Geology*. See Bowler, "Science and the Environment," for this problem in the new field of environmental sciences.

9. Hull, "In Defense of Presentism"; Christie, "Aurora, Nemesis and Clio"; Galison, "History, Philosophy"; Rouse, "Philosophy of Science"; Golinski, "Theory of Practice"; Pickering, *Science as Practice*; Porter, "History of Science."

10. The extensive literature on this topic includes G. Levine, "One Culture," and Serres, *Hermes* (particularly the introduction by Josué Harari and David Bell, pp. ix–xl).

11. Recent key historical analyses of these themes for the eighteenth century include Brewer and Porter, *Consumption and the World of Goods*; Barker-Benfield, *Culture of Sensibility*; Brewer, *Sinews of Power*; Colley, *Britons*; Stewart, *Rise of Public Science*; Golinski, *Science as Public Culture*. These studies are all informed by Jürgen Habermas, *Public Sphere*.

12. R. Smith, *Inhibition*.

13. Thompson, "Eighteenth-Century English Society," p. 151; *Customs in Common*, pp. 87–96.

Chapter One
Mapping Enlightenment England

1. E. Chambers, *Cyclopædia*, vol. 1, pp. ii and ix for geographical metaphors, p. iii for tree (1738 edition); Yeo, "Reading Encyclopedias," pp. 24–9; Darnton, "Great Cat Massacre," pp. 185–207 (this includes reproductions of the trees from Chambers' *Cyclopædia* and the *Encyclopédie*). Fuller, *Mathematical Miscellany*, frontispiece; *Introduction to the Study of Philosophy*, pp. 14–15, 23 (quotation p. 14). For the social construction of classification schemes, see Bloor, "Durkheim and Mauss," and Dean, "Controversy over Classification." Relevant literature on encyclopedias also includes Hughes, "Sciences in English Encyclopaedias," Kafker, *Notable Encyclopaedias*, Layton, "Diction and Dictionaries," Shorr, *Science and Superstition*, and Thorndike, "L'Encyclopédie."

2. There are, of course, excellent studies distinguishing electrical and magnetic topics (for instance, Heilbron, *Electricity*, and Sutton, "Electric Medicine and Mesmerism"). Similarly, there are marvellous analyses of groups such as the Hutchinsons and the Behmenists (notably by Geoffrey Cantor, Simon Schaffer, and Chris Wilde), but their perceptions have had limited impact on more general treatments.

3. Grew, *Musæum Regalis Societatis*, pp. 360–8.

4. Worster, *Compendious and Methodical Account of Natural Philosophy*, p. 234. Magnets were still included in the entry on mechanics in the 1773 edition of the *Encyclopædia Britannica*, vol. 3, p. 36. Although Worster is only briefly referred to by historians, William Whiston consulted Worster as a magnetic expert (Whiston, *Longitude and Latitude*, pp. 19, 55), and his book—published in two editions, one posthumous—appears in several authors' lists of recommended texts and was on two Oxford college reading lists: see Quarrie, "Christ Church Collection Books," pp. 504–5. For Watts and Worster, see Stewart, *Rise of Public Science*, pp. 134–9, 375–7, and Hans, *New Trends in Education*, pp. 82–7.

5. Voltaire, *Letters on England*, p. 68.

6. As four examples: Brewer, *Sinews of Power* (1688–1783); Clark, *English Society* (1688–1832); Colley, *Britons* (1707–1837); Langford, *Polite and Commercial People* (1727–1783).

7. Gilbert, *De Magnete*. Pumfrey, "William Gilbert's Magnetic Philosophy" and "Mechanising Magnetism."

8. Cavallo, *Treatise on Magnetism*; letter from Cavallo to James Lind of 10 July 1784: British Library Additional Manuscripts 22897, fol. 21v. Robison, "Magnetism."

9. Desaguliers, *A Course of Experimental Philosophy*, p. vi: see Stewart, *Rise of Public Science*, pp. 213–54. Daston, "Nationalism and Scientific Neutrality" and "Ideal and Reality." Shapin, "Of Gods and Kings." Atherton, *Political Prints*, pp. 84–105; Colley, *Britons*.

10. Porter, "Enlightenment in England"; Gascoigne, *Joseph Banks*, pp. 29–55; J. H. Brooke, "English Mix Science and Religion."

11. In his *Science as Public Culture*, Golinski explicitly moves south after his study of Scottish chemistry. Other examples include Stewart, *Rise of Public Science*, Porter, *Making of Geology*, and Cantor, *Optics after Newton*. In Scotland, there seems to have been little interest in magnetic phenomena before Robison's important account: "Magnetism." For an exception, see the letter from David Fordyce to John Canton of 30 April 1748: Canton papers, vol. 2, fol. 14. In Dublin, a few magnetic experiments were conducted in the 1720s but were rejected by the London Royal Society: see letters

from William Stephens to James Jurin of 3 and 10 April 1722 (Royal Society of London, Early Letters S.2.42 and S.2.42a), and from Jurin to Stephens of 4 October 1722 (London, Wellcome Institute for the History of Medicine, MSS 6146, Jurin correspondence). I am grateful to Andrea Rusnock for giving me transcripts of these letters. Towards the end of the century, some accounts of Irish investigations were reported more widely. These included Richard Kirwan's "Thoughts on Magnetism" (reproduced in *Nicholson's Journal* from *Transactions of the Royal Irish Academy* and reviewed in *Gentleman's Magazine* and *British Critic*), and Captain O'Brian Drury's comments on compasses (summarized from *Transactions of the Royal Irish Academy* in the *Encyclopædia Britannica* and the *Scots Magazine*). See also M. Young, *Analysis*, pp. 428–50 (Bishop of Clonfert).

12. Templeman, *Curious Remarks and Observations*, p. 173. He was referring to a 1723 text on magnetic experiments by René Réaumur. Descartes, *Principles of Philosophy*; Home, "Introduction," pp. 133–83. Home, "Newtonianism."

13. Rousseau, *Emile*, pp. 135–8: I am grateful to Timothy O'Hagan for this reference. Telescope, *Newtonian System*, p. 11; Magnet, *Newtonian System*, pp. 11–12: see Secord, "Newton in the Nursery," for the changing editions of Tom Telescope. See also Day, *Sandford and Merton*, vol. 2, pp. 161–70.

14. Thackray, *Atoms and Powers*, pp. 101–13; de Pater, "Musschenbroek, a Dutch Newtonian."

15. For example, Réaumur, "Expériences," and Dufay, "Observations," "Suite des observations," and "Troisième mémoire."

16. Home and Connor, *Æpinus's Essay*; Coulomb, "Septième mémoire," pp. 501–5.

17. For the English reception of Aepinus, see Home, "Æpinus and British Electricians." At least one nineteenth-century author wrote in apparent ignorance of Coulomb's results: Bryan, *Lectures on Natural Philosophy*, p. 149. The magnetic research of men like Tobias Mayer and Johann Lambert was scarcely referred to by English writers.

18. Arthur Murphy (in 1793), quoted in Clark, *Samuel Johnson*, p. 34.

19. The 1746 prize was particularly attractive because it had accumulated over three years. For Bernoulli, see Radelet-de-Grave, "Magnetismus." For Euler's magnetic ideas, see *Letters to a German Princess*, vol. 2, pp. 228–310.

20. McClellan, *Science Reorganized.* Gillespie, "Ballooning in France and Britain." Home, "Introduction." Andry and Thouret, "Observations et recherches," pp. 558–653, and "Rapport sur les aimans." Konvitz, *Cartography in France*, pp. 1–31, 63–81.

21. For a summary of French navigation texts and marine inventions, see W. Burney, *Dictionary of the Marine*, pp. 311–12. R. Briggs, "The Académie and Pursuit of Utility," especially pp. 66–81. As a specific magnetic example, the French king's physician Nicolas-Philippe Ledru visited London in 1766 to learn about making compasses. On his return to Paris, he obtained a royal license to make instruments, convert iron to steel by what he called Knight's process, and requisition charts and magnetic observations made by the French navy for his own projects. See Papiers Ledru-Rollin (Bibliothèque historique de la Ville de Paris), no. 2013, vol. 1, fol. 46; entry in the 1859 edition of *Nouvelle biographie générale depuis les temps les plus reculés jusqu'à nos jours*. For Ledru's later involvement in electrical and magnetic therapy, see Sutton, "Electric Medicine."

22. Schofield, *Mechanism and Materialism*, p. 181. Priestley, *History of Electricity*, vol. 1, pp. 501–3: he made a few other brief references to magnetic phenomena.

23. Ben-Chaim, "Social Mobility." Schaffer, "Public Spectacle" and "Consuming Flame." See also Pera, *Ambiguous Frog*, pp. 3–37. The most comprehensive account for this period is Heilbron, *Electricity*.

24. Kryzhanovsky, "Application of Bibliometrics."

25. Cookson, "Account of an Extraordinary Effect of Lightning" and "Further Account"; "Letter to Mr Robins"; Waddell, Effects of Lightning." Letter from Franklin to Peter Collinson of 29 June 1751, reproduced in I. B. Cohen, *Franklin's Experiments*, pp. 242–3; this translation of Anne Robert Turgot's famous epigram (*Eripuit coelo fulmen sceptrumque tyrannis*) is quoted in Pera, *Ambiguous Frog*, p. 25. Such a jar is exhibited at the Boerhaave Museum, Leiden.

26. Experiments at the Royal Society on 6 March 1746 (RSJB, vol. 19, pp. 62–3); letter of 14 March 1746 from Henry Miles to Baker, Baker correspondence, vol. 2, fol. 177. See also B. Wilson, *Treatise on Electricity*, pp. 219–23; "Letter to John Ellicot," p. 99.

27. Substantial comparisons include Musschenbroek, *Elements of Natural Philosophy*, pp. 210–11; Eeles, *Philosophical Essays*, pp. 47–9—a discussion essentially copied in G. Adams, *Lectures on Natural Philosophy* (1799), vol. 4, p. 468; Cavallo, *Treatise on Magnetism*, pp. 126–32; Sulivan, *View of Nature*, vol. 2, pp. 84–6; T. Young, *Course of Lectures*, pp. 448–50.

28. Schenk, "History of Electrotherapy"; Rowbottom and Susskind, *Electricity and Medicine*, pp. 1–30; Schaffer, "Self Evidence," pp. 68–78. Letters from Wesley to Adam Clarke of 25 March 1790 and 9 February 1791, reproduced in Telford, *Letters of John Wesley*, vol. 8, pp. 208, 261–2. Letter from Priestley to J. Bretland of 26 June 1791: Rutt, *Works of Joseph Priestley*, vol. 2, pp. 111–12.

29. Schaffer, "Consuming Flame." Johnson, *Rambler*, vol. 3, pp. 271–6 (quotation from p. 272): he was referring to Knight and John Canton. For Johnson and the longitude problem, see E.G.R. Taylor, "Reward for the Longitude," and A. J. Kuhn, "Dr. Johnson."

30. Schuster and Watchirs, "Natural Philosopy, Experiment and Discourse." For models and mimetic experiments, see Hackmann, "Concept and Instrument Design" and "Scientific Instruments," pp. 45–58, and Galison and Assmus, "Artificial Clouds," pp. 227–32.

31. Cavallo, *Treatise on Magnetism*, p. 41.

32. Compare the entries for "Natural Philosophy" in the 1797 and 1842 editions (the 1823 edition was the same as the one for 1797 on this point).

33. For an overview, see W. P. Jones, *Rhetoric of Science*, pp. 65–78. More detailed studies to be discussed in this book include Goldgar, "Fielding, the Flood Makers, and Natural Philosophy," Christie, "Laputa Revisited," Rogers, "Gulliver and the Engineer," and Schaffer, "Defoe's Natural Philosophy."

34. Adams, *Lectures* (1794), vol. 4, p. 435. Deacon, "Founders of Marine Science." Walters, "Tools of Enlightenment."

35. Desaguliers, *A Course of Experimental Philosophy*, p. viii.

36. J. A. Bennett, "Mechanics' Philosophy," pp. 12–24, and "Challenge of Practical Mathematics." Warner, "Terrestrial Magnetism."

37. J. A. Bennett, "Robert Hooke" and "Hooke's Instruments." Palter, "Early Measurements." RSJB, vol. 9, pp. 257–8 (14 and 21 May 1701).

38. Nairne, "Experiments on Two Dipping-Needles." G. Graham, "Account of the Observations," "Observations of the Dipping Needle," and "Some Observations." Cav-

endish, "Meteorological Instruments." Pumfrey, "Who Did the Work?" McConnell, "Craft Workshop to Big Business."

39. Shapin, "House of Experiment." H. Horne, *Essays concerning Iron and Steel*, p. 148. Cavallo "Magnetical Experiments, and Observations," Royal Society Letters and Papers, vol. 9, cut 147, fols. 8–12 (1789).

40. Shapin and Schaffer, *Leviathan and the Air-Pump*, pp. 30–5 (the frontispiece is reproduced on p. 33 and in many other books on the early Royal Society). Equivalent images by other Societies are reproduced in McClellan, *Science Reorganized.* Letter from Henry Baker to Thomas Walls of 9 March 1745: Baker correspondence, vol. 2, fol. 17; Weld, *History of the Royal Society*, vol. 1, p. 397. See also Bektas and Crosland, "Copley Medal."

41. B. Wilson, "Autobiography," pp. 49–50.

42. N. Ward, *London-Spy Compleat*, pp. 59–60 (dancing the hay referred to a country reel). Zacharias von Uffenbach also found it "remarkable": see Quarrell and Mare, *London in 1710*, p. 99. See J. A. Bennett, *Mathematical Science of Wren*, pp. 44–54. Ward may perhaps have been referring to another of the Society's loadstones, one of which weighed 60 pounds: Grew, *Musæum Regalis Societatis*, p. 317.

43. Fergusson, *Observations on the Art of Navigation*, p. 15. Latour, *Science in Action*, pp. 215–37, and "Force and Reason of Experiment."

44. Cantor, "Eighteenth-Century Problem," pp. 58–62. Stewart, "Texts and Contextualists." Schaffer, "Newtonianism." Yeo, "Genius, Method and Morality." Heimann, "Newtonian Natural Philosophy." Home, "Out of a Newtonian Straitjacket." Cantor and Hodge, "Introduction."

45. Review of Jones, *Essay on First Principles*, in *Critical Review* 13 (1762), 445–52. Letter from Jones to Alexander Catcott of 19 September 1762 in Catcott correspondence. Jones, *Essay on First Principles*, pp. 55–116 (especially pp. 71–2, reproduced in Yolton, *Thinking Matter*, p. 193). For further examples and references, see Wasserman, "Sympathetic Imagination," and Stewart, *Rise of Public Science*, pp. 106–8.

46. The numerous and varied analyses include Jacob, *Newtonians and the English Revolution*, Shapin, "Of Gods and Kings," Stewart, "Samuel Clarke," and Gascoigne, *Cambridge in the Enlightenment*, pp. 142–84.

47. Letter from William Bowman of 3 June 1762, reproduced in Nichols, *Literary History*, vol. 8, pp. 623–4. Paulson, *Hogarth's Graphic Works*, vol. 1, p. 288. See Wagner, *Eros Revived*, pp. 182–91, for this print and scatological literature. For Hebraic glory and gravity, see Cantor, "Revelation and the Cyclical Cosmos of Hutchinson," pp. 12–16, and Caird, *Language and Imagery*, pp. 27–30. More general discussions of the Enlightenment relationship between natural philosophy and religion include J. H. Brooke, *Science and Religion*, pp. 152–91, and Cunningham, "Getting the Game Right" and "How the *Principia* Got Its Name."

48. Guerrini, "Tory Newtonians." Olson, "Tory–High Church Opposition to Science." Wilde, "Hutchinsonianism" and "Matter and Spirit."

49. Gascoigne, *Cambridge in the Enlightenment*, pp. 167–74. Stewart, "Samuel Clarke," especially pp. 63–72. Millard, "Chronology of Roger North."

50. Adrian Wilson, "Politics of Medical Improvement," and P. Wilson, "Out of Sight." Olson, "Tory–High Church Opposition."

51. Ardley, *Berkeley's Philosophy*; Cragg, *Reason and Authority*, pp. 93–124; Cantor, "*Analyst* Revisited." See also the introductions by A. A. Luce and T. E. Jessop in

Berkeley, *Works*, vol. 3, pp. 1–20 (*Alciphron*), and vol. 5, pp. v–ix (*Siris*). Berkeley, *Works*, vol. 5 (*Siris*); Benjamin, "Medicine, Morality and Politics."

52. Mather, *High Church Prophet*, pp. 1–23. Wilde, "Hutchinsonianism." Neve and Porter, "Alexander Catcott." See Cantor, "Revelation," p. 20, for Hutchinson's twelve volumes. For Tory attitudes at Oxford, see Gascoigne, "Mathematics and Meritocracy," pp. 573–6; Sutherland and Mitchell, *University of Oxford*, pp. 9–190; Clark, *Samuel Johnson*, pp. 88–104.

53. Jones, *First Principles of Natural Philosophy* and *Physiological Disquisitions*. The network included Jones, Alexander Catcott, Berkeley's son (George), George Horne, William Stevens (Horne's cousin), and Samuel Glasse: see Park, *Memoirs of William Stevens*, pp. 22–32, 164–80. See letters in the Catcott correspondence and several letters from Jones to the younger Berkeley in British Library Additional Manuscripts 39311, particularly fols. 108–9 (29 January 1762), 160–1 (4 June 1764), and 280–1 (6 July 1774).

54. In his second and major natural philosophy text, Jones was still dissociating himself from Hutchinson: Jones, *Physiological Disquisitions*, pp. iii, ix–x. For his public approbations of Hutchinsonianism, see W. Jones, *Works*, vol. 1, pp. ix–xxx, 300–37; *British Critic* 2 (1793), 208, and 15 (1800), 208. For Horne as a Hutchinsonian, see Horne, *Works*, vol. 1, pp. ix–xxx (by Jones) and 441–505.

55. John Wesley's letter to Mary Bishop of 17 April 1776 (Telford, *Letters of Wesley*, vol. 6, pp. 213–14), letter of 1785 to Dean D—— (vol. 7, pp. 250–2), letter to George Horne of 10 March 1762 (vol. 4, pp. 172–80), letter to Samuel Furly of 26 November 1756 (vol. 3, pp. 206–7). See also Schofield, "John Wesley and Science," especially p. 339, and Rupp, *Religion in England*, p. 411, p. 473.

56. Clark, *English Society*, pp. 216–235; Murray, "Influence of the French Revolution," pp. 44–79.

57. William Stevens' biography of Jones is in W. Jones, *Works*, vol. 1, pp. i–lxiii; Jones helped to promote Adams: see, for example, his letter to Catcott of 16 May 1761, Catcott correspondence, and Jones, *Works*, vol. 10, p. 529.

58. W. Walker, *Life of John Skinner*, pp. 58–70, 148–67; A. J. Kuhn, "Glory or Gravity." John Skinner, the Hutchinsonian Dean of Aberdeen, contributed articles on prophecy and philosophy to the third edition: W. Walker, *Skinner*, pp. 148–52. James Tytler, who was associated with the Glasites, contributed to both editions and wrote many of the articles on natural philosophy—including the one on electricity—for the third edition: Fergusson, *Balloon Tytler*, pp. 26–8, 40–59, 94–5; *Encyclopædia Britannica* (1797), vol. 1, pp. xv–xvi. The influence on Faraday of the electricity article is debated.

59. See Wilde, "Hutchinsonianism" and "Matter and Spirit," for references to Samuel Pike, Andrew Wilson, and others. For a discussion of the possible influence on John Dalton of Hutchinsonian ideas via Adams and Jones, see Thackray, *Atoms and Powers*, pp. 251–3. For Christopher Smart, see D. Greene, "Smart, Berkeley," Williamson, "Smart's *Principia*," and Smart, *Works*, vol. 1, pp. 131–2.

60. A. B. Webster, *Joshua Watson*, pp. 18–32. Corsi, *Science and Religion*, pp. 9–20. Kirby, *Power Wisdom and Goodness*, vol. 1, pp. xvii–cv, especially pp. cii–ciii. For the Sandemanians as Tories, see Cantor, *Michael Faraday*, pp. 11–54 and 89–98.

61. Some historians have drawn continuous links between seventeenth-century Neoplatonic philosophers and nineteenth-century Romantics, in which Freemasonry may have played a vital role in preserving and transmitting occult traditions. Hirst, *Hidden Riches*, especially pp. 180–267; Schuchard, "Freemasons" and "Swedenborg";

Garrett, "Swedenborg and the Mystical Enlightenment." For assessments of Freemasonry's role in eighteenth-century culture, see J. M. Roberts, *Mythology of the Secret Societies*, especially pp. 17–117. For Adams and the Swedenborgians, see Hindmarsh, *New Jerusalem Church*, p. 23, and Danilewicz, "King of the New Israel," p. 61.

62. Hoyles, *Edges of Augustanism*, pp. 79–114; A. J. Kuhn, "Nature Spiritualised"; G. S. Rousseau, "Mysticism and Millenarianism"; Schaffer, "Consuming Flame," pp. 501–12.

63. For example, Horne, *Works*, vol. 1, pp. 209–15, 216–32; letter from Jones to Catcott of 16 May 1761 in the Catcott correspondence; W. Law, *Works*, vol. 5, p. 168, and vol. 7, pp. 191–2. Jonathan Barry, "Piety and the Patient," describes some alliances between individual Hutchinsonians and Behmenists.

64. Clark, *English Society*, pp. 42–276; Hole, *Pulpits, Politics and Public Order*, pp. 11–72; Murray, "Influence of the French Revolution," pp. 44–79. For Filmerian and Lockean ideologies, see Dickinson, *Liberty and Property*, pp. 13–90. See W. R. Ward, *Georgian Oxford*, pp. 205–6, for the political nature of Oxford Hutchinsonian debates, and Colley, *Defiance of Oligarchy*, for the Tory party in the eighteenth century (pp. 105–6 for Horne and the Hutchinsonians). For a specific rousing sermon articulating these views, see W. Jones, *Works*, vol. 5, pp. 274–96 ("Popular Commotions to Precede the End of the World").

65. R. Green, *Truth and Divinity of the Christian Religion*, p. 198.

66. See Schaffer, "Consuming Flame," on the Freke-Martin disputes, and P. Wilson, "Out of Sight," on those between Daniel Turner and James Blondel.

67. For instance, the papers in Frängsmyr, Heilbron, and Rider, *The Quantifying Spirit*. Brewer, *Sinews of Power*, especially pp. 219–49. Burke, *Reflections on the Revolution in France*, p. 285: see Crosland, "Image of Science as a Threat." Thompson, "Moral Economy of the English Crowd." Barrell, *Birth of Pandora*, pp. 41–61. W. Law, *Works*, vol. 5, pp. 89–92 (from an attack on Benjamin Hoadly): see Arthur Walker, *William Law*, pp. 118–26.

68. Quoted on p. 61 of Schaffer, "Self Evidence," from Lichtenberg's *Aphorisms*.

Chapter Two
"A Treasure of Hidden Vertues"

1. E.G.R. Taylor, *Mathematical Practitioners of Tudor and Stuart England*, pp. 292, 416. Crom, *Trade Catalogues*, pp. 66–71. Paulson, *Popular and Polite Art*, pp. 85–114. Langford, *Polite and Commercial People*, especially pp. 1–6.

2. See Tuttell's six of hearts (reproduced in Crom, *Trade Catalogues*, p. 69) and eight of diamonds for compasses in mines.

3. Browne, *Pseudoxica Epidemica*, pp. 107–9; Rees, *Cyclopædia*, vol. 3, entry on Mahometanism. See also Prior, "Alma," p. 229, and Kunz, *Curious Lore of Precious Stones*, pp. 93–4.

4. Hutchison, "Occult Qualities in the Scientific Revolution"; Henry, "Occult Qualities and the Experimental Philosophy"; Schaffer, "Occultism and Reason" and "Godly Men and Mechanical Philosophers." For virtues, see Heilbron, *Electricity*, pp. 19–30, and Schaffer, "Measuring Virtue."

5. Clark, *English Society*. Compare, for example, Langford, *Polite and Commercial People*, or McKendrick, Brewer, and Plumb, *Birth of a Consumer Society*.

6. See McKendrick's "Introduction" (pp. 1–8) in McKendrick, Brewer, and Plumb, *Birth of a Consumer Society*. Many aspects of commercialization are treated from a

worldwide perspective in Brewer and Porter, *Consumption and the World of Goods*. The editors' introduction (pp. 1–15) provides a concise yet comprehensive review of the literature.

7. Compare Plumb's analysis in McKendrick, Brewer, and Plumb, *Birth of a Consumer Society*, pp. 308–9. Hargrave, *History of Playing Cards*, pp. 169–222.

8. Weatherill, *Consumer Behaviour*. Looney, "Cultural Life in the Provinces"; Porter, "Science, Provincial Culture and Public Opinion." Fine and Leopold, "Consumerism and the Industrial Revolution"; de Vries, "Between Purchasing Power and the World of Goods."

9. Colley, *Britons*, especially pp. 55–100. Borsay, *Urban Renaissance*.

10. Sennett, *Fall of Public Man*. Davidoff and Hall, *Family Fortunes*. Barker-Benfield, *Culture of Sensibility*.

11. Stewart, *Rise of Public Science*, and Golinski, *Science as Public Culture*.

12. Deacon, "Founders of Marine Science." Howson, *History of Mathematics Education*, pp. 29–44; E.G.R. Taylor, *Mathematical Practitioners of Tudor and Stuart England*, pp. 114–31.

13. RSJB, vol. 19, p. 366 (30 November 1747, Martin Folkes).

14. Mandeville, *Fable of the Bees*. Goldsmith, *Private Vices, Public Benefits*, pp. 1–27. Copley, *Literature and Social Order*, pp. 1–21. Jack, *Corruption and Progress*. Spadafora, *Idea of Progress*. Barrell, *Political Theory of Painting*, pp. 1–68.

15. Sekora, *Luxury*. Fletcher, *Discourse of Government*, p. 12. Cheyne, *English Malady*, p. 34, discussed in Barker-Benfield, *Culture of Sensibility*, pp. 6–15. For Cheyne's fluctuating health, see also Bowles, "Physical, Human and Divine Attraction," and G. S. Rousseau, "Mysticism and Millenarianism." For Thomas Beddoes and the metaphoric maladies of consumption, see Porter, *Doctor of Society*.

16. David Miller's phrase for the entire century: "Into the Valley of Darkness."

17. Stewart, *Rise of Public Science*, and Golinski, *Science as Public Culture*.

18. McKendrick, Brewer, and Plumb, *Birth of a Consumer Society*, pp. 7–194. Millburn, *Wheelwright of the Heavens*. Doherty, "Anodyne Necklace." Eagleton, *Rape of Clarissa*, pp. 1–39; Raven, *Judging New Wealth*.

19. B. Wilson, "Autobiography," p. 78. The other surviving portrait of Knight is an oil painting, also by Wilson, originally hung in the British Museum and now stored in the archives of London's National Gallery.

20. Lippincott, *Selling Art in Georgian London*. Allibone, *Royal Society and Its Dining Clubs*, pp. 20–80. Solkin, *Painting for Money*. Many of Pond's patrons and friends were close to Thomas Birch, a frequent host to Knight: Birch, Diary (1735–64), passim. In particular, Knight and Pond were both close friends of Daniel Wray, and they both sought patronage from Admiral Anson and the Earl of Hardwicke. Knight's letter to Hardwicke of 22 September 1754 named Wray as a reference: British Library Additional Manuscripts 36269, fol. 29.

21. Jewson, "Medical Knowledge and the Patronage System." Porter, "Before the Fringe" and *Health for Sale*.

22. Porter, "William Hunter." Knight and Hunter were both listed as collectors of papers at the front of the journal *Medical Observations and Inquiries*. John Hunter was an FRS, but his brother William was not.

23. G. Holmes, *Augustan England*. Brewer, *Sinews of Power*.

24. Important examples include Singer, "Sir John Pringle," Crane, "Club of Honest Whigs," and A. E. Gunther, *Life of Thomas Birch*.

25. Crosland, "Qualifications for Membership of the Royal Society." Bektas and Crosland, "Copley Medal." Knight's close friends amongst the mid-century experimenters included Benjamin Wilson, Henry Baker (letter from Baker to William Arderon of 10 March 1748: Baker correspondence, vol. 3, fol. 256, and Arderon correspondence, vol. 2, fol. 6), John Ellicott (letter from J. Collings to William Canton of 13 February 1786: Canton papers, vol. 2, fol. 119), and John Michell (letter from Michell to Wilson of 22 December 1754: Wilson correspondence, fol. 86).

26. McKendrick, "Rôle of Science." For reviews of the literature on the relationship between science and technology in the eighteenth century, see Musson, *Science, Technology and Economic Growth*, pp. 1–68; Mathias, *Transformation of England*, pp. 45–71; Pinch and Bijker, "Social Construction of Facts and Artefacts," pp. 399–408. For reappraisals of the Industrial Revolution, see O'Brien and Quinault, *Industrial Revolution and British Society*.

27. Porter et al., *Science and Profit*. Schaffer, "Natural Philosophy and Public Spectacle." Castle, "Female Thermometer."

28. *Dictionary of National Biography*; Bloxam, *Register of Magdalen College*, vol. 6, pp. 241–3. For Leeds Grammar School, see Smiles, *Lives of the Engineers*, p. 90. Knight never took the final degree which would have enabled him to become a licentiate of the Royal College of Physicians, the major advantage offered by Oxford over the more rigorous medical training institutions at Leiden and Edinburgh. However, this did not prevent him from practicing as a physician.

29. Zacharias von Uffenbach, quoted in Bayne-Powell, *Travellers in Eighteenth-Century England*, p. 96. Gibbon, *Autobiography and Correspondence*, pp. 23–35 (quotation from p. 24). For Oxford education, see Hans, *New Trends in Education*, pp. 18–45; L. Stone, "Oxford Student Body," pp. 37–59; Robb-Smith, "Medical Education at Oxford and Cambridge"; Frank, "Science, Medicine and Universities"; G. V. Bennett, "University, Society and Church," pp. 366–8; C. Webster, "Medical Faculty"; G. L'E. Turner, "Physical Sciences."

30. Surveys of the methods include Euler, *Letters to a German Princess*, vol. 2, pp. 292–310; Sturgeon, *Scientific Researches*, pp. 561–7; W. S. Harris, *Rudimentary Magnetism*, part 1, pp. 84–91.

31. Caswell, "Paper about Magnetism"; Ballard, "Magnetism of Drills"; Derham, "Account of Some Magetical Experiments" and "Farther Observations." For John Payne, see Royal Society Register Book, vol. 13, pp. 144–5. For Restoration Oxford, see Pumfrey, "Mechanising Magnetism," and R. T. Gunther, *Early Science in Oxford*, vol. 4, pp. 18–58. RSJB, vol. 10, p. 308 (23 July 1713). Desaguliers, "Account of Some Magnetical Experiments." See also RSJB, vol. 11, pp. 39 (23 December 1714), 40–1, 46, 50–1 (13 January, 10 and 24 February 1715), 393–4 (5 November 1719); vol. 12, pp. 226, 335 (5 April 1721 and 28 February 1722); vol. 14, pp. 137–8 (11 May 1732); vol. 15, pp. 358–9 (24 June 1736). H. Horne, *Essays concerning Iron and Steel*, especially pp. 147–8. Henry Horne was a smith who provided Canton with steel: see Canton papers, vol. 2, fol. 16. See also Robison, "Magnetism," pp. 132–5, and Imison, *School of Arts*, vol. 2, pp. 161–8.

32. Musschenbroek, *Elements of Natural Philosophy*, pp. 209–10 (John Colson's translation was published in 1744, the same year that Knight first performed at the Royal Society, but the Latin edition dated from 1726). S. Savery, "Magnetical Observations and Experiments." J. Savery, "Account of the Savery Family," fols. 2–3, 83–93. Savery, an Oxford graduate from a distinguished family, was related to Thomas

Savery, inventor of a steam-driven mine pump. He corresponded regularly with several Fellows of the Royal Society about his inventions, which also included barometers and telescopes.

33. Michell, *Artificial Magnets*, pp. 15–16. RSJB, vol. 13, pp. 456, 466, 494 (16 April to 11 June 1730).

34. Jacob Lovelace advertised Savery magnets in Exeter: *Gentleman's Magazine* 75(1) (1785), 135. There were two William Lovelace (father and son): Calvert, *Scientific Trade Cards*, p. 13. Letter from Lovelace to Savery of 3 September 1743, British Library Additional Manuscripts 44058, fol. 91; letter from Henry Horne to John Canton of 15 February 1751, Canton papers, vol. 2, fol. 16; undated letter from Henry Cranke to William Canton, Canton papers, vol. 2, fol. 114. RSJB, vol. 16, p. 464 (5 July 1739); vol. 18, p. 455 (24 October 1745). See Badcock, "Physics at the Royal Society," pp. 251–325, for references to entries on magnetic topics in the RSJB.

35. References include RSJB, vol. 12, pp. 203–4 (15 February 1721); Hamilton, *Calculations and Tables*. (He is variously referred to as William Hamilton, Lord Paisley, and the Earl of Abercorn.)

36. G. Knight, "Account of Magnetical Experiments," "Letter to the President," and "Collection of Magnetical Experiments." The manuscript original of Folkes' 1746 presentation to the Royal Society, printed as Knight, "Collection of Magnetical Experiments," pp. 656–61, is at the London Wellcome Institute for the History of Medicine, Folkes collection 2391(4).

37. RSJB, vol. 19, p. 360 (30 November 1747); Royal Society Council minutes for 11 November 1747. His sponsors for Fellowship included Folkes and Charles Cavendish (Duke of Devonshire). Royal Society Club Minute Book 1748–51. Allibone, *Royal Society Dining Clubs*, pp. 20–39.

38. G. Knight, "Account of the Mariners Compass," "Description of a Mariner's Compass." Mountaine, "Extraordinary Effects of Lightning"; and G. Knight, "Remarks on the Preceding Letter." Lukis, *Memoirs of William Stukeley*, vol. 2, pp. 361–2, comments on Knight, "Account," p. 117.

39. He also invented a patented "Machine Window-Blind": British Library Additional Manuscripts 45871, fol. 197. For his naval sounding device, see PRO: ADM 12/52, cut 59/3 (11 August 1768); a letter from Captain Webster to the Admiralty of 23 September 1768: PRO: ADM 1/2670; and a letter from John Le Roy to Wilson of 6 October 1768: Wilson correspondence, fol. 123.

40. Lippincott, *Selling Art in Georgian London*, p. 50; Morton, "Lectures on Natural Philosophy," p. 417. Quaker practitioner Thomas Hodgkin is quoted in Porter, *Health for Sale*, p. 44. Eminent visitors included Admiral George Anson and Lord Royston (the Lord Chancellor's son): letters from Knight to the Lord Chancellor and the Archbishop of Canterbury of 22 September 1754: British Library Additional Manuscripts 36269, fols. 29–31.

41. G. Knight, "Shock of an Earthquake," p. 603; letter from Knight to the Navy Board of 7 May 1751 (PRO: ADM 106/1092). Nichols, *Literary Anecdotes*, vol. 5, p. 534: the dates of his move given by different sources conflict. For the social geography, see Millburn, *Retailer of the Sciences*, p. 87, and H. Phillips, *Mid-Georgian London*, pp. 187–91.

42. For the use of these initials, see Stearns, "Course of Edmond Halley," p. 294, on John Senex; Millburn, "Benjamin Martin and the Royal Society," on Benjamin Martin; and Wonnacott, "Martin Clare," p. 88, on Martin Clare. For James Simon, see letter from Browning to Baker of 8 April 1748 (Baker correspondence, vol. 4, fol. 81); for

Priestley, see letter from Priestley to Canton of 14 February 1756 (Canton papers, vol. 2, item 58). For Henry Wynne's use of Royal Society prestige to market magnetic instruments, see Bryden, "Magnetic Inclinatory Needles."

43. Letter to Cadwallader Colden of 26 April 1745, reproduced in Colden, *Letters and Papers*, vol. 3, pp. 113–15, quotation from p. 114. Colden passed the information on to John Bartram (p. 159), and Collinson later sent two separate copies of Knight's book to Franklin (Labaree and Willcox, *Papers of Benjamin Franklin*, vol. 4, pp. 114–15, and vol. 5, pp. 232, 331), who subsequently gave one of them to Ezra Stiles (vol. 6, p. 103). Letter from Fothergill to James Logan of 4 May 1750, reproduced in Corner and Booth, *Chain of Friendship*, pp. 137–9.

44. Letter from Baker to William Arderon of 10 March 1748: Baker correspondence, vol. 3, fol. 256, and Arderon correspondence, vol. 2, fol. 6. Letters to Baker about Knight included one from Giuseppe Bruni of Turin in March 1745 (Baker correspondence, vol. 2, fols. 26–7) and one from John Browning of Bristol of 18 May 1749 (Baker correspondence, vol. 4, fols. 93–4).

45. Letter from Henry Miles to Baker of 16 April 1746: Baker correspondence, vol. 2, fols. 221–2. Musschenbroek, *Cours de physique expérimentale*, vol. 1, pp. 431–3, 453–7, 469. Duhamel, "Façon singulière d'aimanter," p. 182; Rivoire, *Traités sur les aimans artificiels*, pp. i–lxxxv.

46. B. Wilson, "Autobiography," pp. 17–18. G. Knight, *Collection of Some Papers*. He presented the copy now in the British Library to Thomas Birch. For the magnets in the Science Museum which are probably Knight's, see Morton and Wess, *Public and Private Science*, pp. 183, 230, 295. For the certificate, see Science Museum photograph no. 5437.

47. G. Adams, "Catalogue" (in *Celestial and Terrestrial Globes*); Turner and Levere, *Van Marum's Scientific Instruments*, p. 187; Lefebvre and de Bruijn, *Martinus van Marum*, vol. 6, pp. 1–2; Magellan, *Collection de différens traités*, pp. 193–5, 215–16.

48. General Meetings of the Board of Trustees of the British Museum, vol. 1, pp. 233–4, 238; British Library Additional Manuscripts 4449, fol. 108. Descriptions of the machine were published only after Knight's death: Fothergill, "Account of the Magnetical Machine," pp. 594–9, and RSJB, vol. 28, pp. 112–13 (23 June 1774). A committee was established to restore it after a fire: RSJB, vol. 28, p. 540 (19 December 1776). The parts used by Michael Faraday at the Royal Military Academy for experiments on galvanometers are now in storage at the London Science Museum: Faraday, "Experimental Researches," p. 135.

49. B. Wilson, "Autobiography," pp. 48–50; Royal Society Council minutes, 18 and 25 January 1759. G. Knight, *Collection of Some Papers*, especially p. 18. Mountaine and Dodson, *Account of the Methods Used*, p. 12; Crosby, *The Mariner's Guide*, p. 182 (edited by William Mountaine); John Robertson, *Elements of Navigation*, vol. 2, pp. 231–2.

50. G. Knight, *Attraction and Repulsion*. G. S. Rousseau, "Science Books and Their Readers." Whiston, *Longitude and Latitude*.

51. Robinson, *Dissertation on Æther*, published initially in Dublin and subsequently in London. Wilson correspondence, fols. 23–8, 55–6; Heilbron, *Electricity*, pp. 302–5. See also Cantor and Hodge, "Introduction," Heimann, "Ether and Imponderables," and Quinn, "Repulsive Force."

52. Feather, "John Nourse." The first edition cost 3 shillings: letter from Collinson to Franklin of 22 February 1751, reproduced in Labaree and Willcox, *Papers of*

Franklin, vol. 4, pp. 114–15. The second edition cost 3 shillings and 6 pence: Watt, *Bibliotheca Britannica*, vol. 2, p. 575o. I am grateful to Alice Walters for pointing out the significance of the change in size: by economizing on paper—more expensive than labor—Nourse reduced his costs. *Monthly Review* 10 (1754), 456–62. The reviewer, William Bewley (see Nangle, *Monthly Review First Series*, pp. 4–5), evidently did not realize that an earlier edition had appeared.

53. Croker, *Experimental Magnetism*, p. 8. Letter from John Browning to Baker of 26 January 1750: Baker correspondence, vol. 4, fols. 166–7.

54. Letter from Franklin to Ezra Styles of 10 July 1755, reproduced in Labaree and Willcox, *Papers of Franklin*, vol. 6, p. 103. Letter from Franklin to James Bowdain of 24 January 1752, reproduced in vol. 4, p. 256. Joseph Black made similar criticisms in *Lectures on Chemistry*, vol. 1, pp. 516–17. Priestley did not consider Knight's book worth reading for his projected history of magnetism: Schofield, *Mechanism and Materialism*, p. 81. Contemporaries showed more interest in the book's treatment of light: Melvill, "Letter," pp. 269–70; Schofield, p. 252. His ideas enjoyed a Victorian revival: W. S. Harris, *Rudimentary Magnetism*, part 3, p. 129; de Morgan, "Dr Gowan Knight" and *Budget of Paradoxes*, p. 90.

55. Letter from John Michell to Wilson of 22 December 1754: Wilson correspondence, fol. 86. Advertisement in *Daily Advertiser* of 12 December 1760; Croker, *Experimental Magnetism*, pp. 8–9. See Stewart, *Rise of Public Science*, p. 363, for Desaguliers's call for a history of magnetic experiments.

56. Royal Society Council minutes for 12 December 1751. RSJB, vol. 21, pp. 28–30 (23 January 1752). Letter from Baker to Arderon of 3 March 1752: Arderon correspondence, vol. 3, fol. 1. Letter from William Bowman to John Nichols, reproduced in Nichols, *Illustrations of Literary History*, vol. 8, pp. 625–6. Nichols, *Literary Anecdotes*, vol. 5, pp. 282–90; A. E. Gunther, *Life of Thomas Birch*, pp. 35–43, 69–78; RSJB, vol. 21, pp. 35–6 (25 January 1752).

57. E. Miller, *Noble Cabinet*, pp. 19–63; A. E. Gunther, "Royal Society", *Founders of Science*, pp. 1–25, and "Matthew Maty." Correspondence between Henry Miles and Baker in 1753 and 1754: Baker correspondence, vol. 5, fols. 248–261, fol. 349, and vol. 6, fols. 7–8.

58. G. S. Rousseau, *Letters and Papers of John Hill*, pp. 54–64; articles in *London Daily Advertiser* of 12 January and 6 December 1754. Birch, Diary, p. 253. Letters from Knight of 22 September 1754 to Lord Hardwicke and the Archbishop of Canterbury: British Library Additional Manuscripts 36269, fols. 29–31. Minutes of Trustees: British Library Additional Manuscripts 4450, p. 204 (3 and 4 June 1756).

59. Charles Morton recorded his death in the Museum library: British Library Additional Manuscripts, fol. 83. Knight received three obituary notices: Musgrave, *Obituary*, vol. 3, p. 385. British Library Additional Manuscripts 4449, fols. 82–109, fol. 171.

60. Fyfe, "Art Exhibitions and Power." British Library Additional Manuscripts 4449, fols. 118–65. A. E. Gunther, *Founders of Science*, pp. 158–9.

61. Letters from Franklin to Rudolph Raspe of 9 September 1766 and 6 July 1767: Labaree and Willcox, *Papers of Franklin*, vol. 13, p. 407, and vol. 14, p. 211. E. Miller, *Noble Cabinet*, pp. 64–90. Extract from a memorandum by Alexander Small on Franklin's view of ventilation: Labaree and Willcox, *Papers of Franklin*, vol. 23, pp. 486–91 (see p. 490); letter from Thomas Gray to James Brown of 8 August 1759: Toynbee and Whibley, *Correspondence of Thomas Gray*, vol. 2, p. 632; British Library Addi-

tional Manuscripts 45868, fols. 26–7. Original letters and papers of the British Museum, vol. 1, fols. 127–72.

62. R. L. Edgeworth, *Memoirs*, p. 74. Nichols, *Literary Anecdotes*, vol. 5, pp. 477–82. I am grateful to C. Gordon for sending me notes on Pringle's references to Knight in the index of his medical annotations at the Library of the Royal College of Physicians, Edinburgh.

63. British Library Additional Manuscripts 30094, fol. 127; Lettsom, *Works of John Fothergill*, p. li; B. Wilson, "Autobiography," pp. 28–9. Fothergill was Knight's executor. Ruggles, "Governor Samuel Wegg."

64. G. Knight, "Recovery from a Fever." A. E. Gunther, *Founders of Science*; Nichols, *Literary Anecdotes*, vol. 3, p. 258. Knight was friendly with many of the club's members. Birch, Diary, pp. 314, 317, and passim.

65. Royal Society Minute Books, 1751–72, passim (especially 1751–2, 1759, 1767–9). Letter from Knight to Wilson of 25 October 1758: Wilson correspondence, fol. 143.

66. Letter from Smeaton to Wilson of 10 July 1748: Wilson correspondence, fol. 69; G. Knight, "Description of a Mariner's Compass," p. 512; Smeaton, "Improvements of the Mariners Compass" and *Reports*, vol. 4, p. 16; Skempton, *John Smeaton*, p. 11; Allibone, *Royal Society Dining Clubs*, p. 85. Millburn, *Wheelwright of the Heavens*, p. 145. Hardin, "Scientific Work of John Michell," p. 28. Canton papers.

67. The lists show him to have been a member from 1759 or 1760 to about 1767. Minutes of the Society, vol. 5, March 1760 and passim; Allan, "Society for the Encouragement of Arts"; and Solkin, *Painting for Money*, pp. 174–90.

68. Letter to the Archbishop of Canterbury, September 1754: British Library Additional Manuscripts 30094, fol. 143. Toynbee and Whibley, *Correspondence of Thomas Gray*, vol. 2, p. 632.

69. Royal Society Register Book (copy) 3 (1674), Petty's "The Discourse made before the Royal Society, the 26 November 1674," 335–61 (quotation from p. 356); Hamilton, *Calculations and Tables*; *Philosophical Transactions* 36 (1730), 245–50.

70. Pomian, *Collectors and Curiosities*. Daston, "Factual Sensibility."

71. Croker, Williams, and Clark, *Dictionary*, entry on magnet; Mottelay, *Bibliographical History*, p. 134, p. 159; E. Stone, *Mathematical Instruments*, p. 305. B. Wilson, "Autobiography," pp. 16–17.

72. Letter to Alexander Pope of 10 October 1716, reproduced in Pope, *Correspondence*, vol. 1, pp. 365–6. Lister, "Journey to Paris," pp. 26–8. Fiennes, *Journeys of Celia Fiennes*, p. 33: I am grateful to Natasha Glaisyer for this reference.

73. Pomet, *Compleat History of Druggs*, pp. 364–5; Templeman, *Curious Remarks and Observations*, p. 173; Andrade, "History of the Permanent Magnet," p. 24; Altick, *Shows of London*, pp. 12–14; Pettus, *Fleta Minor*, book 2, entry on loadstone. H. Wallis, "Geographie Is Better than Divinitie," p. 39.

74. Gunther, *Early Science in Oxford*, vol. 1, pp. 306–7; Ovenell, *Ashmolean Museum*, p. 149.

75. RSJB, vol. 12, pp. 203–4 (15 February 1721). Pettus, *Fleta Minor*, book 2, entry on loadstone. For illustrations of other loadstones, see Andrade, "History of the Permanent Magnet," p. 25; Wynter, *Catalogue of Scientific Instruments*, pp. 2–3; Randier, *Marine Navigation Instruments*, pp. 10–11.

76. Pomet, *History of Druggs*, pp. 364–5. Crawforth, "Evidence from Trade Cards," p. 468. White, *Rich Cabinet*, pp. 9–11. Trade card reproduced in Crom, *Trade Cata-*

logues, p. 25. The long rhyming advertisement for J. Stone of Shepton Mallet is in Lysons II, vol. 2(2), fol. 201.

77. K. Higgins, "Classification of Sundials," pp. 342–5, 350–1, 367; Gouk, *Ivory Sundials*, pp. 9–24.

78. Michell, *Artificial Magnets*, pp. 4–8.

79. G. Adams, "Catalogue," in *Micrographia Illustra*, p. 258; Warter, *Southey's Common-Place Book*, vol. 1, p. 435; advertisement at the foot of an optician's trade card reproduced in Calvert, *Scientific Trade Cards*, no. 132; Wynter, *Catalogue*," pp. 2–3.

80. Johnson, *Rambler*, vol. 3, p. 272; Eamon, *Science and the Secrets of Nature*, especially pp. 310–11. Nicholson, *Introduction to Natural Philosophy*, vol. 2, p. 323. Letter from Baker to Arderon of 6 April 1751: Arderon correspondence, vol. 2, fol. 77.

81. *English Encyclopædia* (1802), entry on magnetism, p. 455. Fothergill, "Magnetical Machine"; B. Wilson, "Dr Knight's Method"; G. Adams, *Natural and Experimental Philosophy* (1794), vol. 4, pp. 448–9.

82. G. Adams, *Essay on Electricity*, p. 392. In addition to Knight's major English rivals, Michell and Canton, for European emulators see Rivoire, *Traités*, pp. i–lxxxv, Duhamel, "Façon singulière," Coulomb, "Septième mémoire," Fuss, *Observations et expériences*, and Radelet-de-Grave, "Magnetismus" (Bernoulli). For a review concentrating on Aepinus, see Home, "Introduction," pp. 137–86.

83. Collins, *Changing Order*, pp. 51–78. Imison, *School of Arts*, vol. 2, pp. 161–8; Robison, "Magnetism," pp. 133–6.

84. Macleod, *Inventing the Industrial Revolution*; Lawrence, "Alexander Monro," pp. 198–200. Letter from Michell, with an addendum by Smeaton, in *Monthly Review* 72 (1785), 478–80.

85. E. Stone, *Mathematical Instruments*, p. 307.

86. Private letters putting indirect pressure on Knight included one from Granville Wheeler to Wilson of 4 February 1748 (Wilson correspondence, fol. 61) and from Arderon to Baker of 12 May 1750 (Baker correspondence, vol. 4, fols. 278–9).

87. Letter from Arderon to Baker of 3 February 1748: Baker correspondence, vol. 3, fols. 239–40; letter from Baker to Henry Miles of 16 February 1748: ibid., fol. 246. See also the letter from John Browning to Baker of 26 January 1750: ibid., vol. 4, fols. 166–7.

88. RSJB, vol. 20, p. 424. Letter from Baker to Arderon, 31 July 1749: Arderon correspondence, vol. 2, fols. 36/18–19. Letter from Miles to Baker of 28 April 1747: Baker correspondence, vol. 3, fol. 83; Canton, "Method of Making Artificial Magnets." Browning later judged Canton foolish to have disseminated his ideas before he established intellectual ownership by publication: letter from Browning to Baker of 4 April 1751: Baker correspondence, vol. 5, fols. 31–2.

89. *Gentleman's Magazine* 50 (1750), 100–2. Michell, *Artificial Magnets*, was reprinted with minor changes in 1751; Rivoire, *Traités*.

90. Kippis, *Biographia Britannica*, vol. 3, pp. 216–17. Many of the relevant letters in volume 2 of the Canton papers are reproduced in de Morgan, "Canton Papers." In particular, see also letters from Priestley to William Canton of July 1985 (Canton papers, vol. 2, fol. 114) and of 20 August 1785 (reproduced in Weld, *History of the Royal Society*, vol. 1, pp. 513–14); *Monthly Review* 72 (1785), 478–80, and 73 (1785), 160; *Gentleman's Magazine* 55 (1785), 511–12, 687. Knight remained on sufficiently good terms with Michell to support his election to the Royal Society: Hardin, "John Michell," p. 28.

91. B. Wilson, "Dr Knight's Method" and "Autobiography," p. 91. Charles Cavendish disliked Wilson and openly favored Canton. Darwin, *Poetical Works*, vol. 1, pp. 90–3: see McNeil, "The Scientific Muse," and Danchin, "Darwin's Scientific and Poetic Purpose."

92. Coulomb, "Septième mémoire."

93. Morton, "Lectures on Natural Philosophy." Burr, *History of Tunbridge Wells*, pp. 145–6. See Inkster, "Public Lecture." Erasmus King was a popular lecturer in natural philosophy.

94. Schaffer, "Natural Philosophy and Public Spectacle" and "Consuming Flame." Golinski, "Utility and Audience."

95. Walters, "Tools of Enlightenment." G. Adams, *Essay on Electricity*, pp. 373–454, and *Natural and Experimental Philosophy* (1794), vol. 4, pp. 435–70. Letter from Cavallo to James Lind of 10 July 1784: British Library Additional Manuscripts 22897, fols. 21–2; Cavallo, *Treatise on Magnetism* (1787). Adams and Cavallo recommended each other, and some editions of Cavallo's books have Jones' catalogues in the back (W. and S. Jones bought the copyright to Adams' books).

96. Stafford, *Artful Science*. Porter, "Before the Fringe," *Health for Sale*, and "Language of Quackery." See also Jewson, "Medical Knowledge" and "Disappearance of the Sick-Man."

97. Quotation from p. 163 of G. Knight, "Account of Some Magnetical Experiments." G. Knight, *Collection of Some Papers*. B. Wilson, "Dr Knight's Method"; Marcel, "Abstract of a Letter."

98. Millburn, *Retailer of the Sciences*, p. 29 (Martin); Calvert, *Trade Cards*, no. 38 (Henry Pyefinch); de la Touche, "Wonderful Property of Magnets"; Lefebvre and de Bruijn, *Martinus van Marum*, p. 2 (Adams); Nairne and Blunt, *Patent Electrical Machine*, pp. 63–8; G. Adams, *Geometrical and Graphical Essays*, p. 496; W. and S. Jones, "Catalogue" (1795), p. 12, and "Catalogue" (1800) (W. and S. Jones bought the copyright to Adams' books, and their catalogues were bound into books by Adams, Cavallo, and others). For auction sales of magnets, see Millburn, *Retailer*, p. 81; Turner and Levere, *Martinus van Marum*, p. 187; Canton papers, vol. 2.

99. Adams, "Catalogue," in *Micrographia Illustra*, p. 258; G. Knight, *Collection of Some Papers*, p. 17; W. and S. Jones, "Catalogue" (1795), p. 12.

100. For example, Desaguliers, *Physico-mechanical Lectures*, pp. 18–19, and *Experimental Philosophy*, pp. 16–21; Worster and Watts, *Experimental Philosophy*, p. 4; Stirling et al., *Mechanical and Experimental Philosophy.*

101. Priestley, *Heads of Lectures*, pp. 155–61 (Hackney College); Enfield, *Institutes of Natural Philosophy*, pp. 337–40 (Warrington Academy); Atwood, *Description of Experiments*, pp. 49–51; Vince, *Plan of a Course of Lectures*, pp. 148–52. Bishop, "Teaching of Physics," pp. 152–387; see especially pp. 242, 254–5, 258, 302, 599–601.

102. Undated advertisement pasted in the British Library copy of Peat, *Mechanical and Experimental Philosophy* (pressmark 8704.aa.10). Moyse's newspaper advertisement of 26 February 1783 is in Lysons I, vol. 3, fol. 177v. Letter from Arden to Baker of 16 March 1750: Baker correspondence, vol. 4, fols. 255–6. See Schofield, *Lunar Society*, pp. 31–2, for a Birmingham advertisement mentioning artificial magnets. For Demainbray, see Morton and Wess, *Public and Private Science*, pp. 123–6, and Morton, "Lectures on Natural Philosophy." Millburn, "London Evening Courses," especially pp. 453–5.

103. Altick, *Shows of London*, pp. 77–86. Porter, "Language of Quackery." Doherty, "Anodyne Necklace."

104. Unsourced copy of a poster reproduced in Bishop, "Teaching of Physics," p. 370. See also A. S. Brown, "Gustavus Katterfelto"; Money, *Experience and Identity*, pp. 140–1; newspaper clippings in Banks. There is a series of advertisements in the *Norfolk Chronicle* (4 December, 11 December, 18 December, and 24 December 1784 and 1 January, 8 January, and 15 January 1785) which he probably repeated as he toured the provinces. For magnetic clocks, see Hankins and Silverman, *Instruments and Imagination*, pp. 14–36.

105. *Morning Post* of 20 February 1778. This is one of several advertisements in Lysons I, vol. 3, fols. 157–73, and items 240/1 and 446 in the Douce Additional MSS at the Bodleian. See Porter, *Health for Sale*, for references, and J. Graham, *Sketch*. In Paris at this time, a commission from the Société royale de médecine was recommending research into therapeutic magnetic chairs, beds, and other equipment: see Fara, "Attractive Therapy."

106. Croker, *Experimental Magnetism*. The instrument was invented by Abraham Mason, a settler in Barbados. *Dictionary of National Biography*; Croker, Williams, and Clark, *Dictionary*; Paulson, *Hogarth: Life, Art and Times*, vol. 2, p. 263; Hindle, *Science in Revolutionary America*, pp. 349–50; Stearns, *Science in the British Colonies*, pp. 362–3. For Croker's compass, see the manuscript notes in Croker's own copy of his *Experimental Magnetism*, British Library pressmark 956.e.33.

107. RSJB, vol. 23, pp. 903–5 (6 November 1760), and vol. 24, pp. 83–4 (26 February 1761). Birch, Diary, p. 387. *Daily Advertiser* of 27 November 1760: I am grateful to Alan Morton for this reference; Croker, *Experimental Magnetism*, pp. 8, 27, 39–40.

108. Kenrick, *Perpetual Motion*; Lukis, *Memoirs of William Stukeley*, vol. 2, p. 372. For Newton's sketch of a magnetic perpetual-motion machine, see Westfall, *Force in Newton's Physics*, pp. 329–32. Schaffer, "Show That Never Ends." Hutton, *Recreations*, vol. 2, pp. 102–6, described and condemned magnetic perpetual-motion. In 1794, John Tulloch submitted a perpetual-motion machine to the Board of Longitude: RGO 14/42, cut 8. John Spence travelled round the country with a magnetic perpetual-motion machine which he claimed had continued for almost three years: undated poster (probably early nineteenth century) in the John Johnson Collection at the Bodleian Library, Oxford, Scientific Instruments box 2.

109. Thomas, *Religion and the Decline of Magic*. Hughes, "Sciences in English Encyclopædias"; Shorr, *Science and Superstition*.

110. Curry, *Prophecy and Power*, pp. 93–152; Money, "Teaching in the Market-Place," pp. 356–60; R. Richardson, *Death, Dissection and the Destitute*, pp. 3–29. MacDonald, "Religion, Social Change, and Psychological Healing"; Clark, *English Society*, pp. 161–73. Letter from Baker to Archibald Blair of 10 February 1749: Baker correspondence, vol. 4, fol. 67. This letter is part of an extensive correspondence on these subjects.

111. Defoe, *System of Magick*, pp. 46–9, 60–1 (quotation from p. 60): see Baine, *Daniel Defoe*, pp. 64–72, and Tavor, *Scepticism*, pp. 7–53. Pemberton, *Newton's Philosophy*, pp. 12–13.

112. Adam Walker, *System of Familiar Philosophy*, vol. 1, p. 60. Bryan, *Lectures on Natural Philosophy*, p. 143. Bloch, *Royal Touch*; Fara, "Attractive Therapy."

113. Bazerman, "Forums of Validation." Edgeworth, *Memoirs*, pp. 74–6. For conjurors, see T. Frost, *Lives of the Conjurors*, pp. 112–56, and for London entertainments, see Altick, *Shows of London*, pp. 34–127. Schofield, *Lunar Society*, p. 45. King-Hele, *Essential Writings of Erasmus Darwin*, p. 130.

114. Breslaw, *Breslaw's Last Legacy*, pp. 35–6.

115. Golinski, *Science as Public Culture*, pp. 93–130, quotation on p. 122.

116. Regnault, *Philosophical Conversations*, vol. 1, p. 213. *Universal Magazine* 33 (1763), 250–4 (plate at pp. 142–3). Torlais, *Un physicien au siècle des lumières*, pp. 77–8. Ozanam, *Recreations*, p. 458. For Watson's and Nollet's contrasting attitudes to electrical deception, see Schaffer, "Self Evidence," pp. 68–78. Rousseau, *Emile*, pp. 135–8. Day, *Sandford and Merton*, vol. 2, pp. 161–70.

117. Breslaw, *Breslaw's Last Legacy*, pp. 33–7, also reproduced in Pinetti, *Conjurer's Repository*, pp. 34–5.

118. Adam Walker, *System of Familiar Philosophy*, vol. 1, pp. 73–4.

119. Hooper, *Recreations*, vol. 3, pp. 113–246: there were four editions in 20 years. For example, H. Clarke, "Prospectus," and Gale, *Cabinet of Knowledge*, pp. 179–88. G. Adams, *Essay on Electricity*, pp. 350–4 (two of these are illustrated in Morton and Wess, *Public and Private Science*, pp. 444–5); W. and S. Jones, "Catalogue" (1792), p. 12, and (1795), p. 11.

120. Lister, "Journey to Paris," p. 27. G. Adams, "Catalogue," in *Micrographia Illustra*, p. 258.

121. Hooper, *Recreations*, vol. 3, p. 185.

122. Ibid., p. 164. J. Godwin, *Athanasius Kircher*, p. 74; Baldwin, "Magnetism and the Anti-Copernican Polemic," pp. 161–5; Hankins and Silverman, *Instruments and Imagination*, pp. 14–36; W. Hine, "Athanasius Kircher and Magnetism." Kircher, *Magnes sive de Arte Magnetica*, fol. 344. For some eighteenth-century discussions of magnetic communication, see Bailey, *Dictionary*, entry on loadstone; Joseph Addison, *Guardian* 2 (1714), 186–7; and *Notes and Queries* 6 (1852), 204–5.

123. Hooper, *Recreations*, vol. 1, pp. iii–viii (quotation from p. vi); Gale, *Cabinet of Knowledge*, pp. iii–vi. *Monthly Review* 52 (1775), 19–29; for Bewley, see Nangle, *Monthly Review First Series*, pp. 4–5. M. Edgeworth and Edgeworth, *Practical Education*, pp. 26–8.

124. *European Magazine, and London Review* 21 (1792), 411–13, quotation from p. 411. Adam Walker, *System of Familiar Philosophy*, vol. 1, pp. 73–4; Breslaw, *Breslaw's Last Legacy*, p. 37; Pinetti, *Conjurer's Repository*, pp. 34–35; *British Critic* 15 (1800), 638–42.

125. Hutton, *Recreations*, vol. 4, pp. 269–312 ("rational amusements" are pp. 305–12).

Chapter Three
The Direction of Invention

1. Boswell, *Life of Johnson*, vol. 1, p. 232; see Brewer, *Sinews of Power*, p. 50 for a critique. Landow, *Images of Crisis*, pp. 35–130 (and pp. 75–84 for *The Shipwreck*). Falconer, *The Shipwreck* (1762 and 1838); W. Burney, *Dictionary of the Marine*, pp. xi–xv (the 1815 update by Dr. William Burney, head of the Gosport Naval Academy, of Falconer's 1769 *Dictionary of the Marine*). Darwin, *Poetical Works*, vol. 1; Danchin, "Darwin's Scientific and Poetic Purpose," and McNeil, "The Scientific Muse." Falconer's poem first appeared in 1762, but there were ten editions by 1800 and numerous reprints through the nineteenth century; Darwin's *Economy of Vegetation* was first published in 1791.

2. Rodger, *Wooden World*, pp. 46–51. May, "Navigational Accuracy." May, "Last Voyage of Cloudisley Shovel" and "Naval Compasses in 1707."

3. F. James, "Davy in the Dockyard." Quill, *John Harrison*; Howse, *Greenwich Time*, pp. 19–80. *Biographical Account of John Hadley*; Martin, *New Construction and Rectification of Hadley's Quadrant*. Allan and Schofield, *Stephen Hales*, pp. 77–99. Lawrence, "Disciplining Disease."

4. Stewart, *Rise of Public Science*, pp. 183–211 (quotation from p. 209); Waters, "Nautical Astronomy"; Howse, *Greenwich Time*, pp. 1–80; E. G. Forbes, *Greenwich Observatory*, pp. 1–156.

5. Variation is now called declination by scientists, although naval and air force personnel still use the older term. In England (unlike Portugal, for instance) "declination" remained mainly an astronomical term, although some writers referred to dip as "inclination," the modern word. Intensity became important for geomagnetic mensuration only in the early nineteenth century: Multhauf and Good, *Brief History of Geomagnetism*, pp. 1–21; Chapman and Bartels, *Geomagnetism*, vol. 2, pp. 898–937; Humboldt, *Cosmos*, vol. 5, pp. 49–162. Robison, "Magnetism," pp. 137–9, summarizes the problems of measuring dip.

6. E.G.R. Taylor, *Haven-finding Art*, pp. 89–263; J. A. Bennett, *Divided Circle*, pp. 51–60; H. Wallis and Robinson, *Cartographical Innovations*, pp. 141–3, 189–91.

7. May, "Binnacle"; Randier, *Marine Navigation Instruments*, pp. 31–6. Blanckley, *Naval Expositor*, p. 41.

8. Hewson, *History of Navigation*, pp. 120–54; May, *History of Marine Navigation*, pp. 43–107; J. A. Bennett, *Divided Circle*, pp. 51–60, 130–42.

9. Falconer, *Dictionary of the Marine*, entry on compass.

10. Grant and Klinkert, *Ship's Compass*, p. 71. See Fanning, *Steady as She Goes*, pp. xx–xxii; Hewson, *History of Navigation*, pp. 55–9; A. Hine, *Magnetic Compasses*, p. 3; Hitchins and May, *From Lodestone to Gyro-Compass*, pp. 24–31; May, *History of Marine Navigation*, pp. 67–9.

11. Schofield, *Mechanism and Materialism*, pp. 175–81; Thackray, *Atoms and Powers*, pp. 141–7; Heimann and McGuire, "Newtonian Forces and Lockean Powers," pp. 296–9; Quinn, "Repulsive Force," pp. 126–8; Home, "Introduction," pp. 158–60.

12. Schuster and Watchirs, "Natural Philosophy, Experiment and Discourse." Bijker, Hughes, and Pinch, *Social Construction of Technological Systems*; MacKenzie and Wajcman, *Social Shaping of Technology*. For technology and Portuguese maritime expansion, see J. Law, "Methods of Long-Distance Control" and "Technology and Heterogeneous Engineering."

13. [Joseph Cottle], "Science Revived, or the Vision of Alfred," reviewed and quoted in *British Critic* 19 (1802), 388–93 (quotations pp. 388, 390). Nieuwentyt, *Religious Philosopher*, vol. 1, p. 567. J. Hervey, *Theron and Aspasio*, vol. 2, pp. 234–6.

14. Mandeville, *Fable of the Bees*, vol. 2, pp. 41–2. *Notes and Queries* 6 (1852), 369.

15. Falconer, *The Shipwreck* (1838), pp. 31, 52. Navigation texts often defined the variation in terms of an arch on the horizon rather than as an angle on the compass.

16. On the appropriateness of the term "administrator" for this period, see Brewer, *Sinews of Power*, pp. 79–85.

17. E.G.R. Taylor, *Mathematical Practitioners of Hanoverian England*, pp. 25–30. E. Stone, *Mathematical Instruments*, p. 307.

18. There were, however, allegations that merchants in league with insurance agents found it financially more advantageous to lose their ships than to spend money on improvements: Bain, *Variation of the Compass*, pp. 135–7. Numerous writers on navi-

gation commented on the inferior equipment of ships owned by merchants other than the large companies.

19. Latour, *Science in Action*; Callon, "Elements of a Sociology of Translation." Pickering, *Science as Practice*, pp. 301–89. Critics cite Latour's account of the Pasteurization of France as a reformulated heroic narrative.

20. Star and Griesemer, "Institutional Ecology"; Fujimura, "Crafting Science." For instance, different participants constructing a natural-history museum invest items to be displayed with different values: a single animal may represent a conservationist's threatened species, a trapper's source of income, a scientist's piece of evidence, or a curator's didactic exhibit.

21. Galison, "History of Physics," especially pp. 31–3. Theoretical and experimental scientists may agree that a particle detector has recorded the passage of an electron yet hold irreconcilable views about how electrons behave and how the detector works.

22. These three sectors of society were not, of course, formally identified as such at the time, and there were important internal distinctions. For example, "navigators" is a loose term covering captains, masters, and other maritime practitioners with contrasting training, skills, and responsibilities.

23. Brock, "Anson." Pond's engraving hangs in London's National Portrait Gallery; it is reproduced in Lippincott, *Selling Art in Georgian London*.

24. Brewer, *Sinews of Power*, pp. 29–63; Baugh, *British Naval Administration*, pp. 93–146; G. Holmes, *Augustan England*, pp. 274–87; Davis, *Rise of the English Shipping Industry*, pp. 117–28.

25. Rodger, *Wooden World*; Rediker, *Devil and the Deep Blue Sea*. Armitage, *Edmond Halley*, pp. 139–42 (quotation from p. 142); Thrower, *Three Voyages of Edmond Halley*, pp. 281–7. A pink was a flat-bottomed three-masted ship.

26. W. Hutchinson, *Treatise on Practical Seamanship*, pp. 85–7, quotation from p. 85. Moore, *Practical Navigator*, p. 189. Konvitz, *Cartography in France*, pp. 63–7; Bravo, "James Rennell," pp. 42–3.

27. G. Holmes, *Augustan England*. Pumfrey, "Who Did the Work?" Golinski, *Science as Public Culture*, especially pp. 10, 15, 55. Russell, *Science and Social Change*, pp. 193–234 (quotation from p. 220). D. Knight, "Science and Professionalism."

28. Macleod, *Inventing the Industrial Revolution*: of those inventors who did choose to take out patents during the eighteenth century, one-fifth described themselves as gentlemen (p. 117). Whiston, *Longitude and Latitude*, pp. 1–7, 94–101, and *Calculation of Solar Eclipses*, pp. 83–91. Michell, *Artificial Magnets*, p. 2 and passim. Letter from Baker to Miles of 2 November 1745: Baker correspondence, vol. 2, fol. 131.

29. *The Patent: a Poem*, p. 9 (an author's footnote informs readers that a loblolly was a ship surgeon's assistant who knew nothing about a seaman's work).

30. G. Holmes, *Augustan England*, pp. 239–61; Brewer, *Sinews of Power*, pp. 25–134; Baugh, *British Naval Administration*, pp. 1–92; Rodger, *Wooden World*, pp. 29–36; R. Middleton, "Naval Administration." Pool, "Navy Contracts," especially p. 162.

31. E. Harrison, *Idea Longitudinis*, pp. 26–49, quotation from p. 30. See also Maitland, *Improvement of Navigation*; Dunn, *Navigator's Guide*, pp. 10–11; John Robertson, *Elements of Navigation*, vol. 2, pp. 231–2, 291; Fergusson, *Observations on the Present State of Navigation*, pp. 1–12; W. Burney, *Dictionary of the Marine*, pp. 98–9.

32. W. Hutchinson, *Treatise on Practical Seamanship*, pp. 102–6, quotation from pp. 105–6. May, "Naval Compasses in 1707." Admiralty instructions to Cook for sailing to Tahiti in 1768: Waters, "Nautical Astronomy," p. 163n7. See Beck, *Folklore and the Sea*, pp. 107–8.

33. I am grateful to Andrew Cook for showing me many logbooks of East India Company ships and for pointing out the errors in May, "Log Books." For the Dutch East India Company and magnetic variation, see Davids, "Finding Longitude at Sea."

34. E. Harrison, *Idea Longitudinis*, pp. 29–33 (quotation from preface and p. 31). Maitland, *Improvement of Navigation*, pp. 29–31, quotation from p. 31.

35. Hitchins and May, *From Lodestone to Gyro-Compass*, p. 33. Norman, *Newe Attractive*, pp. 39–43; Emerson, *Navigation*, p. 4. See May, "True Reading Magnetic Compasses," pp. 144–7.

36. Wakeley, *Mariner's Compass Rectified*, pp. 6, 93–5, 131–5. See also Atkinson, *Epitome of Navigation*, pp. 253–61, Dunn, *Nautical Propositions*, pp. 24–5, and John Adams, *New Much-improved Sinical Quadrant*, p. 22. W. Burney, *Dictionary of the Marine*, pp. 590–1.

37. Nicholas Rodger took his title *Wooden World* from that of a 1708 book by Ned Ward. Rediker, *Devil and the Deep Blue Sea*, pp. 153–204; Beckett, *Naval Customs*; Beck, *Folklore and the Sea*.

38. Joseph Harris, *Treatise of Navigation*, p. 191. Quotation from Captain Livingston's letter: Purdy, *Memoir*, p. 337. May, "Garlic and the Magnetic Compass." This belief was not restricted to mariners: see, for instance, John Cook, *Clavis Naturæ*, p. 100.

39. Moore, *Practical Navigator*, pp. 136–7. This book was a best-seller, running to twenty editions: Cotter, "British Navigation Manuals," pp. 244–5.

40. Lovett, *Philosophical Essays*, pp. 488–514; Churchman, *Magnetic Atlas* (1794), pp. 50–2. Magellan, *Collection de différens traités*, p. 218.

41. May, "Naval Compasses in 1707." The typical quotation is from PRO: ADM 2/511, p. 297 (22 June 1752). Letters from the Deptford Storekeeper to the Navy Board of 4 June 1745 and 27 and 29 January 1746 (PRO: ADM 106/3359) and of 10, 16, 17, and 21 February 1757 (PRO: ADM 106/3362). May, "Compass Makers of Deptford"; and Ranft, "Labour Relations in the Royal Dockyards."

42. J. A. Bennett, "Hooke's Instruments," especially pp. 31–2. For a new compass needle by Hooke, see RSJB, vol. 8, pp. 367–8 (29 July 1696). G. Knight, "Account of Some Magnetical Experiments," "Letter to the President," and "Collection of Magnetical Experiments."

43. G. Knight, "Account of the Mariners Compass"; Lukis, *Stukeley*, vol. 2, pp. 361–2.

44. Addendum by Smeaton to a letter from Michell: *Monthly Review* 72 (1785), 478–80. Woodcroft, *Index of Patents*, vol. 2, pp. 530–1.

45. Hoxton, "Unusual Agitation"; Cookson, "Account of Lightning" and "Further Account of Lightning"; Bremond, "File Rendered Magnetical by Lightning." Letter from John Huxham to James Jurin of 19 January 1728: Royal Society Early Letters H.3.118 (I am grateful to Andrea Rusnock for giving me a transcript of this letter). RSJB, vol. 13, pp. 179, 551 (15 February 1727 and 28 January 1730); vol. 14, pp. 202–3, 247–8 (14 December 1732 and 22 February 1733); vol. 17, p. 157 (27 November 1740).

46. Waddell, "Effects of Lightning"; G. Knight, "Account of the Mariners Compass." For Franklin's response to Waddell's account, see I. B. Cohen, *Franklin's Experiments*, pp. 242–3, 320. G. Knight, "Description of a Mariner's Compass"; Smeaton, "Improvements of the Mariners Compass."

47. In modern terminology, these outer casings are called boxes, but in the eigh-

teenth century the word "box" referred to what is now called the bowl, the inner container of the needle and card. E. Harrison, *Idea Longitudinis*, pp. 29–30 and elsewhere.

48. Magnetic sand had been discovered in several parts of the world. Natural philosophers were interested in analyzing it and using it as a cheap source of iron: Musschenbroek, "Indian Magnetic Sand," H. Horne, "Observations on Sand Iron," Moulen, "Experiments on Black Shining Sand," and elsewhere.

49. Letter from Smeaton to Wilson of 10 July 1748: Wilson correspondence, fol. 69. Smeaton, "Improvements of the Mariners Compass." Magellan, *Traités*, p. 221. Maitland, *Improvement of Navigation*, p. 29 (not specifically referring to Knight and Smeaton).

50. Letter from the Navy Board to the Admiralty Secretary of 15 April 1751—PRO: ADM 106/2185, p. 430. Letter from Admiralty Secretary to Knight of 24 January 1750—PRO: ADM 2/697, pp. 2–3. Mountaine and Dodson, *Methods Used to Describe Lines*, p. 12. Other examples include Crosby, *Mariner's Guide*, p. 182 (edited by Mountaine).

51. W. Hutchinson, *Practical Seamanship*, p. 104.

52. Letters from the Admiralty Secretary to members of the Board of Longitude of 14 May 1752—PRO: ADM 2/511, p. 273. Fellows on the Board included Henry Legge (First Commissioner of the Navy), Folkes, James Bradley, and John Colson.

53. Brock, "Anson as a Naval Reformer"; Anson, *Voyage round the World*, unpaginated introduction. See G. Williams, *Documents Relating to Anson's Voyage*, pp. 230–2, 271–83. Letter from Knight to the Archbishop of Canterbury of 22 September 1754: British Library Additional Manuscripts 36269, fol. 31. Letter from Admiralty Secretary to Navy Board of 28 February 1751—PRO: ADM 106/2077. Letter from the Navy Board to the Admiralty Secretary of 27 March 1751—PRO: ADM 106/2185.

54. Letter from the Admiralty Board to five captains of 4 April 1751—PRO: ADM 2/73; Admiralty Board Minutes of 4 April 1751—PRO: ADM 3/62; letter from the Navy Board to the Admiralty Secretary of 15 April 1751—PRO: ADM 106/2185, p. 446.

55. Letter from Knight to the Navy Board of 7 May 1751—PRO: ADM 106/1092. Letter from Captain John Jermy to the Admiralty Board of 19 April 1751—PRO: ADM 1/1984. Letter from the Admiralty Secretary to Knight of 26 April 1751—PRO: ADM 2/697.

56. Letter from the Navy Board to the Admiralty Board of 29 April 1752—PRO: ADM 106/2186, p. 175.

57. Letter to the Admiralty Board from the Navy Board of 26 January 1754—PRO: ADM 106/2187; letter from the Admiralty Board to the Navy Board of 29 January 1754—PRO: ADM 106/2077. Admiralty Minutes of 11 September 1751—PRO: ADM 3/62. Logbook of the *Fortune*, September 1751—PRO: ADM 51/361.

58. Letter from the Admiralty Secretary to the Treasury of 22 June 1752—PRO: ADM 2/511, p. 297. Admiralty Board Minutes of 24 June 1752—PRO: ADM 3/62. Letters from Admiralty Secretary to the Board of Longitude of 14 May and 9 June 1752—PRO: ADM 2/511, pp. 273, 286–7. Admiralty Board Minutes of 29 July 1752—PRO: ADM 3/62.

59. Letter from the Admiralty Secretary to Knight of 6 March 1756—PRO: ADM 2/703, p. 441. Navy Board Minutes of 20 March 1758—PRO: ADM 106/2567. Correspondence between the Deptford Storekeeper and the Navy Board during February 1757—PRO: ADM 106/3362. Navy Board Minutes of 2 March 1758—PRO: ADM

106/2567. Some orders to Knight are in the Navy Board Minutes of 13 and 18 October 1758—PRO: ADM 106/2568.

60. Letter from Deptford Clerk of the Cheque to the Navy Board of 7 January 1765—PRO: ADM 106/3401, p. 7. Letters from Deptford Clerk of the Cheque to the Navy Board of 29 November 1756 (PRO: ADM 106/3381) and of 12 and 19 August 1768 (PRO: ADM 106/3401, pp. 431–2). Letters from the Navy Board to the Admiralty Board show that between 1758 and 1772 the Navy Board reduced its compass makers from four (PRO: ADM 106/2190, p. 584) to none (PRO: ADM 106/2201, p. 344).

61. Spinney, *Rodney*, p. 107; letter from the Navy Board to the Admiralty of 27 March 1751—PRO: ADM 106/2185, pp. 429–30. Gower, *Theory and Practice of Seamanship*, p. 91. Letter from M. Anthony to the Board of Longitude of 19 June 1799: RGO 14/13, fol. 127.

62. Letter from Matthew Buckle to the Navy Board of 13 February 1753—PRO: ADM 106/1108. Letter from Lucius O'Bryen to the Admiralty Board of 16 January 1757—PRO: ADM 1/2245.

63. G. Robertson, *Discovery of Tahiti*, p. 5.

64. Royal Society Council Book, especially May 1768. Letter from Cook to the Admiralty Board of 25 July 1768—PRO: ADM 1/1609. Admiralty Board Minutes of 27 July 1768—PRO: ADM 3/76, pp. 73–6. Letter from the Admiralty Secretary to the Navy Board of 27 July 1768—PRO: ADM 2/238, p. 90.

65. For example, William Martin (letter of 9 February 1754—PRO: ADM 1/2107). Letters from the Admiralty Board to the Navy Board of 1, 2, and 3 June and 30 August 1756—PRO: ADM 2/220, pp. 297, 298, 303, 499.

66. Letter of 26 February 1758—PRO: ADM 1/2010; Navy Board Minutes of 27 February 1758—PRO: ADM 106/2567. See also letters from the Admiralty Secretary to the Navy Board of 12 August 1752 and 22 February 1753—PRO: ADM 2/511, pp. 329, 441.

67. Letter from Richard Kempenfelt to the Admiralty Board of 10 February 1757—PRO: ADM 1/2010. Letter from John Sayer to the Admiralty Board of 11 October 1758—PRO: ADM 1/2472. Letters from Deptford Clerk of the Cheque to the Navy Board of 29 November 1756 and 8 April 1764—PRO: ADM 106/3381–2.

68. George Rodney, quoted in Spinney, *Rodney*, p. 107. Falconer, *Dictionary of the Marine*, entry on compass.

69. Letter from Cook to the Navy Board of 8 March 1768—PRO: ADM 106/1163. Letter from Cook to the Admiralty of 12 July 1771—PRO: ADM 1/1609. Letter from Cook to the Navy Board of 11 February 1772, reproduced in Beaglehole, *Journals of Captain Cook*, vol. 2, p. 913. Cook seems to have been more satisfied with their performance on land: James Cook, "Variation of the Compass," p. 426.

70. G. Knight, "New Method of Constructing Compasses." Schaffer, "Machine Philosophy," p. 180. May, "History of the Magnetic Compass," pp. 216–17. Magellan, *Traités*, p. 217.

71. Robertson, *Discovery of Tahiti*, pp. 7, 29.

72. Admiralty Board Minutes of 5 April 1769—PRO: ADM 3/76.

73. Beaglehole, *Journals of Captain Cook*, vol. 1, p. 138; Hawkesworth, *Account of the Voyages*, p. 248. Beaglehole, *Journals*, vol. 2, pp. 19–20, 76, 78, 80, 89n, 97, 475n, 525, 563, and vol. 3, p. 287.

74. May, "Deviation of the Compass," pp. 217–22. James Cook and King, *Voyage to the Pacific Ocean*, vol. 2, p. 103. Bligh, *Log of the Bounty*, vol. 2, p. 121. Bain,

Variation of the Compass, p. 140. Gower, *Theory and Practice of Seamanship*, p. 100. G. Knight, "Account of the Mariners Compass," pp. 114–15.

75. W. Burney, *Dictionary of the Marine*, p. 98.

76. Admiralty Board Minutes of 29 July 1752—PRO: ADM 3/62. Standing Order of the Navy Board of 18 December 1778—PRO: ADM 106/2508. W. Burney, *Dictionary of the Marine*, p. 99.

77. Admiralty Indexes (PRO: ADM 12/52–5) go up to 1792 (most of these examples are taken from ADM 12/52, cut 59/3).

78. Letters from Deptford Clerk of the Cheque to the Navy Board of 29 November 1756 and 8 April 1764—PRO: ADM 106/3381–2; and of 7 January 1765—PRO: ADM 106/3401, p. 7. Letters from Deptford to the Navy Board of 12 and 19 August 1768—PRO: ADM 106/3401, pp. 431–2.

79. Letters from the Admiralty Secretary to Knight and Robert Waddington of 26 June and 1 July 1766 and 19 February 1767—PRO: ADM 2/726, pp. 244, 512. Waddington, *Theoretical and Practical Navigation*, pp. 20–1. RGO 14/5, fol. 205.

80. Manuscript notes in Croker's own copy of *Experimental Magnetism*, British Library pressmark 956.e.33.

81. Wheatland, *Apparatus of Science at Harvard*, pp. 155–60, and I. B. Cohen, *Early Tools of American Science*, p. 47.

82. Warner, "What Is a Scientific Instrument?" and "Terrestrial Magnetism." McConnell, "Nineteenth-Century Geomagnetic Instruments." Woodcroft, *Index of Patents*, vol. 2, pp. 530–1. Macleod, *Inventing the Industrial Revolution*.

83. For Kenneth McCulloch, see McCulloch, *New Improved Sea Compasses*, For Benjamin Martin, see Martin, *New Construction and Rectification of Hadley's Quadrant*, pp. 6–8. For Henry Gregory, see Wales and Bayly, *Astronomical Observations*, p. xliv, and a letter from Cook to the Navy Board of 11 February 1772, reproduced in Beaglehole, *Journals of Captain Cook*, vol. 2, p. 913. For Ralph Walker, see Walker, *Treatise on Magnetism*, and May, "Longitude by Variation." For Captain Lane, see May, "Mystery of Captain Lane's Compass." For P. R. Nugent, see Nugent, "New Theory." For O'Brien Drury, see *Encyclopædia Britannica*, supplement to the 3rd edition, vol. 1, pp. 446–7.

84. Lawrence, "Disciplining Disease." D. Knight, "Science and Professionalism," pp. 56–64. Nevil Maskelyne's report to the Board of Longitude of 6 December 1794: RGO 14/4, fol. 17. Varley, "Irregularity in Rate of Time-Pieces." Brooks, "Magnetic Influence on Chronometers"; Fisher, "Errors in Longitude."

85. Bain, *Variation of the Compass*, p. 132n. Mountaine, "Variation of the Magnetic Needle," pp. 218–19 (letter from David Ross, a ship's surgeon). James Cook, "Variation of the Compass," p. 425. Bain, *Variation*, pp. 58–9. Hawkesworth, *Account of the Voyages*, p. 248.

86. Beaglehole, *Life of Captain Cook*, pp. 117, 310–17. RSJB, vol. 30, pp. 465–70 (31 January 1782).

87. Admiralty Board Minutes of 5 April 1769—PRO: ADM 3/76. Letters from Charles Douglas to the Admiralty Board of 23 March and 5 April 1769—PRO: ADM 1/1705. Letter from Charles Douglas to the Admiralty Board of 9 September 1769—PRO: ADM 1/1705. Wales and Bayly, *Astronomical Observations*, pp. 182, 367 et. seq.; Bligh, *Log of the Bounty*, p. 42.

88. Bain, *Variation of the Compass*, pp. 52–66; May, "Deviation of the Compass," pp. 223–4. R. Walker, *Treatise on Magnetism*, pp. 193–210.

89. Flinders, "Differences in the Magnetic Needle." Cotter, "Flinders and Ship

Magnetism"; Mack, *Matthew Flinders*, pp. 246–8; Ingleton, *Matthew Flinders*, pp. 405–16. British Library Additional Manuscripts 32439, fols. 353–69. Flinders, *Voyage to Terra Australis*.

90. W. Burney, *Dictionary of the Marine*, pp. 98–9. F. James, "Davy in the Dockyard." Standing Order of the Navy Board of 15 September 1779—PRO: ADM 106/2508.

91. Quoted in Purdy, *Memoir*, p. 337.

92. Bravo, "James Rennell." Ritchie, *Admiralty Chart*, pp. 3–34. Porter, "Industrial Revolution and the Rise of Geology."

93. Cawood, "Terrestrial Magnetism" and "Magnetic Crusade." D. P. Miller, "Into the Valley of Darkness," "Between Hostile Camps," and "Revival of the Physical Sciences."

Chapter Four
An Attractive Empire

1. Altick, *Shows of London*, pp. 22–98 (quotation from p. 72). Langford, *Polite and Commercial People*, pp. 572–4.

2. Latour, "Force and Reason of Experiment." Friedman, *Appropriating the Weather*. For compasses and Portuguese imperial expansion, see J. Law, "Methods of Long-Distance Control" and "Technology and Heterogeneous Engineering."

3. T. O. Lloyd, *British Empire*, especially pp. 1–137. For varied interpretations of Britain's "Second Empire," see Mackay, "British Imperial Policy," Steven, *Trade, Tactics and Territory*, A. Frost, "Science for Political Purposes," Bayly, *Imperial Meridian*, pp. 1–15, and Baugh, "Seapower and Science," pp. 27–42.

4. Goldsmith, *Private Vices, Public Benefits*, pp. 1–27; Jack, *Corruption and Progress*. Spadafora, *Idea of Progress*. Porter, *Edward Gibbon*, pp. 67–93. Jacobus, *Romanticism Writing and Sexual Difference*, pp. 69–93 (including this quotation from *The Prelude*, book 3, line 63).

5. Lofft, *Eudosia*, pp. 9–13 (quotation from p. 13): see W. P Jones, *Rhetoric of Science*, p. 205. Barthes, "Plates of the *Encyclopedia*." Pratt, *Imperial Eyes*, especially pp. 15–107. Batten, *Pleasurable Instruction*, pp. 84–110. G. S. Rousseau, "Science Books and Their Readers." Porter, "Terraqueous Globe." Rose, "Extract of Two Letters," p. 445.

6. Harley, "Maps, Knowledge and Power" and "Deconstructing the Map." Revel, "Knowledge of the Territory." C.N.G. Clarke, "Taking Possession." Schiebinger, *The Mind Has No Sex?*, pp. 119–59. The first proofs of this chart carried the decorative cartouche but not the dedication: Thrower, *Three Voyages of Halley*, vol. 2 (large-scale reproduction). For French and Dutch versions of Halley's chart, see Konvitz, *Cartography in France*, and Davids, "Magnetic Declination on Dutch East Indiamen."

7. Atherton, *Political Prints*, pp. 89–97; Colley, *Britons*, especially pp. 10–11.

8. H. Wallis and Robinson, *Cartographical Innovations*, pp. 141–3, 189–91. H. Winter, "Pseudo-Labrador and Oblique Meridian." E.G.R. Taylor, *Haven-finding Art*, pp. 151–263.

9. E. Harrison, *Idea Longitudinis*, p. 33. For example, Maitland, *Improvement of Navigation*, and Dunn, *Navigator's Guide*, pp. 10–11.

10. Fergusson, *Present State of the Art of Navigation*, p. 10. D. P. Miller, "Between Hostile Camps." Fergusson is conflated with James Ferguson in E.G.R. Taylor, *Mathematical Practitioners of Hanoverian England*.

11. For example, Sanderson, "Variation of the Needle," Cornwall, "Observations of the Variation," Halley, "Observations of Latitude and Variation," and Hoxton, "Unusual Agitation in the Magnetical Needle" and "Variation of the Magnetic Needle."

12. C. Middleton, "New and Exact Table . . . 1721, to 1725," "New and Exact Table . . . 1721 to 1729," "Observations on the Weather," "Observations of the Variations of the Needle," "Observations Made of the Latitude," "Observation of the Magnetic Needle," "Effects of Cold," and "Use of a New Azimuth Compass." For the allegation that his magnetic research gained him his FRS, see E.G.R. Taylor, *Mathematical Practitioners of Hanoverian England*, p. 186.

13. Joseph Harris, "Astronomical Observations," "Account of Some Magnetical Observations," and *Treatise of Navigation*, pp. 190–3. Stewart, *Rise of Public Science*, pp. 187–8. Maxwell, "Cape of Good Hope," pp. 2423–33, is the printed version of Maxwell's letter to John Harris in British Library Sloane Manuscripts 3399, fols. 49–53. Fol. 48 is a note from Maxwell referring to his measurements of magnetic variation, reproduced on pp. 2433–4.

14. For example, James Cook, "Variation of the Compass," C. Green and Cook, "Observations," Holland, "Astronomical Observations," and Dalrymple, "Journal of a Voyage."

15. P. G. Adams, *Travelers and Travel Liars*, pp. 44–79. R. Holmes, *Coleridge*, pp. 160–2. Ruggles, "Governor Samuel Wegg."

16. Anson, *Voyage round the World*, unpaginated introduction. See G. Williams, *Documents Relating to Anson's Voyage*, pp. 230–2, 271–83, and Baugh, "Seapower and Science," pp. 12–27.

17. J. A. Bennett and Brown, *Compleat Surveyor*, pp. 9–17. Molyneux, "Demonstration of an Error." Churchman, *Magnetic Atlas* (1794), p. v.

18. R. Walker, *Treatise on Magnetism*, pp. 211–26, quotation from p. 222. James Robertson, "Permanency of the Variation."

19. Deacon, "Founders of Marine Science." H. Wallis, "Geographie Is Better than Divinitie."

20. Halley, "Theory of the Variation," pp. 208–9.

21. Golinski, "Utility and Audience." D. Knight, "Science and Professionalism," pp. 56–64.

22. Defoe, *History of Discoveries and Improvements*, pp. 233–307, quotation from p. 303. See Schaffer, "Defoe's Natural Philosophy." RSJB, vol. 19, p. 366 (30 November 1747).

23. Thacker, *Longitudes Examin'd*, p. 2. Whiston, *Longitude and Latitude*, pp. xxiv–xxv.

24. Z. Williams, *Mariners Compass* and *Attempt to Ascertain the Longitude*. E.G.R. Taylor, "Reward for the Longitude" A. J. Kuhn, "Dr Johnson, Zachariah Williams."

25. R. Walker, *Memorial*, pp. 3–9. Lovett, *Philosophical Essays*, pp. 445–520. For biographical information, see J. Chambers, *Biographical Illustrations of Worcestershire*, pp. 363–4: I am grateful to Ronald Stratton, of Worcester Cathedral Library, for this reference.

26. Pringle, "Different Kinds of Air," pp. 27–8.

27. The major magnetic discussions of the aurora were Halley, "Late Surprizing Appearance of the Lights," pp. 421–8; Canton, "Regular Diurnal Variation," pp. 398–404; Dalton, *Meteorological Observations and Essays*, pp. 153–94. See J. M. Briggs, "Aurora and Enlightenment," and Lindqvist, "Spectacle of Science." Jordanova,

"Earth Science and Environmental Medicine." Quotation from *Analytical Review* 6 (1790), 240 (translation from Joseph de Lalande). See also Musschenbroek, "Ephemerides Meterologicæ."

28. Schaffer, "Halley's Atheism." Whiston, *Longitude and Latitude*, pp. 44–75; Churchman, *Explanation of the Magnetic Atlas*, pp. 39–45. G. Knight, *Attraction and Repulsion*, pp. 89–94; John Robison, "Variation," *Encyclopædia Britannica* (1797), vol. 18, pp. 619–25.

29. Derham, *Physico-theology*, p. 278. R. Walker, *Treatise on Magnetism*, p. 43.

30. Marshall and Williams, *Great Map of Mankind*. Passmore, *Man's Responsibility*, pp. 3–40. Priestley, *Experiments and Observations*, vol. 1, pp. xxxi. *Universal Magazine* 6 (1747), 118.

31. Halley, "Remarks on the Variation of the Magnetical Compass," p. 165. See also his "Variation of the Magnetical Compass."

32. Accounts of early geomagnetic measurements include Chapman and Bartels, *Geomagnetism*, vol. 2, pp. 898–937; Humboldt, *Cosmos*, vol. 5, pp. 49–162; E. G. Forbes, *Greenwich Observatory*, pp. 1–24. See Pumfrey, "O Tempora! O Magnes!" and Bryden, "Magnetic Inclinatory Needles," for specific seventeenth-century episodes. Cawood, "Terrestrial Magnetism" and "Magnetic Crusade."

33. McClellan, *Science Reorganized*. Woolf, *Transits of Venus*. Daston, "Ideal and Reality of the Republic of Letters."

34. MacPike, *Correspondence and Papers of Halley*, pp. 55–6. Wargentin, "Variation of the Magnetic Needle." Bergman, "Observations on *Auroræ Borealis*," and "Experiments in Electricity." See Lindqvist, "Spectacle of Science." Heilbron, *Physics at the Royal Society*, pp. 24–5; Appleby, "Daniel Dumaresq," pp. 36–7. Quotation from a letter to George Bulffinger of 25 May 1727: London Wellcome Institute MSS 6146 (I am grateful to Andrea Rusnock for giving me a translation of this letter).

35. *Philosophical Transactions* 18 (1694), 167 (anonymous review of travel accounts). Mountaine and Dodson, "Attempt to Point Out the Advantages," p. 875, and "A Letter concerning the Variation," pp. 330–1.

36. The most comprehensive analysis of his voyages is Thrower, *Three Voyages of Halley*, which includes reproductions of many primary documents. Shorter accounts include Thrower, "Edmond Halley and Thematic Geo-cartography," Armitage, *Halley*, pp. 138–48, Ronan, *Halley*, pp. 161–82, Stearns, "Course of Edmond Halley," and Evans, "Edmond Halley, Geophysicist."

37. Steven, *Trade, Tactics and Territory*, pp. 20–1; Mack, *Flinders*, pp. 75–7.

38. Feldman, "Ancient Climate" and "Late Enlightenment Meteorology." For Halley and the Flood, see Goldgar, "Fielding, the Flood Makers, and Natural Philosophy." Konvitz, *Cartography in France*, pp. 1–31, 71–7.

39. Rooke, "Directions for Sea-Men." Deacon, "Founders of Marine Science." Wallis, "Geographie Is Better than Divinitie."

40. Thrower, *Voyages of Halley*, vol. 1, pp. 29–55. Whiston, *Memoirs*, vol. 1, pp. 253–5, quotation from p. 254.

41. Letter from Charles Burley of 28 September 1757: PRO ADM 14/179, p. 756. Mountaine and Dodson, *Methods Used to Describe Lines*, pp. 4–6 (quotation p. 4). See also their "Letter concerning the Variation," p. 334.

42. Steven, *Trade, Tactics and Territory*. A. Frost, "Science for Political Purposes." Baugh, "Seapower and Science."

43. Ruggles, "Governor Samuel Wegg." Hutchins, "Experiments on the Dipping Needle" and "Success of Some Attempts."

44. Quoted in Gascoigne, *Banks and the English Enlightenment*, p. 61. D. P. Miller, "Into the Valley of Darkness."

45. Mackay, *In the Wake of Cook*; Gascoigne, *Banks and the English Enlightenment*, pp. 185–236. Seymour, *History of the Ordnance Survey*, pp. 1–20; Widmalm, "Accuracy, Rhetoric, and Technology," pp. 184–92. Mack, *Flinders*, pp. 37–152.

46. For example, James Cook, "Variation of the Compass," C. Green and Cook, "Observations," Dalrymple, "Journal of a Voyage," and Pickergill, "Variation of the Compass." For some instructions to Cook, see the Royal Society Council Minutes for May 1768, particularly pp. 316, 344–54. For instructions to William Wales, see Beaglehole, *Journals of Cook*, vol. 2, pp. 724–8. For instructions to William Bayly, see vol. 3, pp. 1500–4. Part of Henry Cavendish's letter to Alexander Dalrymple about dipping needles is reproduced in Chree, "Cavendish's Magnetic Work," p. 462.

47. D. Knight, "Science and Professionalism in England." Batten, *Pleasurable Instruction*. G. S. Rousseau, "Science Books and Their Readers." Pratt, *Imperial Eyes*, pp. 15–107. M. Shelley, *Frankenstein*, p. 14.

48. Macdonald, "Diurnal Variation at Fort Marlborough" and "Diurnal Variation in St Helena." Lorimer, *Concise Essay on Magnetism*, pp. 23 (South Africa), 34 (Quebec); W. Hunter, "Astronomical Observations" (India).

49. Latour, "Force and Reason of Experiment." Mukerji, "Voir le pouvoir."

50. Douglass, "Variation of the Compass," p. 20; see also Mountaine and Dodson, *Methods Used to Describe Lines*, pp. 4–5.

51. J. Law and Whittaker, "On the Art of Representation." See also Lynch and Woolgar, *Representation in Scientific Practice*.

52. L. Roberts, "Setting the Table."

53. For example, compare Canton, "Regular Diurnal Variation," with Gilpin, "Observations on the Variation."

54. Examples include R. Walker, *Treatise on Magnetism*, pp. 54–63; Whiston, *Longitude and Latitude*, p. 76; Churchman, *Explanation of the Magnetic Atlas*.

55. Funkhouser, "Graphical Representation of Statistical Data." Tilling, "Early Experimental Graphs." Hankins and Silverman, *Instruments and Imagination*, pp. 118–28. For a few exceptions, see Boyer, "Early Graph of Statistical Data."

56. Whiston's maps of dip were limited to parts of England and France. In 1768, the Swede Johann Wilcke published a global map showing lines of equal dip. Near the end of the century, the American surveyor John Churchman came to England to promote his magnetic atlas, charts of dip and variation based on calculations and other people's observations. For James Rennell's chart of magnetic variation in the Indian ocean, see Bravo, "James Rennell," pp. 43–5.

57. Daston, "Domestication of Risk" and *Classical Probability*, pp. 112–87. Stewart, *Rise of Public Science*, pp. 271–8. Buck, "People Who Counted," pp. 29–35.

58. Thrower, "Halley and Thematic Geo-cartography." For details of his magnetic charts, see Bauer, "Halley's Earliest Equal Variation Chart," Chapman, "Halley as Physical Geographer," and Thrower, *Voyages of Halley*, vol. 1, pp. 368–70. For the history of maps of natural phenomena, see H. Wallis and Robinson, *Cartographical Innovations*, pp. 135–62.

59. Quotation from John Aubrey in H. Wallis, "Geographie Is Better than Divinitie," p. 6. A. H. Robinson, *Early Thematic Mapping*, pp. 26–67; Konvitz, *Cartography in France*, pp. 63–81. Halley, "Estimate of the Degrees of Mortality." Boyer, "Early Graph of Statistical Data."

60. MacPike, *Correspondence and Papers of Halley*, p. 243. For a modern map of Halley's observations, see Evans, "Edmond Halley, Geophysicist," p. 42.

61. Humboldt, *Cosmos*, vol. 5, pp. 59–60. H. Wallis, "Maps as a Medium of Scientific Communication." A. H. Robinson, "Genealogy of the Isopleth." Halley may have been familiar with Athanasius Kircher's mapping suggestions. John Hutchinson owned an old magnetic map: Whiston, *Longitude and Latitude*, p. 115. For some Portuguese magnetic cartography which Halley would not have encountered, see H. Winter, "Pseudo-Labrador," and Cortesão and da Mota, *Portugaliæ Monumenta*, vol. 3, plate 363. Ravenhill, "Churchman's Contours?"

62. Letters of 30 March and 8 July 1700 to Josiah Burchett at the Admiralty, reproduced in Thrower, *Voyages of Halley*, vol. 1, pp. 306–8.

63. Quoted in H. Wallis, "Geographie Is Better than Divinitie," p. 37.

64. Chapman, "Edmond Halley and Geomagnetism," pp. 231–2 (Mrs. Chapman's English translation).

65. The full text is reprinted in Bauer, "Magnetic Results of Halley's Expedition," pp. 122–3.

66. Halley, "Theory of the Variation," "Account of the Cause of the Change of the Variation," and "Variation of the Magnetical Compass." Quotations from Halley, "Some Remarks on the Variations," pp. 166–7.

67. Mountaine and Dodson, *Methods Used to Describe Lines*, pp. 4–8.

68. Fergusson, *Present State of the Art of Navigation*, pp. 7–8.

69. Nicholson, *Introduction to Natural Philosophy*, vol. 2, p. 330. D. Smith, *Antique Maps of the British Isles*, pp. 19–41; Ritchie, *Admiralty Chart*, pp. 3–6. Stearns, "Course of Halley." Tyacke, "Map-Sellers and the London Map Trade." H. Wallis, "Geographie Is Better than Divinitie."

70. John Robertson, *Elements of Navigation*, vol. 2, p. 291. See also Hutton, *Mathematical and Philosophical Dictionary*, vol. 2, pp. 640–1; James Cook and King, *Voyage to the Pacific Ocean*, vol. 1, pp. 49–50; Dunn, *Navigator's Guide*, pp. 10–11.

71. Rudwick, "Emergence of a Visual Language." B. Hartley, "Artist as Naturalist and Experimenter."

72. T. Young, *Course of Lectures*, vol. 1, pp. 690–1.

73. Ibid., p. 692 and figs. 577–8. Lorimer, *Concise Essay on Magnetism*, pp. 9–18. Cavallo referred extensively to Lorimer's work in the 1800 edition of his *Treatise on Magnetism*.

74. Le Grand, "Is a Picture Worth a Thousand Experiments?" Latour, "Drawing Things Together" and elsewhere.

75. Schuster and Watchirs, "Natural Philosophy, Experiment and Discourse." Golinski, *Science as Public Culture*, pp. 112–28. Schaffer, "Measuring Virtue." A. Walker, *System of Familiar Philosophy*, vol. 1, pp. 52–72.

76. This interpretation differs from that of Gooding, "Magnetic Curves" and *Making of Meaning*, pp. 62–6, 95–113. Gooding falsely stresses nineteenth-century innovation in his accounts, notably concerning the vocabulary of curves and the techniques of mapping. In addition, Barlow did not originate the geometrical construction of magnetic curves: similar arguments and diagrams were used by Young (diagram 569 in Figure 4.7) and by Robison: T. Young, *Course of Lectures*, vol. 1, p. 688 and fig. 569; Robison, "Magnetism," pp. 116–22. Part of Gooding's argument seems to rest on allowing the word "curves" to conflate the significance of isolines such as contours with isogonics: see A. H. Robinson, "Genealogy of the Isopleth."

77. Cawood, "Terrestrial Magnetism" and "Magnetic Crusade." D. P. Miller, "Revival of the Physical Sciences." Widmalm, "Accuracy, Rhetoric, and Technology."

Chapter Five
Measuring Power

1. Canton, "Method of Making Artificial Magnets," p. 35. Examples of keys and swords include Lister, "Journey to Paris," p. 26; RSJB, vol. 8, pp. 154 (22 February 1693), 261 (7 November 1694); vol. 13, p. 454 (9 April 1730). For compass dials, see Gouk, *Ivory Sundials*, and K. Higgins, "Classification of Sundials," pp. 342–5, 350–1, 367.

2. Kippis, *Biographia Britannica*, p. 217. Gardiner, *Memoirs*, p. 80: stroking was a technical term for magnetizing bars, but it did, of course, carry sexual connotations. Desaguliers, *Physico-mechanical Lectures*, p. 18. Robison, "Magnetism," plate 34, figs. 7–10 (and elsewhere).

3. Cajori, *Newton's Mathematical Principles*, p. xv (Motte's translation): see Albury, "Halley's Ode," and Chapman, "Edmond Halley and Geomagnetism," pp. 231–2. Eamon, *Science and the Secrets of Nature*; Keller, "Secrets of God, Nature, and Life." G. Knight, *Attraction and Repulsion*, p. 2. Mason was a settler in Barbados whose experimental apparatus and ideas were disseminated in England by Temple Henry Croker. Mason's work was presented under his own name to the Royal Society on 6 November 1760 (RSJB, vol. 23, pp. 930–5) and under Croker's name on 26 February 1761 (vol. 24, pp. 83–4). There is some information about him in Stearns, *Science in America*, pp. 362–3.

4. Johnson, *Rambler*, vol. 3, p. 271; Z. Williams, *Attempt to Ascertain the Longitude*, pp. 1–2; letters supposedly from Williams to the Admiralty, one of 2 January 1752 and one undated, in *Gentleman's Magazine* 57(2) (1787), 1041. Johnson wrote Williams' *Attempt* (and persuaded his friend Guiseppe Baretti to provide an Italian translation). There is, however, no direct evidence that Johnson was solely responsible for the third quotation. See A. J. Kuhn, "Dr Johnson, Zachariah Williams," and E.G.R. Taylor, "Reward for the Longitude."

5. T. S. Kuhn, *Essential Tension*, pp. 178–224 ("The Function of Measurement in Modern Physical Science"). Frängsmyr, Heilbron, and Rider, *Quantifying Spirit*.

6. Schaffer, "Social History of Plausibility," especially p. 141. This approach stems from the work of Gaston Bachelard: see Tiles, *Bachelard*, pp. 66–119.

7. Schaffer, "Newtonianism," and Yeo, "Genius, Method and Morality." J. A. Bennett, "Mechanics' Philosophy," "Hooke's Instruments," and "Challenge of Practical Mathematics." Schaffer, "Self Evidence."

8. Desaguliers, *Course of Experimental Philosophy*, pp. viii, 21. Cavallo, *Treatise on Magnetism*, p. vi.

9. Letter from Michell to Wilson of 22 December 1754: Wilson correspondence, fol. 86.

10. Schaffer, "Measuring Virtue," p. 313 and passim. Burke, *Reflections on the Revolution*, p. 299; Brewer, *Sinews of Power*, pp. 101–14. See also Buck, "People Who Counted," and Crosland, "Image of Science as a Threat." For an exciseman's life, see Money, "Teaching in the Market-Place." Daston, *Classical Probability*, especially pp. 112–87.

11. Barrell, *Birth of Pandora*, pp. 41–61 (quotation from p. 49), and *Political Theory of Painting*, pp. 1–162.

12. Shuckburgh, "Variation of Temperature," p. 362. P. C. Cohen, *Calculating People*, pp. 15–46. Pedersen, "Philomaths"; P. J. Wallis, "British Philomaths."

13. Quotations from Heilbron, "Mathematicians' Mutiny," pp. 85 (Charles Blagden in a letter to Banks), 87 (Banks' notes): I am grateful to John Heilbron for sending me

a draft version of this paper. D. P. Miller, "Between Hostile Camps." McCormmach, "Henry Cavendish."

14. J. A. Bennett, "Mechanics' Philosophy" and "Challenge of Practical Mathematics." Shapin, "Robert Boyle and Mathematics."

15. Halley, "Theory of the Variation," p. 221.

16. *Monthly Review* 27 (1762), 123: William Kenrick endorsing William Jones' criticism of William Emerson, esteemed by others as the author of widely purchased books on mathematics and navigation for self-taught men. For the identification of Kenrick, see Nangle, *Monthly Review First Series*. For earlier debates, see Shapin, "House of Experiment," and Pumfrey, "Who Did the Work?"

17. RSJB, vol. 10, p. 375 (20 March 1712). G. Knight, "Account of Some Magnetical Experiments," p. 165. Undated report sent from Arderon to Baker "Upon Artificial Magnets lifting wth both poles a great Weight": Baker correspondence, vol. 4, fols. 223–4 (quotation from fol. 224). See also RSJB, vol. 12, p. 204 (15 February 1721). Rivoire, *Traités*, pp. lxvii–lxviii. Canton, "Method of Making Artificial Magnets," p. 35. For replication, see Collins, *Changing Order*, pp. 51–78, and Schaffer, "Self Evidence," pp. 59–78. I am grateful to John Bradley for pointing out that some perplexing results were due to the weakness of magnets, which meant that their polarity was easily reversed. This is why some of Faraday's experiments cannot be replicated with modern magnets.

18. Hamilton, *Calculations and Tables*.

19. Two other techniques were tried. William Stephens' experiment at Dublin was based on a time measurement: see Heilbron, *Physics at the Royal Society*, pp. 75–6. At Worster's suggestion, Whiston measured magnetic force by recording the frequency of the oscillations of a small compass needle: Whiston, *Longitude and Latitude*, pp. 19–20. See also E. Stone, *Construction and Uses of Mathematical Instruments*, p. 304. The best—not entirely comprehensive—account is de Pater, *Petrus van Musschenbroek, een Newtoniaans Natuuronderzoeker*, pp. 122–226 (in Dutch). See also Palter, "Early Measurements," pp. 544–50, and Tilling, "Interpretation of Observational Errors," pp. 28–37.

20. Gilbert, *De Magnete*, pp. 313–58, and Kelly, *The "De Mundo" of William Gilbert*, pp. 25–43, 56–74: see Pumfrey, "Gilbert's Magnetic Philosophy," J. A. Bennett, "Cosmology and the Magnetic Philosophy," and Baldwin, "Magnetism and the Anti-Copernican Polemic." Newton's letters of 28 February 1681 and April 1681, reproduced in Turnbull, *Correspondence of Isaac Newton*, vol. 2, pp. 340–7, 358–62.

21. Schaffer, "Godly Men and Mechanical Philosophers," pp. 65–72, and "Newton's Comets." Letter of 1785 to Johann Sturm, reproduced (in Latin) in MacPike, *Correspondence and Papers of Halley*, pp. 55–6. Halley, "Discourse concerning Gravity," p. 5.

22. Cajori, *Newton's Mathematical Principles*, pp. 4, 301–2. For Newton's use of iron and loadstone to illustrate equilibrium, see Koyré and Cohen, *Newton's Philosophiae Naturalis*, vol. 1, p. 70. Koyré and Cohen, vol. 2, p. 576.

23. For Hooke's experiments of 1666 and 1681 respectively, see Birch, *History of the Royal Society*, vol. 2, pp. 75, 77–8; vol. 4, p. 66. For Halley, see ibid., vol. 4, pp. 518, 526, 527, and MacPike, *Correspondence and Papers of Halley*, pp. 135–7. For the detailed chronology of these and subsequent experiments, see Palter, "Early Measurements," pp. 544–9.

24. RSJB, vol. 10, p. 375; see also p. 376 (20 and 27 March 1712). Hauksbee, "Account of Experiments." B. Taylor, "Account of an Experiment" and "Extract of a

Letter." See also RSJB, vol. 10, pp. 387 (24 April 1712), 398 (15 May 1712), 410 (26 June 1712). Cajori, *Newton's Mathematical Principles*, p. 414.

25. Whiston, *Longitude and Latitude*, p 16. Helsham, *Course of Lectures*, pp. 19–20 (he claimed his measurements were more accurate than Newton's); Michell, *Treatise of Artificial Magnets*, p. 19. Although this was later a Continental research focus, in the first half of the century these experiments were essentially restricted to England, Ireland, and Holland. However, in Italy the Frenchmen Thomas Le Seur and Pierre Prévost described what they claimed to be a successful experimental demonstration of Newton's inverse-cube law: Newton, *Philosophiæ Naturalis*, vol. 3, pp. 39–43. Musschenbroek, *Cours de physique*, vol. 1, p. 436, describes this work but reports he was unable to replicate the results. Palter, "Early Measurements," pp. 551–8, discusses these experiments as the work of Jean-Louis Calandrini in Geneva.

26. Letter from Jurin to Stephens of 4 October 1722: London, Wellcome Institute for the History of Medicine, MSS 6146, Jurin correspondence. I am grateful to Andrea Rusnock for giving me a transcript of this letter and also the letter from Stephens to Jurin of 3 April 1722 (Royal Society, Early Letters S.2.42) and the letter from Stephens to Jurin of 10 April 1722 (ibid. S.2.42a). See also Heilbron, *Physics at the Royal Society*, pp. 75–6. Musschenbroek, "De Viribus Magneticus" and *Physicae Experimentales*, were in Latin. For an English translation of the latter, see Swedenborg, *Principia*, vol. 1, pp. 275–518. Less detailed but widely read accounts include Musschenbroek, *Elements of Natural Philosophy*, pp. 204–11 (the English translation of the 1726 edition), and *Cours de physique*, vol. 1, pp. 430–72 (the French translation of his 1762 book). Musschenbroek adjusted his variables, and possibly edited his readings, to ensure that he obtained simple laws for different configurations. Desaguliers, *Course of Experimental Philosophy*, pp. 21, 449 (as an addendum in the first edition, moved to the main text [p. 41] in the second edition of 1744).

27. Barnes, *T. S. Kuhn and Social Science*, pp. 45–53, describes Kuhnian normal science by restricting the meaning of "paradigm" to exemplary past achievements.

28. Cavallo, *Treatise on Magnetism*, p. iv. T. Young, *Course of Lectures*, vol. 1, pp. 6–10. W. Burney, *Universal Dictionary of the Marine*, p. 269.

29. Tucker (writing as Edward Search), *Light of Nature*, vol. 1, pp. xxvi–xliii (quotation from p. xxxiii). This book was strongly recommended by George Adams and was reissued in three versions in the nineteenth century. See also J. H. Brooke, "English Mix Science and Religion."

30. Howson, *History of Mathematics Education*, pp. 45–85. Gascoigne, "Mathematics and Meritocracy." Richards, "Rigor and Clarity."

31. B. Higgins, *Essay concerning Light*, pp. 183–4. Undated letter from Wilson to Aepinus: Wilson correspondence, fol. 91. For Aepinus and electricity in England, see Home, "Æpinus and the British Electricians." For Wilson's personal mathematical ineptitude, see Heilbron, *Electricity*, p. 304.

32. *Encyclopædia Britannica* (1801), vol. 1, pp. iv (George Gleig's dedication), 559–60 (Robison on electricity), pp. 112–32 (Robison on magnetism), 500–47, quotation from p. 547 (Robison on dynamics). For Robison's own comments on the *Encyclopédie*, see Robison, *Proofs of a Conspiracy*, pp. 519–20. Olson, *Scottish Philosophy and British Physics*, pp. 55–93, 157–61. For Robison's paranoic hostility to the French, see Morrell, "Professors Robison and Playfair," and Christie, "Joseph Black and John Robison." The mathematical tree printed as a frontispiece to Fuller, *Mathematical Miscellany*, shows algebra and bookkeeping as adjacent offshoots from the branch of arithmetic; see also Schaffer, "Defoe's Natural Philosophy." For Faraday, Barlow, and

representations of magnetism, see Gooding, "Magnetic Curves" and *Experiment and the Making of Meaning*, pp. 62–6, 95–113.

33. G. Adams, *Lectures* (1799), vol. 1, p. viii; vol. 4, p. 304; (1794), vol. 4, p. 436.

34. Burke, *Reflections on the Revolution*, especially pp. 284–317. See Crosland, "Image of Science as a Threat," especially pp. 296–8; Buck, "People Who Counted," and Paulson, *Representations of Revolution*. Quotation from pp. 108–9 of Young's review of a book by Laplace, *Quarterly Review* 1 (1809), 107–12. For Young and Burke, see Wood and Oldham, *Thomas Young*, pp. 1–41, 273.

35. C. Middleton, "Observations on the Weather." Musschenbroek, "Ephemerides Meteorologicæ," especially pp. 364–5, 412–13. See Porter, "Terraqueous Globe," and Jordanova, "Earth Science and Environmental Medicine."

36. Hackmann, "Scientific Instruments," pp. 45–58. For Victorian mimetic experimentation, see Galison and Assmus, "Artificial Clouds," pp. 227–32. Halley, "Theory of the Variation," p. 220. Eames, "Magnets Having More Poles than Two." G. Knight, *Attraction and Repulsion*, p. 89.

37. T. Young, *Course of Lectures*, vol. 1, p. 689. A. Walker, *System of Familiar Philosophy*, vol. 1, pp. 71–2. H. Wallis and Robinson, *Cartographical Innovations*, p. 35. For the use of electrical equipment to replicate magnetic patterns of iron filings, see T. Young, *Course of Lectures*, vol. 1, p. 688. Gooding, "Magnetic Curves" especially pp. 202–4, and *Making of Meaning*, especially pp. 64–6.

38. RSJB, vol. 13, pp. 545–6; S. Savery, "Magnetical Observations," pp. 333–40; letters from Savery to the Royal Society of 20 April and 20 May 1732: British Library Additional Manuscripts 4433, fols. 62–70; Whiston, *Longitude and Latitude*, pp. 44–75.

39. Canton, "Regular Diurnal Variation," pp. 400–2. Cavallo, "Magnetical Observations and Experiments," pp. 15–25, and *Treatise on Magnetism* (1800 ed.), pp. 254–9.

40. Particularly G. Graham, "Observations Made of the Horizontal Needle" and "Observations of the Dipping Needle," Canton, "Regular Diurnal Variation," and Cavendish, "Account of the Meteorological Instruments," pp. 385–401.

41. McConnell, "Nineteenth-Century Geomagnetic Instruments," pp. 26–30; Sydenham, *Measuring Instruments*, pp. 227–9. Pinch and Bijker, "Social Construction of Facts and Artefacts." There is no detailed comprehensive analysis of eighteenth-century magnetic instruments, although McConnell, *Geomagnetic Instruments before 1900*, and D. J. Warner, "Terrestrial Magnetism," provide brief reviews. Navigational histories include accounts of compasses: for example, Randier, *Marine Navigation Instruments*, pp. 7–39; Hewson, *History of Navigation*, pp. 45–72, 120–54; May, *History of Marine Navigation*, pp. 43–107. There are also brief descriptions of changes in magnetic surveying instruments: for example, J. A. Bennett and Brown, *The Compleat Surveyor*.

42. J. A. Bennett, "Viol of Water." Warner, "What Is a Scientific Instrument?" and "Terrestrial Magnetism."

43. Grew, *Musæum Regalis Societatis*, p. 364; Calvert, *Scientific Trade Cards*, plate 38 (Henry Pyefinch's card classifies and prices over 100 items).

44. Crawforth, "Evidence from Trade Cards." McConnell, "Craft Workshop to Big Business" and "Nineteenth-Century Geomagnetic Instruments." G. Adams, *Geometrical and Graphical Essays*, pp. 490, 496; Nairne and Blunt, *Nairne's Patent Electrical Machine*, pp. 63–8; W. and S. Jones, "Catalogue" (1795). Calvert, *Scientific Trade*

Cards, plate 31 (George Lee's trade card advertises "Mathematical, Optical and Nautical Instruments").

45. First published in 1581, Norman, *Newe Attractive*, was a reprint organized by Whiston; Gilbert, *De Magnete*, pp. 275–312; Zilsel, "Gilbert's Scientific Method."

46. Robison, "Magnetism," p. 137. For French modifications to suit the requirements of balloonists, see *Nicholson's Journal of Philosophy* 10 (1805), 278–9. *Monthly Review* 25 (1762), 56–7 (William Bewley: see Nangle, *Monthly Review First Series*); RSJB, vol. 10, pp. 308–9 (28 June 1711); Derham, *Physico-theology*, p. 275n.

47. RSJB, vol. 12: see pp. 32–5 (23 June 1720), 49 (27 October 1720), 219 (22 March 1721); Whiston, *Longitude and Latitude*, pp. xxii–xxviii, 1–7, 26–44, 91–101; Whiston, *Calculation of Solar Eclipses*, pp. 83–7; Whiston, *Memoirs*, vol. 1, p. 253. Some other inventors suggested using dip to measure longitude: R. Walker, *Treatise on Magnetism*, pp. 7–8; Nugent, "New Theory," pp. 381–92.

48. G. Graham, "Observations of the Dipping Needle." Nairne, "Experiments on Two Dipping-Needles." Hutchins, "Experiments on the Dipping Needle" (quotation from p. 132) and "Attempts to Freeze Quicksilver," pp. 179–81. Wheatland, *Apparatus of Science at Harvard*, pp. 155–60. Nugent, "New Theory," pp. 378–81.

49. R. Walker, *Treatise on Magnetism*, pp. 204–6. His petition to the Longitude Board received a sympathetic hearing, despite Maskelyne's dogmatic and possibly self-interested rejection of magnetic methods. Another important compass of this period, described in McCulloch, *New Improved Sea Compasses*, was designed and patented by the maritime specialist Kenneth McCulloch.

50. G. Graham, "Observations Made of the Horizontal Needle" and "Observations of the Dipping Needle." See also "Observations, Made during the Last Three Years." Archinard, *Collection de Saussure*, p. 31 (I am grateful to Allen Simpson for this reference). Canton, "Regular Diurnal Variation."

51. Nairne, "Experiments on Two Dipping-Needles." McCormmach, "Henry Cavendish." Chree, "Cavendish's Magnetic Work," pp. 465–92; Cavendish, "Account of the Meteorological Instruments," pp. 385–401; Gilpin, "Observations on the Variation."

52. Ingenhousz, "New Methods of Suspending Magnetical Needles," and Bennet, "New Suspension of the Magnetic Needle."

53. Pettus, *Fleta Minor*, part 1, pp. 317–18. For example, see R. T. Gunther, *Early Science in Oxford*, vol. 4, pp. 29, 31, 36, 42 (I am grateful to Ken Arnold for pointing out this Oxford activity). Ballard, "Magnetism of Drills"; Caswell, "Paper about Magnetism"; Derham, "Account of Some Magnetical Experiments" and "Farther Observations."

54. Lesch, "Systematics and the Geometrical Spirit"; R. Laudan, *From Mineralogy to Geology*, pp. 70–86. Bloor, "Durkheim and Mauss Revisited"; Dean, "Controversy over Classification."

55. Albury and Oldroyd, "From Renaissance Mineral Studies to Historical Geology." Porter, *Making of Geology*, pp. 170–6. Golinski, *Science as Public Culture*, pp. 269–83.

56. F. D. Adams, *Birth and Development of Geological Sciences*, pp. 77–136. Hooson, *Miners Dictionary*, entry "Tyth".

57. RSJB, vol. 19, p. 365 (30 November 1747: Folkes). Rudwick, *Meaning of Fossils*, pp. 1–100. In his *Fossils*, the Linnean propagandist John Hill divided fossils into several genera, including metals and their ores, such as loadstone. Woodward, *Fossils*

of All Kinds, pp. 1–5, iii–xvi, 50–2 (quotation p. v). See J. R. Levine, *Dr. Woodward's Shield*.

58. Lister, "Journey to Paris," p. 27. G. Knight, *Attraction and Repulsion*, p. 92. Derham, *Physico-theology*, p. 63. Pettus, *Fleta Minor*, part 2, entry on loadstone (unpaginated).

59. Some schemes seen as important at the time are summarized in Linnaeus, *General System of Nature*, vol. 7, pp. 10–59. See, for example, Hill, *Fossils*, pp. 402–17; Kirwan, *Elements of Mineralogy* (1784), pp. 269–91.

60. Preface to W. Phillips, *Elementary Introduction to Mineralogy*. Kirwan, *Elements of Mineralogy* (1794–6), vol. 2, pp. 155–94: this edition is very different from the 1784 edition.

61. Linnaeus, *General System of Nature*, vol. 7, p. 54; Townson, *Philosophy of Mineralogy*, p. 164; *Nicholson's Journal of Philosophy* 1 (1797), 100–1.

62. The word "brass" referred to alloys of copper and zinc (the modern definition) and also to alloys of copper and tin (now called bronze): Piggott, *Ancient Britons*, p. 95. Letter from Arderon to Baker of 24 July 1758, Baker correspondence, vol. 8, fols. 33–8; letter from Baker to Arderon of 2 October 1758, Arderon correspondence, vol. 4, fol. 33. For Baker's editorship, see letter from Baker to Henry Miles of 16 February 1748, Baker correspondence, vol. 3, fol. 246. See also Arderon, "Abstract of a Letter." In the Arderon correspondence, see also letter from Arderon to Baker of 9 September 1758 (vol. 4, fol. 32) and letter from Arderon to Baker of 26 October 1758 (vol. 4, fol. 34). In the Baker correspondence, see letter from Arderon to Baker of 26 July 1758 (vol. 8, fols. 62–3).

63. Cavallo, "Magnetical Experiments and Observations" and "Magnetical Observations and Experiments." The third was unpublished—Cavallo, "Magnetical Experiments, and Observations": Royal Society Letters and Papers, vol. 9, cut 147, fols. 7–12 (1789); see also RSJB, vol. 34, p. 3. Bennet, "New Suspension of the Magnetic Needle," who used a needle suspended by a spider's thread, maintained that only brass containing iron could be made magnetic.

64. Cavallo, "Magnetical Experiments, and Observations," fol. 5 (see n. 63).

65. Letter of 30 April 1763 from Baker to William Borlase: Baker correspondence, vol. 8, fol. 34. Moulen, "Experiments on a Black Shining Sand"; Musschenbroek, "Experiments on the Indian Magnetic Sand"; H. Horne, "Observations on Sand Iron"; Ingenhousz, "Experiments on Platina."

66. *Nicholson's Journal of Philosophy* 2 (1802), 143–4 (quotation from title); 5 (1802), 287–9 (Richard Chevenix).

67. Ibid., 10 (1805), 265 (Charles Hatchett). T. Young, *Course of Lectures*, vol. 1, pp. 686–7. For a general discussion, see Cavallo, *Treatise on Magnetism*, pp. 283–306.

68. Shapin and Schaffer, *Leviathan and the Air-pump*, especially pp. 2–79. D. P. Miller, "Revival of the Physical Sciences."

69. E. Chambers, *Cyclopædia*, vol. 2, p. 485. In Rees' *Cyclopædia*, Knight's experiments were added at this point in the entry.

70. For example, Gilpin, "Observations on the Variation," p. 398.

71. For example, E. Chambers, *Cyclopædia*, vol. 2, pp. 484–7; Rees, *Cyclopædia*, vol. 3, entry on magnetism; and Hutton, *Mathematical and Philosophical Dictionary*, vol. 2, pp. 71–3.

72. E. Chambers, *Cyclopædia*, vol. 2, pp. 484–7; Rees, *Cyclopædia*, vol. 3, entries on magnet, magnetism, etc.

73. Yeo, "Reading Encyclopaedias"; Layton, "Diction and Dictionaries." For eigh-

teenth-century English encyclopedias, see Hughes, "Sciences in English Encyclopaedias," pp. 340–50, and Kafker, *Notable Encyclopaedias*. J. A. Bennett and Brown, *Compleat Surveyor*, especially pp. 9–15. Richeson, *English Land Measuring*, pp. 142–88.

74. Magnets retained this location in the *Encyclopædia Britannica* until at least 1773 (vol. 3, p. 36).

75. For Harris, see Kafker, *Notable Encyclopaedias*, pp. 107–21, and Stewart, *Rise of Public Science*, pp. 108–19. The *Lexicon Technicum* was unpaginated. The entries are in vol. 1 of the 1710 and 1736 editions.

76. *Encyclopædia Britannica* (1771), vol. 3, p. 3; (1778–83), vol. 6, pp. 4373–83. For Tytler, see Fergusson, *Balloon Tytler*, pp. 40–59, 94–5. Particularly in the first edition, entries got progressively shorter through the alphabet. The entries on magnet are in vol. 3 of the three-volume first edition, vol. 6 of the ten-volume second edition, and vol. 10 of the eighteen-volume third edition.

77. Compare *Encyclopædia Britannica* (1778–83), vol. 10, pp. 8691–2, with (1797), vol. 18, pp. 619–25.

78. Robison, "Magnetism," edited by David Brewster and reprinted in Robison, *System of Mechanical Philosophy*, vol. 4. Morrell, "Professors Robison and Playfair," pp. 44–51; Olson, *Scottish Philosophy*, pp. 157–61.

79. Quoted in Weld, *History of the Royal Society*, vol. 2, p. 431. The 1778 edition of Chambers' *Cyclopædia* provides evidence of an earlier stage in this process, since the entry on variation refers the reader to the article on magnets, with its own subheading "Magnet, variation of."

80. D. Knight, "Science and Professionalism," pp. 56–64. A Senior Wrangler was a Cambridge graduate who had gained top place in his final Mathematical Tripos examinations

81. For example, John Robertson, *Elements of Navigation*, vol. 1, pp. ix–xii; vol. 2, pp. 232–5. For changes in navigational texts, see Cotter, "British Navigation Manuals," especially pp. 244–5.

82. Fergusson, *Present State of the Art of Navigation*, pp. 7, 2–3. W. Hutchinson, *Treatise on Practical Seamanship*, p. 86. Similar scattered comments were commonplace.

83. D. P. Miller, "Between Hostile Camps," pp. 1–19. Fergusson, *Present State of Navigation*, p. 15 and passim. I have found no further evidence of this Society. W. Hutchinson, *Practical Seamanship*, pp. 85–6.

84. R. Walker, *Treatise on Magnetism*, p. 9. Such books include Haselden, *Seaman's Daily Assistant* (used by Cook), and Moore, *Practical Navigator* (which became the most popular navigational text of the period). For learning by rule, see P. G. Cohen, *Calculating People*, pp. 15–46, 116–49.

85. Burney, *Dictionary of the Marine*, pp. xi–xv. Falconer, *Shipwreck* (1762): this much-reprinted poem includes detailed footnotes explaining nautical terms and incorporates such technical terms as azimuth, index, and magnetic variance in the verse.

86. *Gentleman's Magazine* 87 (1) (1817), 76; *Scots Magazine* 63 (1801), 179, 338–9. See also the letter from Indagator (Canton) in *Gentleman's Magazine* 31 (1761), 499–500. May, "Garlic and the Magnetic Compass."

87. There was an extended correspondence between Canton (writing as Indagator) and Chapple (see Pedersen, "Philomaths," p. 249), who claimed that tallow attracted a compass needle: *Gentleman's Magazine* 31 (1761), 357–9, 397–8, 459–60, 499–500, 569–70. This debate also appeared in *London Magazine* 30 (1761), 483–4. According

to Canton's own records, his letters also appeared in *Lloyd's Evening Post* from 7 to 9 September 1761 (Canton papers, vol. 1, fols. 21–31). This correspondence stimulated Arderon to experiment with tallow: undated letter from Arderon to Baker, Baker correspondence, vol. 8, fol. 49v; another copy of this letter, dated 7 (month illegible) 1761 is in the Arderon correspondence, vol. 4, fol. 34. Letter from Robert Mason (a naval lieutenant) to James Clarke of 5 February 1801, reproduced in J. S. Clarke, *Progress of Maritime Discovery*, pp. 254–5; Bain, *Essay on the Variation*, p. 132n.

88. *Monthly Review* 27 (1762), 123. *British Critic* 14 (1799), 337–41 (quotation on p. 340); 5 (1795), 487–90 (quotation on p. 489). See O. Smith, *Politics of Language*, pp. 1–34.

89. *Quarterly Review* 1 (1809), 108.

Chapter Six
God's Mysterious Creation

1. Coleridge, *Lectures 1795*, p. 157.

2. Ray, *Wisdom of God*, p. 70. R. Walker, *Treatise on Magnetism*, pp. 42–3.

3. Hale, *Magnetismus Magnus*, pp. 1–64. J. Hervey, *Theron and Aspasio*, vol. 2, pp. 233–6 (quotation on pp. 235–6): see Russell, *Science and Social Change*, pp. 66–8.

4. J. H. Brooke, "English Mix Science and Religion" and "Science and Secularisation of Knowledge"; Cunningham, "Getting the Game Right" and "How the *Principia* Got Its Name." Cantor, "Theological Significance of Ethers." Stewart, *Rise of Public Science*, pp. 31–59.

5. Bryan, *Lectures on Natural Philosophy*, p. 143.

6. For Kircher's magnetic texts and their influence, see J. Godwin, *Athanasius Kircher*, W. Hine, "Athanasius Kircher and Magnetism," Hankins and Silverman, *Instruments and Imagination*, pp. 14–36, and Daly, *Literature in the Light of the Emblem*, pp. 77–9. Cavallo was still referring approvingly to Kircher at the end of the century.

7. Reynolds, *View of Death*, p. 34. Trenchard, *Natural History of Superstition*, pp. 26–7, 30; see Schwartz, *Knaves, Fools, Madmen*, especially p. 53, and Keith, *Magick of Quakerism*, pp. 60–9. E. Chambers, *Cyclopædia* (1728), vol. 2, p. 161 (entry on sympathy). See Hughes, "Sciences in English Encyclopaedias."

8. Herwig, *Art of Curing Sympathetically*; Boerhaave (allegedly), *Virtue and Efficient Cause of Magnetical Cures*; *Athenian Oracle* (1703), 71–2. Dictionary references include Bailey, *Universal Dictionary* (entry on sympathetic), and E. Chambers, *Cyclopædia* (1728), vol. 2, p. 487 (entry on magnet—also in the 1741 edition). Campbel, *Friendly Dæmon*; *Wonderful Magazine* 1 (1793), 87–91 (quotation p. 87); see Baine, *Defoe and the Supernatural*, pp. 137–80. Sibly, *Key to Physic*, pp. 276–80: see Debus, "Scientific Truth and Occult Tradition."

9. For example, Roger North: British Library Additional Manuscripts 32546, fols. 73r–78v. Goad, *Astro-meteorologica*, p. 124. T. Robinson, *Philosophical and Theological Exposition*, pp. 99–104; Hobbs, *Earth Generated and Anatomised*, pp. 63–4. For an illustration of a magnetic tidal indicator, see Pettus, *Fleta Minor*, part 2, unpaginated entry on loadstone. Worster, *Principles of Natural Philosophy*, p. 234. For Worster on Oxford reading lists, see Yolton, "Schoolmen, Logic and Philosophy," pp. 580–2, and Quarrie, "Christ Church Collection Books," pp. 504–5.

10. Grew, *Anatomy of Plants*, pp. 16, 22, and *Cosmologia Sacra*, p. 12. Pettus, *Fleta Minor*, part 2, unpaginated entry on loadstone. For Ovid's account of the chryso-

magnet (which attracts gold), see Addison's *Guardian* 122 (31 July 1713). Dobbs, *Foundations of Newton's Alchemy*, especially pp. 146–61, 187–90. T. Robinson, *Natural History of Westmorland*, unpaginated preface and p. 16: see Rossi, *Dark Abyss of Time*, pp. 9–11.

11. R. James, *Medicinal Dictionary*, vol. 1 (entry for arsenic); Pomet, *Compleat History of Druggs*, p. 370; *magnes arsenicalis* was added to the entry on magnetism in the 1741 edition of Chambers' *Cyclopædia*. See also the entry on *magnes arsenical* in Bailey, *Universal Dictionary*, and several similar definitions in mid-century dictionaries. James, *Dictionary*, vol. 2, entry Magneticus. Several other dictionaries carried similar descriptions.

12. Virtually any encyclopedia entry for magnet or loadstone. It was also called *lapis siderites*. Stukeley maintained that Hercules had led the Druids to Britain: *Stonehenge*, pp. 57–8.

13. Salmon, *Compleat English Physician*, pp. 189–90; Pomet, *Compleat History of Druggs*, p. 367; Barrow, *Dictionarium Polygraphicum*, vol. 2, double entry for manganese and magnese. John Hill included manganese, Latinized as *magnesia*, under iron, after the entry for *ferrum magnes*, or loadstone: *Fossils*, pp. 402–17. Browne, *Pseudoxica Epidemica*, pp. 102–6. For *magnesia alba*, see Multhauf, "History of Magnesia Alba."

14. *Adamas* means "invincible," and *adamare* means "to have an attraction for." For an example of their interchangeability, see Swift, *Gulliver's Travels*, p. 179. See also Stukeley, *Stonehenge*, p. 60. For Shelley and adamantine, see Dawson, "Sort of Natural Magic," pp. 29–30. Erasmus Darwin referred to "adamantite" steel for making artificial magnets: *Poetical Works*, vol. 1, p. 93. Browne, *Pseudoxica Epidemica*, p. 103. According to Samuel Clarke, this belief derived from Pliny: Rohault, *System of Natural Philosophy*, vol. 2, p. 186n1 (see Hoskin, "Mining All Within," for Clarke's editorial role).

15. Dobbs, "Natural Philosophy of Kenelm Digby." Werenfels, *Superstition in Natural Things*, pp. 5–21; Boerhaave, *Magnetical Cures*, is a collection of magnetic cures and recipes including ingredients like the moss from the skull of a hanged man. Reynolds, *View of Death*, p. 32. See also Campbel, *Friendly Dæmon*, pp. 3–12, 27–31, and *Wonderful Magazine* 1 (1793), 87–91.

16. Salmon, *Compleat English Physician*, p. 188. De la Touche, "Wonderful Property of Magnets." The 1778 edition of Chambers' *Cyclopædia* included a new entry on the medical uses of magnets. The foot of W. Dowling's 1828 trade card, reproduced in Calvert, *Scientific Trade Cards*, no. 132; Warter, *Southey's Common-Place Book*, vol. 1, pp. 434–5. For historical surveys, see Fernie, *Precious Stones*, pp. 316–22; W. G. Black, *Folk-Medicine*, pp. 52–5; Walsh, *Cures*, pp. 77–87; E. H. Frei, "Medical Applications of Magnetism." For contrasting summaries at the end of the century, see Cavallo, *Treatise on Magnetism*, pp. 101–3, and Barrett, *Magus*, book 2, pp. 8–13 (see Heisler, "Behind 'The Magus'").

17. John Harris, *Lexicon Technicum* (1736), entry on magnetism. Harvey, *Works*, p. 575 (*On Conception*); Erikson, "Books of Generation," p. 84.

18. G. L'E. Turner, "Defence," pp. 133–4; see also Blondel, *Power of the Mother's Imagination*, p. 9; P. Wilson, "Out of Sight, Out of Mind" (especially p. 74n70); and Boucé, "Imagination, Pregnant Women, and Monsters."

19. Chamberlen, *Actions on Distant Subjects*, especially pp. 6, 18 (and in other texts). For Chamberlen's commercial magnetic medicine, see Doherty, "Anodyne Necklace." Browne, *Pseudoxica Epidemica*, p. 102, and several eighteenth-century

encyclopedias. F. D. Adams, *Birth and Development of Geological Sciences*, pp. 98–102.

20. Leonardus, *Mirror of Stones*, pp. 209–10; Browne, *Pseudoxica Epidemica*, pp. 111–13; Kunz, *Curious Lore of Precious Stones*, pp. 93–7. Boucé, "Sexual Beliefs and Myths." Croker, Williams, and Clark, *Dictionary*, vol. 2, entry on magnet (and elsewhere). Kunz, *Rings for the Finger*, pp. 326–7.

21. Sargent, *The Mine*, pp. 9–10; letter from Frances Burney to Dr. Burney of 2 March 1794, reproduced in Hemlow, *Fanny Burney*, p. 30; Pasquin, *Memoirs of the Royal Academicians*, pp. 118–20; *Journal Britannique* (1752), 264–6; Johnson, *Rambler*, vol. 3, p. 274.

22. Manuel, *Eighteenth Century Confronts the Gods*; J. R. Levine, *Dr. Woodward's Shield* and *Humanism and History*; Rossi, *Dark Abyss of Time*; Aarsleff, *Study of Language in England*; Frei, *Eclipse of Biblical Narrative*; Prickett, *Words and the Word*; Porter, *Gibbon*.

23. J. Hutchinson, *Philosophical and Theological Works*, vol. 11, p. 349.

24. Halley, "Change of the Variation of the Magnetical Needle." See Clark, *Johnson*, pp. 11–58, for Anglo-Latin culture. Lovett, *Electrical Philosopher*, pp. 455–8.

25. Halley, "Theory of the Variation" and "Cause of the Change of the Variation," pp. 563–9.

26. Burnet, *Sacred Theory of the Earth*, vol. 1, p. 267 (also in the 1691 edition); vol. 2, p. 367 (added as a response to Erasmus Warren's criticisms): see Rossi, *Dark Abyss of Time*, pp. 33–41, and Force, *William Whiston*, pp. 32–62.

27. Schaffer, "Halley's Atheism." Halley, "Cause of the Change of the Variation," pp. 569, 572–8. Zirkle, "Theory of Concentric Spheres." Halley later obliquely referred to this model, drawing on it to suggest explanations of the aurora borealis: "Late Surprizing Appearance of the Lights," especially pp. 422–3 and 427–8.

28. Halley, "Cause of the Change of the Variation," pp. 573, 575. Crowe, *Extraterrestrial Life Debate*, pp. 3–37. Halley's paper discussing life on other planets, was published before Richard Bentley's *Confutation of Atheism*, which Crowe mistakenly accords priority in eighteenth-century debates about the plurality of worlds. For contemporary ideas about primeval light, see Collier, *Cosmogonies of Our Fathers*, pp. 338–50. It was this light which Halley linked to the aurora borealis in 1716 (see Figure 7.1). See also Harrington, *A New System*.

29. RSJB, vol. 8, pp. 130–1 (19 October 1692): reproduced with minor transcription errors in MacPike, *Correspondence and Papers of Halley*, p. 229.

30. Whiston, *New Theory of the Earth* (1722), pp. 73, 109–12. These sections were added to the text of the 1696 edition. Whiston, *Longitude and Latitude*, pp. 44–75. See Farrell, *Life and Work of Whiston*, and Force, *William Whiston*.

31. Goldgar, "Fielding, Flood Makers, and Natural Philosophy." Although the opening pages have not survived, Catcott surely copied all of Hutchinson, *Philosophical and Theological Works*, vol. 11, pp. 344–52, into a notebook: the Catcott papers at the Bristol Central Library include a collection of papers (ref. no. 149.3H) which contains fols. 29–34 of Catcott's copy. Catcott, *Treatise on the Deluge*, p. 280. See Neve and Porter, "Alexander Catcott," and Collier, *Cosmogonies of Our Fathers*, pp. 230–41.

32. Cockburn, *Enquiry into the Mosaic Deluge*, p. 311: see Collier, *Cosmogonies of Our Fathers*, pp. 230–4. Churchman, *Explanation of the Magnetic Atlas*, pp. 39–45; he did not include these ideas in his *Magnetic Atlas* of 1794.

33. Whiston, *Astronomical Principles*, pp. 93–7, and *Eternity of Hell Torments*, pp.

1–3. D. P. Walker, *Decline of Hell*; Almond, "Contours of Hell" (I am grateful to Philip Almond for sending me a draft version of this article). Genuth, "Devils' Hells and Astronomers' Heavens." G. Knight, *Attraction and Repulsion*, p. 58. For pluralism, deism, and hell, see Crowe, *Extraterrestrial Life Debate*, pp. 36–7, 67.

34. Shelley, *Frankenstein*, pp. 13–14.

35. Trapp, *Lectures on Poetry*, pp. 189–90: see Guest, *Form of Sound Words*, pp. 44–8. Derham, *Physico-theology*, pp. 263–78, "Magnetical Experiments," and "Farther Observations." See W. P. Jones, *Rhetoric of Science*. *Vanity of Philosophick Systems*, pp. 7, 16.

36. Nieuwentyt, *Religious Philosopher*, vol. 2, p. 566.

37. Coleridge, *Lectures 1795*, p. 97; G. Adams, *Lectures* (1794), vol. 4, p. 436. Burns, *Great Debate on Miracles*. Pemberton, *View of Newton's Philosophy*, pp. 12–13; Defoe, *System of Magick*, pp. 60–1. D. Hartley, *Observations on Man*, vol. 2, pp. 142–3. Coleridge, *Lectures 1795*, p. 112. Bishop Butler is quoted in Burns, *Great Debate*, p. 130.

38. J. Hervey, *Theron and Aspasio*, vol. 2, p. 234; Edwards, *Existence and Providence of God*, p. 146. Other examples include Boyle in *Style of the Holy Scriptures*, quoted in Howell, *British Logic and Rhetoric*, p. 479, Telescope, *Newtonian System*, p. 81, and *British Critic* 19 (1802), 390 (a poem "Science Revived, or the Vision of Alfred"). Some of these writers were undoubtedly familiar with Claudian's lines on this theme, quoted by Johnson in *Rambler*, vol. 3, p. 271.

39. Gascoigne, *Cambridge in the Enlightenment*, pp. 167–74. Edwards, *Demonstration of the Existence and Providence of God*, p. 145. See also Baker, *Reflections upon Learning*, pp. 76–86, R. Green, *Principles of Natural Philosophy*, §15 of unpaginated preface and p. 24; for Green(e)'s opposition to reason, see Green, *Demonstration of Truth and Divinity*, and Guest, *Form of Sound Words*, pp. 209–18.

40. Hobson, *Christianity the Light of the Moral World*, pp. 55, 56. Blackmore, *Creation*, pp. 40–5, quotation from p. 42. See Solomon, *Sir Richard Blackmore*, especially pp. 120–44. For Lucretius on magnetic atoms, see Lucretius, *On the Nature of Things*, pp. 283–91 (book 6, vv. 908–1089).

41. Smart, *Poetical Works*, vol. 4, p. 276 (*On the Power of the Supreme Being*). Ibid., p. 207 (*On the Omniscience of the Supreme Being*): the asterisk indicates Smart's note saying that he is referring to the longitude. Guest, *Form of Sound Words*; Williamson, "Smart's *Principia*."

42. N. Ward, *London-Spy Compleat*, pp. 59–61. For satires from 1660–1760, see W. P. Jones, *Rhetoric of Science*, pp. 65–78.

43. T. Brown, *Third Volume of the Works*, p. 100. See H. J. Cook, *Decline of the Old Medical Regime*, pp. 210–62, and J. R. Levine, *Dr. Woodward's Shield*, especially p. 126. Wagner, "Discourse on Sex" and *Eros Revived*, especially pp. 182–200. Browne, *Pseudoxica Epidemica*, pp. 111–12; Derham, *Physico-theology*, pp. 301–3.

44. Johnson, *Rambler*, vol. 3, pp. 271–6, quotations from pp. 274–6 (11 February 1752) (previously discussed in Chapters 2 and 5); for Anglo-Latin culture, see Clark, *Johnson*, pp. 11–58. *Journal Britannique* (1752), 243–74 (quotations pp. 264–6): A. E. Gunther, "Matthew Maty," pp. 11–28 (see p. 22 for Johnson's animosity towards Maty).

45. Johnson, *Rambler*, vol. 3, pp. 271–6, quotations from pp. 274–6 (11 February 1752).

46. Swift, *Gulliver's Travels*, pp. 163–205 (quotation p. 179). Nicolson and Mohler, "Swift's 'Flying Island'" and "Scientific Background." Korshin, "Intellectual Context

of Swift's Flying Island." Rogers, "Gulliver and the Engineers." D. Todd, "Laputa." Christie, "Laputa Revisited."

47. Lucretius, *On the Nature of Things*, book 6, pp. 283–91. Plato, *Ion*. Addison, *Guardian*, 28 July 1713 (no. 119).

48. Weisinger, "Rise of Science and the Renaissance." For printing, see Johns, "Wisdom in the Concourse," pp. 102–34.

49. Madden, *Memoirs of the Twentieth Century*, p. 512.

50. J. R. Levine, *Humanism and History*, pp. 155–87; Spadafora, *Progress*, pp. 21–84. Temple, *Works*, pp. 516–17. Wotton, *Reflections upon Ancient and Modern Learning*, pp. 247–57, quotation from p. 257.

51. Locke quoted in Hoyles, *Edges of Augustanism*, p. 64 (1785 edition of the *Dictionary*). Locke, "Mr Locke's History of Navigation," pp. 79, 84–7 (this account was attributed to Locke but may be by another author). Defoe, *Discoveries and Improvements*, pp. 226–307 (quotation pp. 298–9); see Schaffer, "Defoe's Natural Philosophy," and Tavor, *Scepticism*, pp. 7–53.

52. Priestley, *Lectures on History*, pp. 268–442, especially pp. 294, 407. Weisinger, "Rise of Science and the Renaissance," pp. 251–7; John Harris, "Philological Enquiries," p. 573. Playfair, *Causes of Decline*, p. 73, quotation from title page. Anderson, *Historical and Chronological Deduction*, vol. 1, pp. 144–5, quotation from title page.

53. *Universal Magazine* 1 (1747), pp. 117–19 (quotation p. 118). J. Wallis, "Second Letter to Dr Wallis," pp. 1035–8, and "Letter to Edmond Halley," p. 1111; Defoe, *Discoveries and Improvements*, p. 251.

54. RSJB, vol. 8, p. 144 (21 December 1692). The Fellows admired a Chinese compass (vol. 10, p. 153—26 March 1707) and discussed whether ancient Muslims used compasses to orient their prayer mats (vol. 9, p. 316—17 June 1702). Appleton, *Cycle of Cathay* (for Defoe, see pp. 55–60). Defoe, *Discoveries and Improvements*, pp. 226–32; *Universal Magazine* 1 (1747), 117–19; *Wonderful Magazine* 2 (1793), 47–52 (especially p. 50) (but compare 4 [1794], 231–3). Marshall, "Oriental Studies." Vincent, *Commerce and Navigation of the Ancients*, vol. 2, pp. 200, 285–8, 656–60. Lucretius, *Nature of Things*, book 2, p. 557 (note by the translator, John M. Good). J. S. Clarke, *Progress of Maritime Discovery*, pp. ii–ix. Maurice, *History of Hindoostan*, vol. 1, pp. 435–7.

55. *Bee* 13 (1793), 82–8; Vincent, *Commerce and Navigation*, vol. 2, pp. 285–8. Hackmann, "Van der Straet and the Origins of the Mariner's Compass."

56. Cheyne, *English Malady*, p. 34; Barker-Benfield, *Culture of Sensibility*, pp. 6–15. Fletcher, *Discourse of Government*, pp. 11–12: see Schaffer, "Defoe's Natural Philosophy," p. 37, Goldsmith, *Private Vices, Public Benefits*, pp. 4–21, and Sekora, *Luxury*, pp. 78–9.

57. Derham, *Physico-theology*, pp. 267–75 (quotation p. 268). Other examples include Hale, *Magnetismus Magnus*, pp. 47–51, Howard, *New Philosophy of Descartes*, p. 309, and Z. Williams, *Mariners Compass Compleated*, part 1, pp. 25–35. For similar French debates, see Rossi, *Dark Abyss of Time*, pp. 166–7; Regnault, *Philosophical Conversations*, vol. 1, p. 219n; G. Adams, *Lectures* (1799), vol. 4, pp. 459–60. Collinson, *Birthpangs of Protestant England*, pp. 1–27; Colley, *Britons*, pp. 11–54. J. S. Clarke, *Progress of Maritime Discovery*, pp. 399–400 (seconding the opinion of William Monson). *Universal Magazine* 6 (1750), 117–18 (quotation from p. 117).

58. Rossi, *Dark Abyss of Time*, pp. 121–92, especially pp. 152–3; J. R. Levine, *Dr. Woodward's Shield*, pp. 18–79; Piggott, *Ancient Britons*.

59. Hyde, *Historia Religionis Veterum Persarum*, pp. 495–7 (in Latin). Marshall, "Oriental Studies." Protagonists often cited Samuel Bochart, a seventeenth-century French scholar who denied ancient knowledge of compasses.

60. J. Hutchinson, *Philosophical and Theological Works*, vol. 4, pp. 120–4 (quotation from p. 121), and vol. 5, pp. 208–13. Jeremiah 31.18, cited amongst several others in Hutchinson, vol. 11, pp. 349–51, and vol. 4, p. 123. See also Hale, *Magnetismus Magnus*, pp. 67–8, and G. Horne, *Works*, vol. 3, p. 34. D. Forbes, *Letter to a Bishop*, p. 29.

61. Stukeley, *Stonehenge*, pp. 56–66 (quotation p. 60). W. Cooke, *Enquiry into the Patriarchal and Druidical Religion*, pp. 1 (quotation), 23–8.

62. Stukeley, *Stonehenge*, pp. 56–66; Piggott, *William Stukeley*, especially pp. 79–109, and *Ancient Britons*, especially pp. 40, 132. Like other contemporary writers, Stukeley did not use the B.C./A.D. dating convention.

63. Letter from Arderon to Baker of 31 July 1749: Arderon correspondence, vol. 2, item 36/18–19. Manuel, *Eighteenth Century Confronts the Gods*, pp. 85–102, and *Isaac Newton: Historian*. Stukeley, *Stonehenge*, pp. 57–64; W. Cooke, *Enquiry into Religion*, pp. 23–8; G. Adams, *Lectures* (1799), vol. 4, pp. 459–60; J. S. Clarke, *Progress of Maritime Discovery*, pp. ii–ix; Lucretius, *Nature of Things*, book 2, p. 557n (by John M. Good); Vincent, *Commerce and Navigation of the Ancients*, vol. 2, pp. 199–200; Maurice, *History of Hindoostan*, vol. 1, pp. 435–7.

64. Quotations from Desaguliers's introduction to Nieuwentyt, *Religious Philosopher*, vol. 1, pp. ii, xxvi–xxvii. The censoring editors overlooked Nieuwentyt's expressions of gratitude for God's recent decision to release magnetic knowledge: ibid., vol. 2, p. 567. *Monthly Review* 38 (1768), 373–83, quotation from pp. 381–2.

65. Billingsley, *Longitude at Sea*, pp. 11–12: see Stewart, *Rise of Public Science*, pp. 272–8. G. Knight, *Attraction and Repulsion*, p. 94.

66. Bryan, *Lectures on Natural Philosophy*, p. 158. G. Adams, *Lectures on Natural and Experimental Philosophy* (1794), vol. 4, pp. 433–6, 468–70 (quotation from p. 470): see Wilde, "Hutchinsonianism."

67. From Hume's 1752 essay, first called "On Luxury," then retitled "Of Refinement in the Arts," quoted and discussed in Solkin, *Painting for Money*, p. 157.

Chapter Seven
A Powerful Language

1. Johnson, *Dictionary*, entry on magnet (Milton quotation from book 3, lines 582–3, of *Paradise Lost*), and unpaginated preface. Griffin, *Regaining Paradise*. Clark, *Johnson*, pp. 184–7; O. Smith, *Politics of Language*, pp. 1–34.

2. M. Cohen, *Sensible Words*, pp. 43–136. Land, *Signs to Propositions* and *Philosophy of Language in Britain*. Spadafora, *Idea of Progress*, pp. 179–240. Aarsleff, *From Locke to Saussure* and *Study of Language in England*. Locke, *Essay concerning Human Understanding*, book 3, especially chs. 10 and 11. Howell, *British Logic and Rhetoric*, pp. 439–691. Quotation from G. Knight, *Attraction and Repulsion*, p. 2—repeating a well-worn image; see Shapin, "Boyle and Mathematics," pp. 23–6.

3. Berkeley, *Works*, vol. 3 (*Alciphron*—pp. 13–14 of the editors' introduction is particularly relevant); vol. 5, pp. 1–164 (*Siris*). See Ardley, *Berkeley's Philosophy of Nature*, pp. 38–50; Land, *Philosophy of Language*, pp. 79–130; Cantor, "Berkeley, Reid, and Optics" and "*Analyst* Revisited"; and Benjamin, "Medicine, Morality and Politics."

4. Wesley quoted in A. J. Kuhn, "Nature Spiritualised," p. 411. Wasserman, "Nature Moralised." William Jones, *Physiological Disquisitions*, p. 148. See also Jones, *Lectures on Figurative Language*, especially p. 302.

5. William Jones, *Works*, vol. 5, pp. 274–96 (*Popular Commotions to Precede the End of the World*), especially p. 286, of September 1789: Jones was concerned about the American Revolution. See Korshin, *Typologies in England*, especially pp. 101–85, and Froom, *Prophetic Faith*, especially vol. 1, pp. 31–166, and vol. 2, pp. 640–796. Paulson, *Representations of Revolution*; Hole, *Pulpits, Politics and Public Order* and "English Sermons." For millenarianism, see J.F.C. Harrison, Second Coming.

6. Aarsleff, *From Locke to Saussure*, especially pp. 3–31, 278–92. Rossi, *Dark Abyss of Time*, pp. 193–270. Guest, *Form of Sound Words*, pp. 167–95 (on poetry). Essick, *Blake and the Language of Adam*, pp. 6–103.

7. Cantor, "Light and Enlightenment" and "Weighing Light." For a parallel analysis of sound, see Hankins and Silverman, *Instruments and Imagination*, pp. 86–112.

8. G. Levine, "One Culture." The extensive literature on these themes is reviewed in Beer and Martins, "Introduction," and Golinski, "Language, Discourse and Science."

9. In addition to previous references in this chapter, accessible reviews include Leatherdale, *Role of Analogy*, Weyant, "Protoscience, Pseudoscience," and Bono, "Science, Discourse, and Literature."

10. Westfall, *Force in Newton's Physics*, pp. 330–2, 374–7. Leatherdale, *Role of Analogy*, pp. 223–44. Paul Taylor, *Independent*, 22 March 1991, p. 17 (on Strindberg's *Miss Julie*).

11. De Man, "Epistemology of Metaphor." Howell, *British Logic and Rhetoric*, pp. 439–691 (Boyle quotation on p. 479, from *Style of the Holy Scriptures*). For Boyle and mathematical language, see Shapin, "Boyle and Mathematics."

12. Hutchison, "Occult Qualities in the Scientific Revolution"; Henry, "Occult Qualities and the Experimental Philosophy." See also Schaffer, "Occultism and Reason," "Godly Men and Mechanical Philosophers," and "Newton's Comets and the Transformation of Astrology." Pumfrey, "Gilbert's Magnetic Philosophy," pp. 334–45. See Heilbron, *Electricity*, pp. 19–30, for the terminology of virtues, and Schaffer, "Measuring Virtue," for eudiometric measurements of virtue. For metaphorical implications of credit, see Schaffer, "Defoe's Natural Philosophy."

13. From "An Inquiry concerning Virtue, or Merit," partially reproduced in Copley, *Literature and the Social Order*, pp. 174–5 (quotation p. 174). Grew, *Musæum Regalis Societatis*, p. 318.

14. Quarles, *Quarles' Emblems, Divine and Moral*, p. 202. The verse in this ca. 1790 edition is entirely different in the 1764 edition. In her magnetic poem, Bryan wrote, "Thus, may the force of virtue charm my soul": Bryan, *Lectures on Natural Philosophy*, p. 143. See also Telescope, *Newtonian System of Philosophy*, p. 81: "But there are other stones, which tho' void of beauty, may, perhaps, have more virtue . . . such as the *loadstone*." For Tom Telescope's moral lessons, see Secord, "Newton in the Nursery."

15. Quarles, *Quarles' Emblems, Divine and Moral*, p. 202.

16. Bate, *Experimental Philosophy*, p. 72. The word "sensible" referred far more strongly to perception than it does nowadays. Locke quoted in entry on attraction in Johnson, *Dictionary*. Andrew Wilson, *Principles of Natural Philosophy*, pp. 32–3. William Jones, *Works*, vol. 10, p. 592.

17. Desaguliers, "Account of Some Magnetical Experiments," p. 387 (based on a French suggestion). G. Knight, *Attraction and Repulsion*, pp. 66–77 (quotation p. 67), 89–94. Sennett, *Flesh and Stone*, pp. 255–316. Marcovich, "Image of Society and Image of the Body."

18. Relevant analyses include Home, "Introduction," pp. 137–88; Cantor and Hodge, "Introduction"; Heimann, "Ether and Imponderables"; L. Laudan, "The Medium and Its Message."

19. Descartes, *Principles of Philosophy*, pp. 133–83. Halley, "Late Surprizing Appearance of the Lights," p. 421 (his model was also based on the passage from Newton's *Opticks* discussed below).

20. Edwards, *Existence and Providence of God*, p. 143; see also Howard, *New Philosophy of Descartes*, p. 311; Desaguliers, *Course of Experimental Philosophy*, p. viii; Blackmore, *Creation*, book 1, lines 256–309; Templeman, *Curious Remarks and Observations*, p. 173.

21. Rohault, *System of Natural Philosophy*, vol. 2, pp. 163–87. Samuel Clarke made only three minor footnotes to the magnetic material (identical in the editions of 1723, 1729, and 1735): see Hoskin, "Mining All Within." Regnault, *Philosophical Conversations*, vol. 1, pp. 197–232; *Universal Magazine* 33 (1763), 250–4. Letter from John Browning to Baker of 6 January 1750: Baker correspondence, vol. 4, fol. 167.

22. Langrish, *Essay on Muscular Motion*, pp. 81–2 (see also p. 57). A. Smith, *Essays on Philosophical Subjects*, pp. 40–3 (I am grateful to John Christie for this reference). When the French Camisard prophets arrived in London, younger Whig critics vouched for their sincerity by explaining that their conduct was governed by the same types of effluvia as those causing the attraction of loadstone and iron: Schwartz, *Knaves, Fools, Madmen*; Trenchard, *Natural History of Superstition*, pp. 26–30; Keith, *Magick of Quakerism*, pp. 51–9.

23. Home, "'Newtonianism' and the Theory of the Magnet." McGuire, "Force, Active Principles and Newton's Invisible Realm." Westfall, *Force in Newton's Physics*.

24. Newton, *Opticks*, p. 267; Thackray, *Atoms and Powers*, pp. 18–32. For magnetic effluvia in Newton's unpublished "De Aere et Aethere," see Hall and Boas, *Unpublished Scientific Papers*, p. 228, and Hawes, "Newton and the 'Electrical Attraction Unexcited.'" Greene, *Expansive and Contractive Forces*, pp. 3–6, 21–4; Baxter, *Nature of the Human Soul*, p. 327n.

25. Newton, *Opticks*, pp. 375–6 (Query 31), 351–2 (Query 21). For relevant draft versions of Query 22, see McGuire, "Force, Active Principles," pp. 160–1. For other manuscript references, see McGuire, pp. 180–1, and Westfall, *Force in Newton's Physics*, pp. 393–5, 465.

26. North: British Library Additional Manuscripts 32548, fol. 52; 32456, fol. 73. Other examples include Berkeley, *Siris*, pp. 108–15; Sulivan, *View of Nature*, vol. 2, pp. 393–4; vol. 1, pp. 113–28. Bate, *Experimental Philosophy*, pp. 9, 10. See also J. Hutchinson, *Works*, vol. 5, p. 231.

27. Yolton, *Thinking Matter*; Shapin, "Of Gods and Kings." Prior, "Alma," p. 230: see Rippy, *Matthew Prior*, pp. 90–4, and Spears, "Prior's Attitude toward Natural Science," pp. 496–7.

28. Quinn, "Repulsive Force in England." Desaguliers, "Cause of Electricity," pp. 180–5. Langrish, *Essay on Muscular Motion*, pp. 1–45 (quotation p. 31); Martin Clare was amongst those using similar arguments.

29. G. Knight, *Attraction and Repulsion*, pp. 66–95, quotation p. 79.

30. Schofield, *Mechanism and Materialism*, p. 180. Thackray, *Atoms and Powers*, p. 142. Heimann and McGuire, "Newtonian Forces and Lockean Powers," pp. 296–9. Heimann, "Ether and Imponderables," p. 71. Home, "Introduction," p. 160. Quinn, "Repulsive Force in England," p. 127.

31. G. Knight, *Attraction and Repulsion*, pp. 80–7, quotations pp. 82–3, 83. Sisco, *Réaumur's Memoirs on Steel and Iron*. Metzger, *Newton, Stahl, Boerhaave*, pp. 165–9. Sudduth, "Identifications of Phlogiston with Electricity." Golinski, *Science as Public Culture*, pp. 130–7.

32. Of the numerous references to magnetic fluids, the more substantial British accounts include Cavallo, *Treatise on Magnetism*, pp. 132–9, Darwin, *Poetical Works*, vol. 3, pp. 289–95, and Robison, "Magnetism," pp. 145–53. Continental translations include Winkler, *Natural Philosophy Delineated*, vol. 1, pp. 307–32, and Euler, *Letters to a German Princess*, vol. 2, pp. 261–82.

33. Dalton, *Meteorological Observations*, pp. 159–87; he claimed previous ignorance of Halley's magnetic explanation. Hillary, *Nature of Fire*, especially pp. 80–1; for Boerhaave, see Thackray, *Atoms and Powers*, pp. 106–13. Adam Walker, *System of Familiar Philosophy*, pp. 52–64, especially p. 53. Harrington, *New System on Fire and Planetary Life*; see Golinski, *Science as Public Culture*, pp. 151–2.

34. John Cook, *Clavis Naturæ*, p. 56; for Cook's magnetic arguments, see pp. 99–103. Wilde, "Hutchinsonianism"; Cantor, "Theological Significance of Ethers." Freke, *Nature and Property of Fire*, pp. 145–70 (quotation p. 145); Schaffer, "Consuming Flame," pp. 494–7, 506–9.

35. Penrose, *Essay on Magnetism* and *Animadversions on a Late Sermon*. Penrose's books on medicine and natural philosophy were not viewed favorably even by the Hutchinsonians: William Stevens' letters to Catcott of 28 March 1752 and 27 April 1752, Catcott correspondence. Penrose's books on natural philosophy were criticized by reviewers normally dealing with religious topics: *Monthly Review* 6 (1752), 438–41; 8 (1753), 439–41 (by John Ward and Abraham Dawson: see Nangle, *Monthly Review First Series*).

36. John Hutchinson, *Works*, vol. 11, pp. 305–52 (quotation p. 305); vol. 2, p. 210. Cantor, "Revelation and the Cyclical Cosmos" and "Light and Enlightenment," pp. 83–93. Wilde, "Matter and Spirit."

37. John Hutchinson, *Works*, vol. 11, pp. 351–2; vol. 4, pp. 121–2. Job 28.18. Preston, "Biblical Criticism." Cooke, *Patriarchal and Druidical Religion*, p. 26. See also D. Forbes, *Letter to a Bishop*, p. 29.

38. Parkhurst, *Hebrew and English Lexicon*, pp. 276–7. Parkhurst's discussion was reproduced in William Dodd's *Christians' Magazine* 3 (1762), 260–2, under the heading "An enquiry whether the loadstone be mentioned in scripture." Sir William Jones made copious Latin and Arabic notes in the margins of his personal copy, British Library pressmark 12904.f.23. For the Hebrew controversy at Oxford, see Patterson, "Hebrew Studies." For Jones and philology, see Aarsleff, *Study of Language in England*.

39. Cavallo, *Treatise on Magnetism*, pp. 132–9, 192–6; T. Young, *Course of Lectures*, vol. 1, pp. 685–6; Nicholson, *Introduction to Natural Philosophy*, vol. 2, p. 325; Priestley, *Heads of Lectures*, pp. 157–8. See also *Encyclopædia Britannica* (1797), vol. 10, pp. 433–5 (taken from Cavallo). For resistance to Aepinus' theories, see Home, "Æpinus and the British Electricians" and "Introduction," pp. 189–224.

40. Robison, "Magnetism," pp. 112–25. Olson, *Scottish Philosophy and British Physics*, especially pp. 157–61. For Lambert's magnetic research, see Tilling, "Inter-

pretation of Observational Errors," pp. 37–46. Barlow's diagrams are reproduced in Gooding, "Magnetic Curves" p. 188, and *Experiment and the Making of Meaning*, pp. 105–6. For French influence on English magnetic theories of the early nineteenth century, see Crosland and Smith, "Transmission of Physics," pp. 19–30.

41. *Encyclopædia Britannica* (1801 supplement), vol. 1, pp. 500–5, quotation from p. 501 (on dynamics). Shapin, "Boyle and Mathematics."

42. Quarles, *Quarles' Emblems, Divine and Moral*, pp. 202–3. This edition of ca. 1790 reproduces the original verses, but the plate is slightly different: the original probably wore out. Compare his *Emblems and Hieroglyphicks*, pp. 134–5: the magnetic verses in this 1764 edition were completely rewritten—by Dr. Watt, according to a handwritten note in the preface of the British Library copy, 11623.b.24. Moseley, *Century of Emblems*, pp. 3–6, 17, 92, 95; Daly, *Literature in the Light of the Emblem*, pp. 104–5.

43. Norris, *Poems*, p. 174. Dubois, *The Magnet*, p. 29. For Johnson, see Chapter 6.

44. *Political Magnet*, p. 3. J. Hutchinson, *Works*, vol. 5, pp. 211–12.

45. Prickett, *Words and the Word*, pp. 196–242. Bono, "Science, Discourse, and Literature." Cantor, "Light and Enlightenment" and "Weighing Light." G. Horne, *Works*, vol. 3, p. 34 (from *The Necessity of Rising with Christ*, pp. 30–43).

46. W. Law, *Works*, vol. 5, pp. 89–91 (quotations pp. 89, 90). Law wrote this Behmenist tract, originally published in 1737, in response to Bishop Hoadly; see Arthur Walker, *William Law*, pp. 118–26, and Wormhoudt, "Newton's Natural Philosophy."

47. Freke, *Nature and Property of Fire*, pp. 60–1, 145–70 (quotations pp. 150, 65); Schaffer, "Consuming Flame." For Law and Freke, see Walton, *Notes and Materials*, pp. 405n-407n; Talon, *Journals and Papers of John Byrom*, p. 222n; Arthur Walker, *Law*, pp. 200–3. See also letter from Freke to Byrom of 20 May 1754 in Byrom, "Private Journal and Literary Remains," vol. 34, p. 538, and *Monthly Review* 7 (1752), 387–94 (by William Bewley).

48. Smart, *Works*, vol. 1, pp. 39–40; Guest, *Form of Sound Words*, especially pp. 232–8. See also D. Greene, "Smart, Berkeley," and Williamson, "Smart's *Principia*." Smart was evidently familiar with Newton's magnetic analogy for doubly refracting crystals (Newton, *Opticks*, p. 373—in the 29th Query), for in the same poem he reinterpreted this relationship to provide elaborate parallels between rainbows and God's covenant with humanity, between material aids for terrestrial voyagers and divine guidance for spiritual travellers. Compass roses were originally labelled with the names of winds rather than the four cardinal points, and East was often marked by a cross to indicate the direction of Jerusalem and paradise: see Figure 3.2.

49. H. Brooke, "Universal Beauty," p. 349; Griffin, *Regaining Paradise*, pp. 109–10. Commentators have pointed to this poem as an influence on Darwin's *Botanic Garden*; compare the mysticism of Darwin's magnetic celebration quoted in Chapter 2.

50. Hirst, *Hidden Riches*. S. Richardson, *Sir Charles Grandison*, vol. 3, p. 67. Hoyles, *Edges of Augustanism*, pp. 79–114; Kuhn, "Nature Spiritualised"; G. S. Rousseau, "Mysticism and Millenarianism"; Schaffer, "Consuming Flame." Bechler, "Triall by What Is Contrary"; Tavor, *Scepticism, Society*, pp. 54–107.

51. Plato, *Ion*. Wasserman, "Sympathetic Imagination." P. B. Shelley, *Works*, vol. 7, p. 380 ("Epipsychidion"): see Dawson, "A Sort of Natural Magic," and Leask, "Shelley's 'Magnetic Ladies.'" For Shelley's translation of the *Ion*, see *Works*, vol. 7, pp. 249–8.

52. Dryden, *Poems*, vol. 2, p. 513. Zwicker, *Politics and Language in Dryden's Poetry*, pp. 56–60, 124–58. For Dryden's celebration of Gilbert, see *Poems*, vol. 1,

p. 33 (in his ode to Dr. Charleton). In addition to Johnson's entries, good sources of quotations include the Oxford English Dictionary, Stevenson, *Home Book of Quotations*, p. 306 (on constancy), and *Notes and Queries* 6 (1852), 127, 207, 368–9, 566.

53. Norris, *Discourses upon Divine Subjects*, vol. 3, p. 12: note that "incline" is a magnetic term. Some other examples are Ken, *Works*, vol. 4, pp. 63–5 (*Preparatives for Death*); vol. 3, p. 335 (*Hymnotheo*); for other examples from Ken, see Hoyles, *Edges of Augustanism*, pp. 7–77, especially pp. 59–65. Just as there were few references to magnetic evil, there were virtually no metaphoric references to repulsion.

54. Hutcheson quoted and discussed in Solkin, *Painting for Money*, p. 233 (from *An Inquiry into the Original of Our Ideas of Beauty and Virtue*). See Lawrence, "Nervous System and Society," especially p. 31.

55. Cleland, *Memoirs of a Woman of Pleasure*, p. 38. Adam Walker, *System of Familiar Philosophy*, vol. 1, pp. 52–74. P. B. Shelley, *Works*, vol. 2, p. 257 (*Prometheus Unbound*): for further examples, see Dawson, "Sort of Natural Magic."

56. F. Burney, *Evelina*, vol. 3, p. 53. E. Hervey, *Mourtray Family*, vol. 2, p. 64. Burney, *Evelina*, vol. 1, p. 186; vol. 3, p. 124.

57. Daly, *Literature in the Light of the Emblem*, pp. 104–5. Gay quoted in Stevenson, *Home Book of Quotations*, p. 306 (from "Sweet William's Farewell to Black-eyed Susan"). John Harris, *Astronomical Dialogues*, p. 22: Harris was a Fellow of the Royal Society and a Doctor of Divinity. Mullan, "Gendered Knowledge, Gendered Minds"; Myers, "Science for Women and Children." Reynolds, *View of Death*, p. 33.

58. Quarles, *Quarles' Emblems*, p. 261. Smollett, *Humphry Clinker*, p. 213 (see also p. 236). Castle, "Female Thermometer."

59. C. Williams, "Changing Face of Change." Barker-Benfield, *Culture of Sensibility*; J. Todd, *Sensibility*. F. Brooke, *Emily Montague*, vol. 1, p. 83. Greville quoted and discussed in Todd, p. 61 (from "A Prayer for Indifference").

60. From Darwin's *Economy of Vegetation*, lines 309–10, 313–14, quoted in Danchin, "Darwin's Scientific and Poetic Purpose," pp. 139–40. Byron, *Works*, vol. 5, pp. 72, 71 (*Don Juan*, canto I, 1579–80, lines 1557–60). Daston, "Naturalised Female Intellect"; Keller, "Secrets of God, Nature, and Life."

61. O. Smith, *Politics of Language*. Barrell, *English Literature in History*, pp. 110–75. Barker-Benfield, *Culture of Sensibility*. Shiach, *Popular Culture*, pp. 1–34.

62. *Encyclopædia Britannica* (1803 supplement), vol. 1, p. 501 (article on dynamics). Darwin, *Poetical Works*, vol. 2, pp. 64–5. Shapin, "Robert Boyle and Mathematics."

63. Curry, *Prophecy and Power*, pp. 93–152. Well into the century, readers could learn about sending magnetic messages from Paris to London or magnetically curing diseases by transferring them to vegetables: Bailey, *Dictionary*, under the entries for loadstone and transplantation (and elsewhere). See Hughes, "Sciences in English Encyclopaedias," Shorr, *Science and Superstition*, and Layton, "Diction and Dictionaries." Leonardus, *Mirror of Stones*, preface and pp. 206–10 (as one example).

64. Cavallo, *Treatise on Magnetism*, pp. 101–3 (quotations pp. 103, 102). See also Werenfels, *Dissertation upon Superstition*, W. G. Black, *Folk-Medicine*, pp. 52–5, Walsh, *Cures*, pp. 77–87, and E. H. Frei, "Medical Applications of Magnetism."

65. See especially Chapters 2, 3, and 5. F. Burney, *Diary and Letters*, vol. 2, pp. 180–1.

66. John Harris, *Astronomical Dialogues*, p. 81. Griffin, *Regaining Paradise*, pp. 62–71. Myers, "Science for Women and Children." Mullan, "Gendered Knowledge, Gendered Minds."

67. Cooter, "Deploying 'Pseudoscience'" and "History of Mesmerism." Darnton, *Mesmerism and the End of the Enlightenment*. McCalman, *Radical Underworld*, pp. 1–94. L. Wilson, *Women and Medicine*, pp. 104–24.

68. Andry and Thouret, "Observations et recherches" and "Rapport sur les aimans." Weyant, "Protoscience, Pseudoscience." English people most commonly referred to the practice as animal magnetism. For details of the literature on Mesmer, see Darnton, *End of the Enlightenment*, Gauld, *History of Hypnotism*, pp. 1–123, and Crabtree, *From Mesmer to Freud*. This account is based on Fara, "An Attractive Therapy," which includes copious references to the primary and secondary literature.

69. W. Godwin, *Report of Dr Benjamin Franklin*. For Godwin's authorship, see Schaffer, "Self Evidence," p. 80. Inchbald, *Animal Magnetism*, is virtually identical to Dumaniant, *Le médecin malgré tout le monde*.

70. Porter, *Health for Sale*, "Before the Fringe," and "Language of Quackery"; Jewson, "Medical Knowledge and the Patronage System." For a good selection of rival advertisements, see Lysons I, vol. 3, fols. 157–73.

71. Letter of 4 July 1786 from George Elliot to his wife, reproduced in Minto, *Life and Letters*, vol. 1, p. 111. Bell, *New System of the World* and *General and Particular Principles*. For the social geography, see H. Phillips, *Mid-Georgian London*, especially pp. 210–12, 236–9, 295. Sibly, *Key to Physic*, p. 256: see Debus, "Scientific Truth and Occult Tradition." *Memoir of the Late Thomas Holloway*, pp. 5–32. For further details of these and other magnetizers, see Fara, "Attractive Therapy."

72. De Mainauduc, *Lectures*: about 150 subscribers for this posthumous book are listed. There is a handwritten list of 195 students and patients at London's Royal College of Surgeons, MS 42.e.1. Porter, "William Hunter." Rosenberg, "Medical Text and Social Context."

73. Letter to William Withering of 5 December 1787: Royal Society of Medicine, Withering correspondence, fols. 92–4 (quotation from fol. 93r). See also her letter to Withering of 25 December 1786, fols. 88–90. Lefanu, *Betsy Sheridan's Journal*, pp. 123–4; see also Minto, *Life and Letters*, vol. 1, pp. 113, 285–6.

74. De Almeida, *Romantic Medicine*, pp. 59–134: see p. 100 for John Haighton lecturing on this topic at Guy's Hospital. J. Hunter, *Works*, vol. 1, especially pp. 221–8 (quotations from pp. 222, 213). For his vitalism, see Duchesnau, "Vitalism," pp. 279–93, and Cross, "John Hunter." Darwin, *Zoonomia*, vol. 1, p. 64. De Mainauduc, *Lectures*, pp. 66–75 (quotation p. 69); Darwin, vol. 1, pp. 30–3.

75. De Mainauduc, *Lectures*, pp. 1–116. J. Hunter, *Works*, vol. 1, pp. 211–8. Rosenberg, "Medical Text and Social Context," pp. 31–7. Sennett, *Flesh and Stone*, pp. 255–316. Cross, "John Hunter," pp. 68–77. Marcovich, "Image of Society."

76. Lawrence, "The Nervous System and Society." J. Hunter, *Works*, vol. 1, pp. 317–37, especially pp. 336–7. Macalpine and Hunter, *George III and the Mad-Business*. Porter, "Barely Touching." G. S. Rousseau, "Science and the Discovery of the Imagination." De Mainauduc, *Lectures*, pp. 41–75.

77. Foucault, *Birth of the Clinic*, pp. 107–73. Stafford, *Body Criticism*, especially pp. 450–63. Macalpine and Hunter, *George III and the Mad-Business*, pp. 270–2. De Almeida, *Romantic Medicine*, pp. 182–96. Quoted in *Philosophical Magazine* 19 (1804), 383.

78. Letter from Cavallo to James Lind of 10 July 1784: British Library Additional Manuscripts 22897, fols. 21–2. For British anti-Catholicism, see Colley, *Britons*, pp. 11–54. Letter from Cavallo to Prince Hoare of 8 February 1788: Yale University Beinecke Rare Book and Manuscript Library, Osborn Files/Cavallo. S. Lloyd, "The Accomplished Maria Cosway." Bloch, *The Royal Touch*, pp. 214–28.

79. Inchbald, *Animal Magnetism*. This seems to have been a real event, although I have been unable to find any written references to it. Certainly Thomas Ribright (the optician named above the door in the cartoon) existed and practiced as an optician at 40 The Poultry from 1783 to 1796: I am grateful to Gloria Clifton for supplying detailed information about Ribright from the Project Simon database.

80. W. Godwin, *Report of Dr Benjamin Franklin*, pp. iii–xx (quotation p. xviii). Letter from More to Horace Walpole of September 1788 reproduced in Lewis, *Walpole's Correspondence*, vol. 31, pp. 279–81 (quotations p. 280); see also his reply of 22 September 1788: pp. 282–4. Letter from Walpole to Lady Ossory of 1 July 1789: vol. 34. pp. 50–1; see also his letter of 26 November 1789: p. 83.

81. *Monthly Review* 73 (1785), 41–5. Wollstonecraft, *Vindication of the Rights of Woman*, pp. 414–52, especially pp. 419–25 (quotations pp. 420, 422). Schiebinger, *The Mind Has No Sex*, pp. 160–213. Gilman et al., *Hysteria beyond Freud*, pp. 3–221. Colley, *Britons*, pp. 250–62. Barker-Benfield, *The Culture of Sensibility*, pp. 1–36. See also M. Edgeworth, *Letters for Literary Ladies*, p. 47.

82. *World* for 22, 23, 24, and 26 December 1788 and 5 January 1789 (quotation from 22 December 1788): I am grateful to Keith Schuchard for this reference. Cone, *Burke and the Nature of Politics*, vol. 2, pp. 257–82. C. Reid, "Burke, the Regency Crisis."

83. Coleridge, *Lectures 1795*, p. 328: I am grateful to Reeve Parker for this reference.

84. Relevant literature includes Paulson, *Representations of Revolution*, Hole, *Pulpits, Politics and Public Order*, and Philp, *The French Revolution. Supernatural Magazine* (1809), 8–9. Robison, *Proofs of a Conspiracy*, p. 6, and W. H. Reid, *Rise and Dissolution of the Infidel Societies*, p. 91. For Reid, see McCalman, *Radical Underworld*, p. 1 and passim; for Robison, see Morrell, "Professors Robison and Playfair," and Christie, "Joseph Black and John Robison." *Times*, 24 March 1790 and 4 March 1795. Southey, *Letters from England*, pp. 304–19.

85. For relevant literature on Priestley and Beddoes, see Golinski, *Science as Public Culture*, pp. 153–87, and Porter, *Doctor of Society. The Sceptic*, pp. 1–11 (quotations pp. 4–6). For porcine symbolism, see O. Smith, *Politics of Language*, pp. 68–109.

86. F. Burney, *Diary and Letters*, vol. 3, pp. 271–2 (19 June 1787). Sibly, *Key to Physic*, pp. 256–76: see Debus, "Scientific Truth and Occult Tradition," and Duchesnau, "Vitalism in Late Eighteenth-Century Physiology." Several encyclopedias of the early nineteenth century discussed Bell's rather than de Mainauduc's animal magnetism. Cooper, "Power of the Eye"; quotation from an early draft of *The Ancient Mariner*. Letter of 1 January 1787, reproduced in Boyd, *Papers of Thomas Jefferson*, vol. 11, pp. 3–4. St. Clair, *The Godwins and the Shelleys*, p. 243 (letter to his second wife, Mary Jane Godwin). W. Godwin, *Caleb Williams*, p. 112.

87. F. Burney, *The Wanderer*, pp. 70, 663; compare *Evelina*, vol. 1, pp. 186–7; vol. 3, pp. 53, 124 (quoted earlier in this chapter). P. B. Shelley, *Works*, vol. 7, p. 120 (from "Defence of Poetry," written 1821); the relevant passage in the *Ion* is in vol. 7, p. 263: see Dawson, "A Sort of Natural Magic," and Leask, "Shelley's Magnetic Ladies.'" Tatar, *Spellbound*; A. Winter, "The Island of Mesmeria."

CONCLUSION

1. Gerard, *An Essay on Taste*, pp. 168–9. Barrell, *Political Theory of Painting*, pp. 1–68.

2. Barrell, *Political Theory of Painting*, pp. 192–9 (Burke quotation from p. 197). Golinski, *Science as Public Culture*, and Stewart, *Rise of Public Science*. Solkin, *Painting for Money*. Raven, *Judging New Wealth*.

3. Barry, *Series of Pictures*, pp. 60–1. Schaffer, "Scientific Discoveries." Daston, "Naturalised Female Intellect." Barrell, *Birth of Pandora*, pp. 41–61. McNeil, "Scientific Muse."

4. Dryden, *Poems*, vol. 1, p. 33 (from "To My Honour'd Friend, D[r] Charleton"). Robison, "Magnetism," pp. 112–13. William Jones, *Physiological Disquisitions*, pp. x–xii. T. S. Kuhn, *Essential Tension*, pp. 31–65; relevant critiques include J. A. Bennett, "Mechanics' Philosophy," and Schuster and Watchirs, "Natural Philosophy," pp. 1–14.

5. John Robertson, *Elements of Navigation*, vol. 2, p. 232.

6. Letter from Jones to Alexander Catcott of 25 April 1759: Catcott correspondence. Schaffer,"Newtonianism"; Yeo, "Genius, Method and Morality." Feyerabend, "Classical Empiricism"; Schaffer, "Machine Philosophy"; Schuster, "Methodologies as Mythic Structures."

7. William Jones, *First Principles of Natural Philosophy*, pp. 55–116 (the dialogue on pp. 71–2 is reproduced in Yolton, *Thinking Matter*, p. 193). Letter from Jones to Catcott of 22 October 1759: Catcott correspondence. Quotation from *Monthly Review* 20 (1759), 300–1 (on Lovett); see also *Monthly Review* 27 (1762), 122–7 (on Jones). For the nineteenth century, see Cantor, *Michael Faraday*, pp. 162–8; A. B. Webster, *Joshua Watson*, pp. 18–32; Corsi, *Science and Religion*, pp. 18–32; W. Walker, *Life and Times of Skinner*, pp. 58–70, 148–67; Kirby, *Power Wisdom and Goodness of God*, vol. 1, pp. xvii–cv, especially pp. cii–ciii.

8. Aarsleff, *Study of Language in England*. Prickett, *Words and "The Word,"* pp. 105–23. Jacobus, *Romanticism*, pp. 69–93 (including this quotation from Wordsworth's *The Prelude*, book 3, line 61); Greenberg, "Eighteenth-Century Poetry."

9. *Quarterly Review* 18 (1818), 457, 458 (from his critical review of James Burney's "Memoir" in the *Philosophical Transactions* about the Northwest Passage).

10. Lawrence, "Disciplining Disease." Mullan, "Gendered Knowledge, Gendered Minds." Myers, "Science for Women and Children." Daston, "Naturalized Female Intellect." Keller, "Secrets of God, Nature, and Life."

11. Burney, *The Wanderer*, p. 544. M. Edgeworth, *Letters for Literary Ladies*, pp. 20–1 (misquoting—probably unintentionally—Darwin's advertisement for *The Loves of the Plants* in Darwin, *Poetical Works*, vol. 2). Barrell, *Birth of Pandora*, pp. 132–43.

12. M. Shelley, *Frankenstein*, p. 11. Darwin, *Zoonomia*, vol. 1, p. 64. W. Godwin, *Caleb Williams*, p. 112. For references to the literature on *Frankenstein*, see Bann, *Frankenstein, Creation and Monstrosity*, especially pp. 77–9, 210–12.

13. Shelley, *Frankenstein*, pp. 14 (Walton), 203 (Frankenstein), 213 (creature), 215 (final words of the book).

BIBLIOGRAPHY

Aarsleff, Hans. *The Study of Language in England, 1780–1860*. Princeton: Princeton University Press, 1967.

———. *From Locke to Saussure: Essays on the Study of Language and Intellectual History*. London: Athlone, 1982.

Adams, Frank D. *The Birth and Development of the Geological Sciences*. London: Baillière, Tindall & Cox, 1938.

Adams, George. "A Catalogue of Mathematical, Philosophical, and Optical Instruments, as Made and Sold by George Adams, at Tycho Brahe's Head, in Fleet-Street, London." In *Micrographia Illustra*, pp. 243–63. London, 1746.

———. "A Catalogue of Mathematical, Philosophical, and Optical Instruments." In *New Celestial and Terrestrial Globes*, by G. Adams. London, 1777.

———. *An Essay on Electricity, Explaining the Theory and Practice of That Useful Science; and the Mode of Applying It to Medical Purposes*. London, 1787.

———. *Geometrical and Graphical Essays*. London, 1791.

———. *Lectures on Natural and Experimental Philosophy, Considered in It's Present State of Improvement*, 5 vols. London, 1794.

———. *Lectures on Natural and Experimental Philosophy, Considered in It's Present State of Improvement*, 2d ed., 5 vols. London, 1799.

Adams, John. *The Description and Use of a New Much-improved Sinical Quadrant*. Ratcliff, 1781.

Adams, Percy G. *Travelers and Travel Liars, 1660–1800*. 1962. Reprint. New York: Dover Publications, 1980.

Albury, W. R. "Halley's Ode on the *Principia* of Newton and the Epicurean Revival in England," *Journal of the History of Ideas* 39 (1978), 24–43.

Albury, W. R., and D. R. Oldroyd. "From Renaissance Mineral Studies to Historical Geology, in the Light of Michel Foucault's *The Order of Things*," *British Journal for the History of Science* 10 (1977), 187–215.

Allan, David G. C. "The Society for the Encouragement of Arts, Manufactures and Commerce: Organisation, Membership and Objectives in the First Three Decades." Ph.D. thesis, University of London, 1979.

Allan, David G. C., and Robert E. Schofield. *Stephen Hales: Scientist and Philanthropist*. London: Scolar Press, 1980.

Allibone, T. E. *The Royal Society and Its Dining Clubs*. Oxford: Pergamon Press, 1976.

Almond, Philip. "The Contours of Hell in English Thought, 1660–1750," *Religion* 22 (1992), 197–311.

Altick, Richard D. *The Shows of London*. Cambridge: Harvard University Press, 1978.

Anderson, Adam. *An Historical and Chronological Deduction of the Origin of Commerce, from the Earliest Accounts to the Present Time*, 2 vols. London, 1764.

Andrade, E. N. da C. "The Early History of the Permanent Magnet," *Endeavour* 17 (1958), 22–30.

Andry, Charles, and Thouret, Michel. "Observations et recherches sur l'usage de l'aimant en médecine, ou mémoire sur le magnétisme médecinal," *Histoires et mémoires de la Société royale de médecine* 3 (1782), 531–688.

Andry, Charles, and Thouret, Michel. "Rapport sur les aimans présentés par M. l'abbé Le Noble." *Extrait des registres de la Société royale de médecine*, 1783.

Anson, George. *A Voyage round the World, in the Years MDCCXL, I, II, III, IV. By George Anson, Esq; Commander in Chief of a Squadron of His Majesty's Ships, Sent upon an Expedition to the South-Seas*. London, 1748.

Appleby, John H. "Daniel Dumaresq, DD, FRS (1712–1805), as a Promoter of Anglo-Russian Science and Culture," *Notes and Records of the Royal Society of London* 44 (1990), 25–50.

Appleton, William. *A Cycle of Cathay: The Chinese Vogue in England during the Seventeenth and Eighteenth Centuries*. New York: Columbia University Press, 1951.

Archinard, Margarida. *Collection de Saussure*. Geneva: Musée d'art et d'histoire, 1979.

Arderon, William. "Abstract of a Letter . . . on the Giving Magnetism and Polarity to Brass," *Philosophical Transactions* 50 (1758), 774–7.

Ardley, G. W. *Berkeley's Philosophy of Nature*. Auckland: University of Auckland, Pelorus Press, 1962.

Armitage, Angus. *Edmond Halley*. London: Nelson, 1966.

Atherton, Herbert M. *Political Prints in the Age of Hogarth*. Oxford: Clarendon Press, 1974.

Atkinson, James. *Epitome of the Art of Navigation*. London 1718.

Atwood, George. *A Description of the Experiments Intended to Illustrate a Course of Lectures, on the Principles of Natural Philosophy, Read in the Observatory at Trinity College, Cambridge*. London, 1776.

Badcock, Allan W. "Physics at the Royal Society 1687–1800." Ph.D. thesis, University of London, 1962.

Bailey, Nathan. *The Universal Etymological English Dictionary*. London 1731.

Bain, William. *An Essay on the Variation of the Compass, Shewing How Far It Is Influenced by a Change in the Direction of the Ship's Head, with an Exposition of the Dangers Arising to Navigators from Not Allowing for This Change of Variation*. Edinburgh, 1817.

Baine, R. M. *Daniel Defoe and the Supernatural*. Athens: University of Georgia Press, 1968.

Baker, Thomas. *Reflections upon Learning, Wherein Is Shown the Insufficiency Thereof, in Its Several Particulars*. London, 1699.

Baldwin, Martha R. "Magnetism and the Anti-Copernican Polemic," *Journal for the History of Astronomy* 16 (1985), 155–74.

Ballard, Richard. "Concerning the Magnetism of Drills," *Philosophical Transactions* 20 (1698), 417–21.

Bann, Stephen (ed.). *Frankenstein, Creation and Monstrosity*. London: Reaktion Books, 1994.

Barker-Benfield, G. J. *The Culture of Sensibility: Sex and Society in Eighteenth-Century Britain*. Chicago and London: University of Chicago Press, 1992.

Barnes, Barry. *T. S. Kuhn and Social Science*. London: Macmillan, 1982.

Barrell, John. *English Literature in History 1730–80: An Equal, Wide Survey*. London: Hutchinson, 1983.

———. *The Political Theory of Painting from Reynolds to Hazlitt: "The Body of the Public*." New Haven and London: Yale University Press, 1986.

———. *The Birth of Pandora and the Division of Knowledge*. Basingstoke: Macmillan, 1992.

Barrett, Francis. *The Magus; or, The Celestial Intelligencer; Being a Complete System of Occult Philosophy*. London 1801.

Barrow, John. *Dictionarium Polygraphicum; or, The Whole Body of Arts Regularly Digested*, 2 vols. London, 1735.

Barry, James. *An Account of a Series of Pictures, in the Great Room of the Society of Arts, Manufactures and Commerce, at the Adelphi*. London, 1783.

Barry, Jonathan. "Piety and the Patient: Medicine and Religion in Eighteenth Century Bristol." In *Patients and Practitioners: Lay Perceptions of Medicine in Pre-industrial Society*, ed. Roy Porter, pp. 145–75. Cambridge: Cambridge University Press, 1985.

Barthes, Roland. "The Plates of the *Encyclopedia*." In *A Barthes Reader*, ed. Susan Sontag, pp. 218–35. London: Jonathan Cape, 1982.

Bate, Julius. *Experimental Philosophy Asserted and Defended, against Some Late Attempts to Undermine It*. London, 1740.

Batten, Charles. *Pleasurable Instruction*. Berkeley and Los Angeles: University of California Press, 1978.

Bauer, L. A. "Halley's Earliest Equal Variation Chart," *Terrestrial Magnetism* 1 (1896), 28–31.

———. "Magnetic Results of Halley's Expedition, 1698–1700," *Terrestrial Magnetism and Atmospheric Electricity* 18 (1913), 113–26.

Baugh, David. *British Naval Administration in the Age of Walpole*. Princeton: Princeton University Press, 1965.

———. "Seapower and Science: The Motives for Pacific Exploration." In *Background to Discovery: Pacific Exploration from Dampier to Cook*, ed. Derek Howse, pp. 1–55. Berkeley and Los Angeles: University of California Press, 1990.

Baxter, Andrew. *An Enquiry into the Nature of the Human Soul; Wherein the Immateriality of the Soul Is Evinced from the Principles of Reason and Philosophy*. London, 1733.

Bayly, Christopher A. *Imperial Meridian: The British Empire and the World 1780–1830*. London: Longman, 1989.

Bayne-Powell, Rosamund. *Travellers in Eighteenth-Century England*. London: John Murray, 1951.

Bazerman, Charles. "Forums of Validation and Forms of Knowledge: The Magical Rhetoric of Otto von Guericke's Sulfur Globe," *Configurations* 1 (1993), 201–28.

Beaglehole, John C. (ed.). *The Journals of Captain James Cook*, 5 vols. Cambridge: Hakluyt Society, 1967–9.

———. *The Life of Captain James Cook*. Cambridge: Hakluyt Society, 1973.

Bechler, Rosemary. "'Triall by What Is Contrary': Samuel Richardson and Christian Doctrine." In *Samuel Richardson: Passion and Prudence*, ed. Valerie Myer, pp. 93–113. London: Vision Press, 1986.

Beck, Horace. *Folklore and the Sea*. Middletown, Conn.: Marine Historical Association, 1973.

Beckett, Walter N. T. *A Few Naval Customs, Expressions, Traditions and Superstitions*. Portsmouth: Gieves, 1932.

Beer, Gillian, and Herminio Martins. "Introduction," *History of the Human Sciences* 3 (1990), 163–75.

Bektas, M. Yakup, and Maurice Crosland. "The Copley Medal: The Establishment of a Reward System in the Royal Society, 1731–1839," *Notes and Records of the Royal Society of London* 46 (1992), 43–76.

Bell, John. *New System of the World; and the Laws of Motion; in Which Are Explained Animal Electricity and Magnetism, Both Natural and Artificial.* London, 1788.

———. *The General and Particular Principles of Animal Electricity and Magnetism &c.* London, 1792.

Ben-Chaim, Michael. "Social Mobility and Scientific Change: Stephen Gray's Contribution to Electrical Research," *British Journal for the History of Science* 23 (1990), 3–24.

Benjamin, Marina. "Medicine, Morality and the Politics of Berkeley's Tar-Water." In *The Medical Enlightenment of the Eighteenth Century*, ed. Andrew Cunningham and Roger French, pp. 165–93. Cambridge: Cambridge University Press, 1990.

Bennet, Abraham. "A New Suspension of the Magnetic Needle, Intended for the Discovery of Minute Quantities of Magnetic Attraction . . . with New Experiments on the Magnetism of Iron Filings," *Philosophical Transactions* 82 (1792), 81–98.

Bennett, G. V. "University, Society and Church 1688–1714." In *The History of the University of Oxford.* Vol. V, *The Eighteenth Century*, ed. L. Sutherland and L. Mitchell, pp. 359–400. Oxford: Clarendon Press, 1986.

Bennett, J. A. "Robert Hooke as Mechanic and Natural Philosopher," *Notes and Records of the Royal Society of London* 35 (1980), 33–48.

———. "Cosmology and the Magnetic Philosophy 1640–80," *Journal for the History of Astronomy* 12 (1981), 165–77.

———. *The Mathematical Science of Christopher Wren.* Cambridge: Cambridge University Press, 1982.

———. "The Mechanics' Philosophy and the Mechanical Philosophy," *History of Science* 24 (1986), 1–28.

———. *The Divided Circle: A History of Instruments for Astronomy, Navigation and Surveying.* Oxford: Phaidon, 1987.

———. "Hooke's Instruments for Astronomy and Navigation." In *Robert Hooke: New Studies*, ed. Michael Hunter and Simon Schaffer, pp. 21–32. Woodbridge, Suffolk: Boydell Press, 1989.

———. "A Viol of Water or a Wedge of Glass." In *The Uses of Experiment: Studies in the Natural Sciences*, ed. David Gooding, Trevor Pinch, and Simon Schaffer, pp. 105–14. Cambridge: Cambridge University Press, 1989.

———. "The Challenge of Practical Mathematics." In *Science, Culture and Popular Belief in Renaissance Europe*, ed. Stephen Pumfrey, Paolo Rossi, and Maurice Slawinski, pp. 176–90. Manchester: Manchester University Press, 1991.

Bennett, J. A., and Olivia Brown. *The Compleat Surveyor.* Cambridge: Whipple Museum of the History of Science, 1982.

Bergman, Torbern. "Observations on *Auroræ Boreales* in Sweden," *Philosophical Transactions* 52 (1762), 479–86.

———. "A Letter Containing Some Experiments in Electricity," *Philosophical Transactions* 54 (1764), 84–8.

Berkeley, George. *Siris: A Chain of Philosophical Reflexions and Inquiries concerning the Virtues of Tar Water, and Divers Other Subjects Connected Together and Arising One from Another.* Dublin, 1744.

———. *The Works of George Berkeley, Bishop of Cloyne*, 9 vols., ed. A. A. Luce and T. E. Jessop. London: Thomas Nelson & Sons, 1948–57.

Bijker, Wiebe, Thomas Hughes, and Trevor Pinch. *The Social Construction of Technological Systems: New Directions in the Social Study of Technology*. Cambridge: MIT Press, 1987.

Billingsley, Case. *The Longitude at Sea, Not to Be Found by Firing Guns, Nor by the Most Curious Spring-Clocks or Watches*. London, 1714.

———. *Biographical Account of John Hadley, Esq VPRS, the Inventor of the Quadrant, and of His Brothers George and Henry*. 1840.

Birch, Thomas. Diary. (1735–64). British Library Additional Manuscripts 4478C.

———. *The History of the Royal Society*, 4 vols. London, 1756–7.

Bishop, George D. "A History of the Teaching of Physics in England from 1650 to 1850." Ph.D. thesis, University of London, 1959.

Black, Joseph. *Lectures on the Elements of Chemistry, Delivered in the University of Edinburgh*, 2 vols. Edinburgh, 1803.

Black, William G. *Folk-Medicine: A Chapter in the History of Culture*. London, 1883 (for the Folk-Lore Society).

Blackmore, Richard. *Creation: A Philosophical Poem, in Seven Books*. London, 1797.

Blanckley, Thomas R. *The Naval Expositor*. London, 1750.

Bligh, William. *The Log of the Bounty*, 2 vols. London: Golden Cockerel Press, 1937.

Bloch, Marc. *The Royal Touch: Sacred Monarchy and Scrofula in England and France*. 1924 (in French). Trans. J. E. Anderson. London: Routledge & Kegan Paul, 1973.

Blondel, James A. *The Power of the Mother's Imagination over the Foetus Examin'd*. London, 1729.

Bloor, David. "Durkheim and Mauss Revisited: Classification and the Sociology of Knowledge," *Studies in History and Philosophy of Science* 13 (1982), 267–92.

Bloxam, John R. *A Register of . . . Magdalen College*, 7 vols. Oxford: Magdalen College, 1879.

Boerhaave, Hermann. *An Essay on the Virtue and Efficient Cause of Magnetical Cures*. London, 1743.

Bono, James J. "Science, Discourse, and Literature: The Role/Rule of Metaphor in Science." In *Literature and Science: Theory and Practice*, ed. Stuart Peterfreund, pp. 59–89. Boston: Northeastern University Press, 1990.

Borsay, Peter. *The English Urban Renaissance: Culture and Society in the Provincial Town 1660–1770*. Oxford: Clarendon Press, 1989.

Boswell, James. *Boswell's Life of Johnson*, 2 vols. London: Oxford University Press, 1927.

Boucé, Paul-Gabriel. "Some Sexual Beliefs and Myths in Eighteenth-Century Britain." In *Sexuality in Eighteenth-Century Britain*. ed. Paul-Gabriel Boucé, pp. 28–46. Manchester: Manchester University Press, 1982.

———. "Imagination, Pregnant Women, and Monsters, in Eighteenth-Century England and France." In *Sexual Underworlds of the Enlightenment*, ed. George S. Rousseau and Roy Porter, pp. 86–100. Manchester: Manchester University Press, 1987.

Bowler, Peter. "Science and the Environment: New Agendas for the History of Science?" In *Science and Nature: Essays in the History of the Environmental Sciences*, ed. Michael Shortland, pp. 1–21. British Society for the History of Science, 1993.

Bowles, Geoffrey. "Physical, Human and Divine Attraction in the Life and Thought of George Cheyne," *Annals of Science* 31 (1974), 473–88.

Boyd, Julian P. (ed.). *The Papers of Thomas Jefferson*, 25 vols. Princeton: Princeton University Press, 1950–92.

Boyer, Carl B. "Note on an Early Graph of Statistical Data (Huygens 1669)," *Isis* 37 (1947), 148–9.

Bravo, Michael. "James Rennell: Antiquarian of Ocean Currents," *Ocean Challenge* 4 (1993), 41–50.

Bremond,?. "Extract of a Letter to Dr Mortimer, Concerning a File Rendered Magnetical by Lightning." *Philosophical Transactions* 41 (1741), 614–16.

Breslaw, Phillip. *Breslaw's Last Legacy; or, The Magical Companion: Containing All That Is Curious, Pleasing, Entertaining, and Comical; Selected from the Most Celebrated Masters of Deception; as Well with Slight of Hand, as with Mathematical Inventions*. London, 1784.

Brewer, John. *The Sinews of Power*. London: Unwin Hyman, 1989.

Brewer, John, and Roy Porter (eds.). *Consumption and the World of Goods*. London: Routledge, 1993.

Briggs, J. Morton. "Aurora and Enlightenment: Eighteenth-Century Explanations of the Aurora Borealis," *Isis* 58 (1967), 491–503.

Briggs, Robin. "The Académie Royale des Sciences and the Pursuit of Utility," *Past and Present* 131 (1991), 38–88.

Brock, P. W. "Anson and His Importance as a Naval Reformer," *Naval Review* 17 (1929), 497–528.

Brooke, Frances. *The History of Emily Montague*, 2 vols. London, 1769.

Brooke, Henry. "Universal Beauty." In *The Works of the English Poets, from Chaucer to Cowper*, ed. Alexander Chalmers, vol. 17, pp. 337–65. London, 1810.

Brooke, John H. "Why Did the English Mix Their Science and Their Religion?" In *Science and Imagination in XVIIIth-Century British Culture*, ed. Sergio Rossi, pp. 57–78. Milan: Edizioni Unicopoli, 1987.

———. "Science and the Secularisation of Knowledge: Perspectives on Some Eighteenth-Century Transformations," *Nuncius* 4 (1989), 43–65.

———. *Science and Religion*. Cambridge: Cambridge University Press, 1991.

Brooks, Randall C. "Magnetic Influence on Chronometers, 1798–1834: A Case Study," *Annals of Science* 44 (1987), 245–64.

Brown, A. Stuart. "Gustavus Katterfelto: Mason and Magician," *Ars Quatuor Coronatorum* 69 (1956), 136–8.

Brown, Thomas. *The Third Volume of the Works of Mr Thomas Brown, Serious and Comical in Prose and Verse*. London, 1720.

Browne, Thomas. *Pseudoxica Epidemica; or, Enquiries into Very Many Received Tenents and Commonly Presumed Truths*. Vol. 2 of *The Works of Thomas Browne*, 4 vols., ed. Geoffrey Keynes. London: Faber & Faber, 1964.

Bryan, Margaret. *Lectures on Natural Philosophy: The Result of Many Years' Practical Experience of the Facts Elucidated*. London, 1806.

Bryden, D. J. "Magnetic Inclinatory Needles: Approved by the Royal Society?" *Notes and Records of the Royal Society of London* 47 (1993), 17–31.

Buck, Peter. "People Who Counted: Political Arithmetic in the Eighteenth Century," *Isis* 73 (1982), 28–45.

Burke, Edmund. *Reflections on the Revolution in France, and on the Proceedings in Certain Societies in London Relative to That Event*, ed. Conor C. O'Brien. Harmondsworth: Penguin, 1968.

Burnet, Thomas. *The Sacred Theory of the Earth: Containing an Account of the Origi-*

nal of the Earth, and of All the General Changes Which It Hath Already Undergone, or Is to Undergo, till the Consummation of All Things, 2 vols. London, 1722.

Burney, Frances. *Evelina; or, A Young Lady's Entrance into the World*, 3 vols. London, 1778.

———. *Diary and Letters of Madame d'Arblay (1778–1840)*, 6 vols., ed. Charlotte Barrett. London: Macmillan, 1904–5.

———. *The Wanderer*. 1814. Reprint. Oxford: Oxford University Press, 1991.

Burney, William. *A New Universal Dictionary of the Marine*. London, 1815.

Burns, R. M. *The Great Debate on Miracles: From Joseph Glanvill to David Hume*. Lewisburg: Bucknell University Press, 1981.

Burr, Thomas B. *The History of Tunbridge Wells*. London, 1766.

Byrom, John. *The Private Journal and Literary Remains of John Byrom*, ed. Richard Parkinson. 2 vols. in 4. *Remains Historical and Literary Connected with the Palatine Counties of Lancaster and Chester, Published by the Chetham Society*, 32, 34, 40, 44 (Manchester, 1854–7).

Byron, George Gordon. *The Complete Poetical Works*, 7 vols., ed. Jerome McGann. Oxford: Clarendon Press, 1980.

Caird, George B. *The Language and Imagery of the Bible*. London: Duckworth, 1980.

Cajori, Florian. *Sir Isaac Newton's Mathematical Principles of Natural Philosophy and His System of the World*. Berkeley and Los Angeles: University of California Press, 1960.

Callon, Michel. "Some Elements of a Sociology of Translation: Domestication of the Scallops and the Fishermen of St Brieuc Bay." In *Power, Action and Belief: A New Sociology of Knowledge?*, ed. John Law, pp. 196–233. London: Routledge & Kegan Paul, 1986.

Calvert, Henry R. *Scientific Trade Cards in the Science Museum Collection*. London: HMSO, 1971.

Campbel, Duncan. *The Friendly Dæmon; or, The Generous Apparition; Being a True Narrative of a Miraculous Cure, Newly Perform'd Upon That Famous Deaf and Dumb Person*. London, 1726.

Canton, John. "A Method of Making Artificial Magnets without the Use of Natural Ones," *Philosophical Transactions* 47 (1751), 31–8.

———. "An Attempt to Account for the Regular Diurnal Variation of the Horizontal Magnetic Needle; and Also for Its Irregular Variation at the Time of an Aurora Borealis," *Philosophical Transactions* 51 (1759), 398–445.

Cantor, Geoffrey. "Berkeley, Reid, and Optics," *Journal of the History of Ideas* 38 (1977), 429–48.

———. "Revelation and the Cyclical Cosmos of John Hutchinson." In *Images of the Earth: Essays in the History of the Environmental Sciences*, ed. Ludmilla Jordanova and Roy Porter, pp. 3–22. St. Giles: British Society for the History of Science, 1978.

———. "The Theological Significance of Ethers." In *Conceptions of Ether: Studies in the History of Ether Theories, 1740–1900*, ed. Geoffrey Cantor and Michael Hodge, pp. 135–56. Cambridge: Cambridge University Press, 1981.

———. "The Eighteenth-Century Problem," *History of Science* 20 (1982), 44–63.

———. *Optics after Newton: Theories of Light in Britain and Ireland, 1704–1840*. Manchester: Manchester University Press, 1983.

———. "*The Analyst* Revisited," *Isis* 75 (1984), 668–83.

———. "Light and Enlightenment: An Exploration of Mid-Eighteenth-Century Modes of Discourse." In *The Discourse of Light from the Middle Ages to the Enlightenment*, ed. David Lindberg and Geoffrey Cantor, pp. 67–106. Los Angeles: William Andrews Clark Memorial Library, University of California, 1985.

Cantor, Geoffrey. "Weighing Light: The Role of Metaphor in Eighteenth-Century Optical Discourse." In *The Figural and the Literal: Problems of Language in the History of Science and Philosophy, 1630–1800*, ed. Andrew Benjamin, Geoffrey Cantor, and John Christie, pp. 124–46. Manchester: Manchester University Press, 1989.

———. *Michael Faraday: Sandemanian and Scientist.* Basingstoke: Macmillan, 1991.

Cantor, Geoffrey, and Michael Hodge. "Introduction." In *Conceptions of Ether: Studies in the History of Ether Theories, 1740–1900*, ed. Geoffrey Cantor and Michael Hodge, pp. 1–60. Cambridge: Cambridge University Press, 1981.

Castle, Terry. "The Female Thermometer," *Representations* 17 (1987), 1–27.

Caswell, John. "A Paper about Magnetism, or Concerning the Changing and Fixing the Polarity of a Piece of Iron," *Philosophical Transactions* 18 (1694), 257–62.

Catcott, Alexander. *A Treatise on the Deluge*. London, 1768.

Cavallo, Tiberius. "Magnetical Experiments and Observations," *Philosophical Transactions* 76 (1786), 62–80.

———. *A Treatise on Magnetism, in Theory and Practice, with Original Experiments.* London, 1787.

———. "Magnetical Observations and Experiments," *Philosophical Transactions* 77 (1787), 6–25.

Cavendish, Henry. "An Account of the Meteorological Instruments Used at the Royal Society's House," *Philosophical Transactions* 66 (1776), 375–401.

Cawood, John. "Terrestrial Magnetism and the Development of International Collaboration in the Early Nineteenth Century," *Annals of Science* 34 (1977), 551–87.

———. "The Magnetic Crusade: Science and Politics in Early Victorian Britain," *Isis* 70 (1979), 493–518.

Chamberlen, ?Paul. *A Philosophical Essay upon Actions on Distant Subjects*. London, 1715.

Chambers, Ephraim. *Cyclopædia; or, An Universal Dictionary of Arts and Sciences,* 2 vols. London, 1728.

Chambers, John. *Biographical Illustrations of Worcestershire: Including Lives of Persons, Natives or Residents, Eminent Either for Piety or Talent: to Which Is Added, a List of Living Authors of the County*. Worcester, 1820.

Chapman, Sidney. "Edmond Halley as Physical Geographer, and the Story of His Charts," *Occasional Notes of the Royal Astronomical Society* 1 (1941), 122–34.

———. "Edmond Halley and Geomagnetism," *Nature* 152 (1943), 231–7.

Chapman, Sidney, and Julius Bartels. *Geomagnetism*, 2 vols. Oxford: Clarendon Press, 1940.

Cheyne, George. *The English Malady*. 1733. New York: Scholars' Facsimiles & Reprints, 1976.

Chree, Charles. "Cavendish's Magnetic Work." In *The Scientific Papers of the Honourable Henry Cavendish, FRS*. Vol. 2, *Chemical and Dynamical*, ed. Edward Thorpe, pp. 438–92. Cambridge: Cambridge University Press, 1921.

Christie, John R. R. "Joseph Black and John Robison." In *Joseph Black 1728–1799: A Commemorative Symposium*, ed. Alan Simpson, pp. 47–52. Edinburgh: Royal Scottish Museum, 1982.

———. "Laputa Revisited." In *Nature Transfigured: Science and Literature, 1700–*

1900, ed. John Christie and Sally Shuttleworth, pp. 45–60. Manchester: Manchester University Press, 1989.

———. "Aurora, Nemesis and Clio," *British Journal for the History of Science* 26 (1993), 391–405.

Churchman, John. *An Explanation of the Magnetic Atlas, or Variation Chart, Hereunto Annexed, Projected on a Plan Entirely New, by Which the Magnetic Variation on Any Part of the Globe May Be Precisely Determined, for Any Time, Past, Present, or Future: and the Variation and Latitude Being Accurately Known, the Longitude Is of Consequence Truly Determined*. Philadelphia, 1790.

———. *The Magnetic Atlas, or Variation Charts of the Whole Terraqueous Globe; Comprising a System of the Variation and Dip of the Needle, by Which, the Observations Being Truly Made, the Longitude May Be Ascertained*. London, 1794.

Clark, J.C.D. *English Society 1688–1832: Ideology, Social Structure and Political Practice during the Ancien Régime*. Cambridge: Cambridge University Press, 1985.

———. *Samuel Johnson: Literature, Religion and English Cultural Politics from the Restoration to Romanticism*. Cambridge: Cambridge University Press, 1994.

Clarke, C.N.G. "Taking Possession: The Cartouche as Cultural Text in Eighteenth-Century American Maps," *Word and Image* 4 (1988), 455–74.

Clarke, Henry. "Prospectus." In *The Seaman's Desiderata*. Bristol, 1800.

Clarke, James S. *The Progress of Maritime Discovery, from the Earliest Period to the Close of the Eighteenth Century, Forming an Extensive System of Hydrography*. London, 1803.

Cleland, John. *Memoirs of a Woman of Pleasure: Fanny Hill*. 1750. Ware, Herts: Wordsworth Classics, 1993.

Cockburn, Patrick. *An Enquiry into the Truth and Certainty of the Mosaic Deluge*. London, 1750.

Cohen, I. Bernard. *Benjamin Franklin's Experiments: A New Edition of Franklin's Experiments and Observations on Electricity*. Cambridge: Harvard University Press, 1941.

———. *Some Early Tools of American Science*. Cambridge: Harvard University Press, 1950.

Cohen, Murray. *Sensible Words: Linguistic Practice in England 1640–1785*. Baltimore and London: Johns Hopkins University Press, 1977.

Cohen, Patricia Cline. *A Calculating People: The Spread of Numeracy in Early America*. Chicago: University of Chicago Press, 1982.

Colden, Cadwallader. *The Letters and Papers of Cadwallader Colden*, 9 vols. New York: New-York Historical Society, 1918–37.

Coleridge, Samuel Taylor. *Lectures 1795 on Politics and Religion*, ed. Lewis Patton and Peter Mann. London: Routledge & Kegan Paul, 1971.

Colley, Linda. *In Defiance of Oligarchy: The Tory Party 1714–60*. Cambridge: Cambridge University Press, 1982.

———. *Britons: Forging the Nation 1707–1837*. New Haven: Yale University Press, 1992; London: Pimlico, 1994.

Collier, Katharine B. *Cosmogonies of Our Fathers: Some Theories of the Seventeenth and Eighteenth Centuries*. New York: Columbia University Press, 1934.

Collins, Harry M. *Changing Order: Replication and Induction in Scientific Practice*. London and Beverly Hills, Calif.: Sage, 1985.

Collinson, Patrick. *The Birthpangs of Protestant England*. London: Macmillan, 1988.

Cone, Carl. *Burke and the Nature of Politics*, 2 vols. Lexington: University of Kentucky Press, 1964.

Cook, Harold J. *The Decline of the Old Medical Regime in Stuart London*. Ithaca: Cornell University Press, 1986.

Cook, John. *Clavis Naturæ; or, The Mystery of Philosophy Unvail'd*. London, 1733.

Cook, James. "Variation of the Compass, as Observed on Board the Endeavour Bark, in a Voyage round the World," *Philosophical Transactions* 61 (1771), 422–32.

Cook, James, and James King. *A Voyage to the Pacific Ocean: Undertaken by the Command of His Majesty, for Making Discoveries in the Northern Hemisphere; Performed under the Direction of Captains Cook, Clerke, and Gore, in His Majesty's Ships the Resolution and Discovery in the Years 1776, 1777, 1778, 1779 and 1780*, 3 vols. London, 1784.

Cooke, William. *An Enquiry into the Patriarchal and Druidical Religion, Temples, &c*. London, 1755.

Cookson, Dr. "An Account of an Extraordinary Effect of Lightning in Communicating Magnetism," *Philosophical Transactions* 39 (1735), 74–5.

———. "A Further Account of the Extraordinary Effects of the Same Lightning at Wakefield," *Philosophical Transactions* 39 (1735), 75–8.

Cooper, Lane. "The Power of the Eye in Coleridge." In *Late Harvest*, by Lane Cooper, pp. 65–96. Ithaca: Cornell University Press, 1952.

Cooter, Roger. "Deploying 'Pseudoscience': Then and Now." In *Science, Pseudo-Science and Society*, ed. Marsha Hanen, Margaret Osler, and Robert Weyant, pp. 237–72. Waterloo, Ont.: Wilfred Laurier University Press, 1980.

———. "The History of Mesmerism in Britain: Poverty and Promise." In *Franz Anton Mesmer und die Geschichte des Mesmerismus*, ed. Heinz Schott, pp. 152–62. Stuttgart: Franz Steiner Verlag Wiesbaden, 1985.

Copley, Stephen. *Literature and the Social Order in Eighteenth-Century England*. London: Croom Helm, 1984.

Corner, Betsy C., and Christopher C. Booth. *Chain of Friendship: Selected Letters of Dr. John Fothergill of London, 1735–1780*. Cambridge: Harvard University Press, 1971.

Cornwall, Captain. "Observations of the Variation on Board the Royal African Pacquet, in 1721," *Philosophical Transactions* 32 (1722), 55–6.

Corsi, Pietro. *Science and Religion: Baden-Powell and the Anglican Debate*. Cambridge: Cambridge University Press, 1988.

Cortesão, A., and A. da Mota. *Portugaliæ Monumenta*, 5 vols. Lisbon, 1960.

Cotter, Charles H. "Matthew Flinders and Ship Magnetism," *Journal of Navigation* 29 (1976), 123–34.

———. "A Brief Historical Survey of British Navigation Manuals," *Journal of Navigation* 36 (1983), 237–49.

Coulomb, Charles A. "Septième mémoire sur l'électricité et le magnétisme," *Mémoires de mathématique et de physique de l'Académie royale des sciences* 7 (1789), 455–505.

Crabtree, Adam. *From Mesmer to Freud: Magnetic Sleep and the Roots of Psychological Healing*. New Haven and London: Yale University Press, 1993.

Cragg, Gerald. *Reason and Authority in the Eighteenth Century*. Cambridge: Cambridge University Press, 1964.

Crane, V. W. "The Club of Honest Whigs: Friends of Science and Liberty," *William and Mary Quarterly* 23 (1966), 210–33.

Crawforth, M. A. "Evidence from Trade Cards for the Scientific Instrument Industry," *Annals of Science* 42 (1985), 453–554.

Croker, Temple H. *Experimental Magnetism; or, The Truth of Mr Mason's Discoveries in That Branch of Natural Philosophy, That There Can Be No Such Thing in Nature, as an Internal Central Loadstone, Proved and Ascertained.* London, 1761.

Croker, Temple H., T. Williams, and S. Clark. *The Complete Dictionary of Arts and Sciences*, 3 vols. London, 1773.

Crom, Theodore. *Trade Catalogues—1542 to 1842*. Florida, 1989.

Crosby, Thomas. *The Mariner's Guide*. London, 1762.

Crosland, Maurice. "Explicit Qualifications as a Criterion for Membership of the Royal Society: A Historical Review," *Notes and Records of the Royal Society of London* 37 (1983), 167–87.

———. "The Image of Science as a Threat: Burke versus Priestley and the 'Philosophic Revolution,'" *British Journal for the History of Science* 20 (1987), 277–307.

Crosland, Maurice, and Crosbie Smith. "The Transmission of Physics from France to Britain: 1800–1840," *Historical Studies in the Physical Sciences* 9 (1978), 1–61.

Cross, Stephen J. "John Hunter, the Animal Oeconomy, and Late Eighteenth-Century Physiological Discourse," *Studies in the History of Biology* 5 (1981), 1–110.

Crowe, Michael J. *The Extraterrestrial Life Debate 1750–1900*. Cambridge: Cambridge University Press, 1986.

Cunningham, Andrew. "Getting the Game Right: Some Plain Words on the Identity and Invention of Science," *Studies in History and Philosophy of Science* 19 (1988), 365–89.

———. "How the *Principia* Got Its Name; or, Taking Natural Philosophy Seriously," *History of Science* 29 (1991), 377–92.

Curry, Patrick. *Prophecy and Power: Astrology in Early Modern England*. Cambridge: Polity Press, 1989.

Dalrymple, Alexander. "Journal of a Voyage to the East Indies, in the Ship Grenville, Captain Burnett Abercrombie, in the Year 1775," *Philosophical Transactions* 68 (1778), 389–418.

Dalton, John. *Meteorological Observations and Essays*. London, 1793.

Daly, Peter M. *Literature in the Light of the Emblem*. Toronto: University of Toronto Press, 1979.

Danchin, Pierre. "Erasmus Darwin's Scientific and Poetic Purpose in *The Botanic Garden*." In *Science and Imagination in XVIIIth-Century British Culture*, ed. Sergio Rossi, pp. 133–50. Milan: Edizioni Unicopoli, 1987.

Danilewicz, N. L. "'The King of the New Israel': Thaddeus Grabianka (1740–1807)," *Oxford Slavonic Papers* 1 (1968), 49–73.

Darnton, Robert. *Mesmerism and the End of the Enlightenment in France*. Cambridge: Harvard University Press, 1968.

———. *The Great Cat Massacre and Other Episodes in French Cultural History*. Harmondsworth: Penguin, 1985.

Darwin, Erasmus. *Zoonomia; or, The Laws of Organic Life*, 2 vols. London, 1794–6.

———. *The Poetical Works of Erasmus Darwin, Containing The Botanic Garden, in Two Parts; and The Temple of Nature*, 3 vols. London, 1806.

Daston, Lorraine J. "The Domestication of Risk: Mathematical Probability and Insurance 1650–1830," In *The Probabilistic Revolution*, 2 vols., ed. Lorenz Krüger, Lorraine J. Daston, and Michael Heidelberger, vol. 1, pp. 237–60. Cambridge: MIT Press, 1987.

Daston, Lorraine J. "The Factual Sensibility," *Isis* 79 (1988), 452–67.
———. *Classical Probability in the Enlightenment*. Princeton: Princeton University Press, 1988.
———. "Nationalism and Scientific Neutrality under Napoleon." In *Solomon's House Revisited: The Organisation and Institutionalisation of Science*, ed. Tore Frängsmyr, pp. 95–119. Canton, Ohio: Science History Publications, 1990.
———. "The Ideal and Reality of the Republic of Letters in the Enlightenment," *Science in Context* 4 (1991), 367–86.
———. "The Naturalised Female Intellect," *Science in Context* 5 (1992), 209–35.
Davidoff, Leonore, and Catherine Hall. *Family Fortunes: Men and Women of the English Middle Class, 1780–1850*. London: Hutchinson, 1987.
Davids, Karel A. "Finding Longitude at Sea by Magnetic Declination on Dutch East Indiamen, 1596–1795." *American Neptune* 50 (1990), 281–90.
Davis, Ralph. *The Rise of the English Shipping Industry in the Seventeenth and Eighteenth Centuries*. London: Macmillan, 1962.
Dawson, Paul M. S. "'A Sort of Natural Magic': Shelley and Animal Magnetism," *Keats-Shelley Review* 1 (1986), 15–34.
Day, Thomas. *The History of Sandford and Merton, a Work Intended for the Use of Children*, 3 vols. London, 1783–9.
de Almeida, Hermione. *Romantic Medicine and John Keats*. Oxford: Oxford University Press, 1991.
de la Touche, H. Boesnier. "Wonderful Property of Magnets to Cure the Tooth-ach," *Annual Register* 8 (1765), 112.
de Mainauduc, John B. *The Lectures of J. B. de Mainauduc MD*. London, 1798.
de Man, Paul. "The Epistemology of Metaphor," *Critical Inquiry* 5 (1978), 13–30.
de Morgan, Augustus. "The Canton Papers," *Athenaeum* (1849), 5–7, 162–4, 375.
———. "Dr Gowan Knight," *Notes and Queries* 10 (1860), 281–2.
———. *A Budget of Paradoxes*. London: Longmans, Green, 1872.
de Pater, C. "Petrus van Musschenbroek (1692–1761), a Dutch Newtonian," *Janus* 64 (1977), 77–87.
———. *Petrus van Musschenbroek (1692–1761), een Newtoniaans Natuuronderzoeker*. Utrecht: Museum Boerhaave, 1979.
de Vries, Jan. "Between Purchasing Power and the World of Goods: Understanding the Household Economy in Early Modern Europe." In *Consumption and the World of Goods*, ed. John Brewer and Roy Porter, pp. 85–132. London and New York: Routledge, 1993.
Deacon, Margaret. "Founders of Marine Science in Britain: The Work of the Early Fellows of the Royal Society," *Notes and Records of the Royal Society of London* 20 (1965), 28–50.
Dean, John. "Controversy over Classification: A Case Study from the History of Botany." In *Natural Order: Historical Studies of Scientific Culture*, ed. Barry Barnes and Steven Shapin, pp. 211–28. Beverly Hills, Calif.: Sage, 1979.
Debus, Allen G. "Scientific Truth and Occult Tradition: The Medical World of Ebenezer Sibly (1751–1799)," *Medical History* 26 (1982), 259–78.
Defoe, Daniel. *A System of Magick; or, A History of the Black Art*. London, 1727.
———. *A General History of Discoveries and Improvements, in Useful Arts, Particularly in the Great Branches of Commerce, Navigation, and Plantation, in All Parts of the Known World*, 4 pts. in 1. London, 1727–8.

Derham, William. "An Account of Some Magnetical Experiments and Observations," *Philosophical Transactions* 24 (1705), 2136–8.

———. "Farther Observations and Remarks on the Same Subject," *Philosophical Transactions* 24 (1705), 2138–44.

———. *Physico-theology; or, A Demonstration of the Being and Attributes of God, from the Works of Creation*. London, 1723.

Desaguliers, John T. *Physico-mechanical Lectures*. London, 1717.

———. *A Course of Experimental Philosophy*. London, 1734.

———. "An Account of Some Magnetical Experiments Made before the Royal Society," *Philosophical Transactions* 40 (1738), 385–7.

———. "Some Thoughts and Conjectures concerning the Cause of Electricity," *Philosophical Transactions* 41 (1739), 175–85.

Descartes, René. *Principles of Philosophy*. Trans. V. R. Miller and R. P. Miller. Dordrecht: Reidel, 1983.

Dickinson, H. T. *Liberty and Property: Politics and Ideology in Eighteenth-Century Britain*. London: Methuen, 1977.

Dobbs, Betty Jo Teeter. "Studies in the Natural Philosophy of Sir Kenelm Digby. Part I," *Ambix* 18 (1971), 1–25.

———. *The Foundations of Newton's Alchemy or "The Hunting of the Greene Lyon."* Cambridge: Cambridge University Press, 1975.

Doherty, Francis. "The Anodyne Necklace: A Quack Remedy and Its Promotion," *Medical History* 34 (1990), 268–93.

Douglass, Robert. "The Variation of the Compass; Containing 1719 Observations to, in, and from, the East Indies, Guinea, West Indies and Mediterranean, with the Latitudes and Longitudes at the Time of Observation," *Philosophical Transactions* 66 (1776), 18–72.

Dryden, John. *The Poems of John Dryden*, ed. James Kinsley, 4 vols. Oxford: Clarendon Press, 1958.

Dubois, Dorothea. *The Magnet: A Musical Entertainment*. London, 1771.

Duchesnau, François. "Vitalism in Late Eighteenth-Century Physiology: The Cases of Barthez, Blumenbach and John Hunter." In *Medical Fringe and Medical Orthodoxy, 1750–1850*, ed. William Bynum and Roy Porter, pp. 259–95. London: Croom Helm, 1987.

Dufay, Charles F. "Observations sur quelques expériences de l'aimant," *Mémoires de mathématique et de physique de l'Académie royale des sciences*, 1728, pp. 355–69.

———. "Suite des observations sur l'aimant," *Mémoires de mathématique et de physique de l'Académie royale des sciences*, 1730, pp. 142–57.

———. "Troisième mémoire sur l'aimant," *Mémoires de mathématique et de physique de l'Académie royale des sciences*, 1731, pp. 417–32.

Duhamel, Henri. "Façon singulière d'aimanter un barreau d'acier," *Mémoires de mathématique et de physique de l'Académie royale des sciences*, 1745, pp. 181–93.

Dumaniant, ?. *Le médecin malgré tout le monde: comédie en trois actes, en prose*. Paris, 1786.

Dunn, Samuel. *The Navigator's Guide, to the Oriental or Indian Seas; or, The Description and Use of a Variation Chart of the Magnetic Needle*. London, 1776.

———. *Nautical Propositions and Institutes*. London, 1781.

Eagleton, Terry. *The Rape of Clarissa: Writing, Sexuality and Class Struggle in Samuel Richardson*. Oxford: Basil Blackwell, 1982.

Eames, John. "An Extract from the Journal Books of the Royal Society, Concerning Magnets Having More Poles than Two; with Some Observations by Dr Desaguliers on the Same Subject," *Philosophical Transactions* 40 (1738), 383–4.

Eamon, William. *Science and the Secrets of Nature: Books of Secrets in Medieval and Early Modern Culture*. Princeton: Princeton University Press, 1994.

Edgeworth, Maria. *Letters for Literary Ladies*. 1795. Ed. Claire Connolly. London: Dent, 1993.

Edgeworth, Maria, and Richard Lovell Edgeworth. *A Practical Education*. London, 1798.

Edgeworth, Richard Lovell. *Memoirs of Richard Lovell Edgeworth Esq Begun by Himself, and Concluded by His Daughter, Maria Edgeworth*. London, 1844.

Edwards, John. *A Demonstration of the Existence and Providence of God, from the Contemplation of the Visible Structure of the Greater and the Lesser World*. London, 1696.

Eeles, Henry. *Philosophical Essays: In Several Letters to the Royal Society*. London, 1771.

Emerson, William. *Navigation; or, The Art of Sailing upon the Sea*. London, 1764.

Enfield, William. *Institutes of Natural Philosophy, Theoretical and Experimental*. London, 1785.

Erikson, Robert. "'The Books of Generation': Some Observations on the Style of the British Midwife Books, 1671–1764." In *Sexuality in Eighteenth-Century Britain*, ed. Paul-Gabriel Boucé, pp. 74–94. Manchester: Manchester University Press, 1982.

Essick, Robert N. *William Blake and the Language of Adam*. Oxford: Clarendon Press, 1989.

Euler, Leonhard. *Letters of Euler to a German Princess, on Different Subjects in Physics and Philosophy*, 2 vols. London, 1795.

Evans, Michael E. "Edmond Halley, Geophysicist," *Physics Today* 41(2) (1988), 41–5.

Falconer, William. *The Shipwreck. A Poem. In Three Cantos. By a Sailor*. London, 1762.

———. *An Universal Dictionary of the Marine*. London, 1769.

———. *The Shipwreck and Other Poems . . . with a Life of the Author*. London, 1838.

Fanning, Anthony E. *Steady as She Goes: A History of the Compass Department of the Admiralty*. London: HMSO, 1986.

Fara, Patricia. "An Attractive Therapy: Animal Magnetism in Eighteenth-Century England," *History of Science* 33 (1995), 127–77.

Faraday, Michael. "Experimental Researches in Electricity," *Philosophical Transactions* 122 (1832), 125–94.

Farrell, Maureen. *The Life and Work of William Whiston*. New York: Arno Press, 1981.

Feather, John P. "John Nourse and His Authors," *Studies in Bibliography* 34 (1981), 205–26.

Feldman, Theodore S. "Late Enlightenment Meteorology." In *The Quantifying Spirit in the Eighteenth Century*, ed. Tore Frängsmyr, John Heilbron, and Robin Rider, pp. 143–77. Berkeley and Los Angeles: University of California Press, 1990.

———. "The Ancient Climate in the Eighteenth and Early Nineteenth Century." In *Science and Nature: Essays in the History of the Environmental Sciences*, ed. Michael Shortland, pp. 23–40. British Society for the History of Science, 1993.

Fergusson, James. *Observations on the Present State of the Art of Navigation, with a Short Account of the Nature and Regulations of a Society Now Forming for Its Effectual Improvement*. London, 1787.

Fergusson, James. *Balloon Tytler*. London: Faber & Faber, 1972.

Fernie, William T. *Precious Stones: For Curative Wear; and Other Remedial Uses: Likewise the Nobler Metals*. Bristol: John Wright, 1907.

Feyerabend, Paul. "Classical Empiricism." In *The Methodological Heritage of Newton*, ed. Robert Butts and John Davis, pp. 150–70. Oxford: Basil Blackwell, 1970.

Fiennes, Celia. *The Journeys of Celia Fiennes*, ed. Christopher Morris. London: Cresset Press, 1949.

Fine, B., and E. Leopold. "Consumerism and the Industrial Revolution," *Social History* 15 (1990), 151–79.

Fisher, George. "On the Errors in Longitude . . . Arising from the Action of the Iron in the Ships upon the Chronometers," *Philosophical Transactions* 110 (1820), 196–208.

Fletcher, Andrew. *A Discourse of Government with Relation to Militia's*. Edinburgh, 1698.

Flinders, Matthew. "Concerning the Differences in the Magnetic Needle . . . Arising from an Alteration in the Direction of the Ship's Head," *Philosophical Transactions* 95 (1805), 186–97.

———. *A Voyage to Terra Australis; Undertaken for the Purpose of Completing the Discovery of That Vast Country, and Prosecuted in the Years 1801, 1802, and 1803, in His Majesty's Ship the Investigator, and Subsequently in the Armed Vessel Porpoise and Cumberland Schooner*, 2 vols. London, 1814.

Forbes, Duncan. *A Letter to a Bishop, concerning Some Important Discoveries in Philosophy and Theology*. London, 1732.

Forbes, Eric G. *Greenwich Observatory*. Vol. 1, *Origins and Early History (1675–1835)*. London: Taylor & Francis, 1975.

Force, James E. *William Whiston: Honest Newtonian*. Cambridge: Cambridge University Press, 1985.

Fothergill, John. "An Account of the Magnetical Machine Contrived by the Late Dr Gowan Knight, FRS," *Philosophical Transactions* 66 (1776), 591–9.

Foucault, Michel. *The Birth of the Clinic*, trans. A. M. Sheridan. London: Tavistock Publications, 1973.

Frängsmyr, Tore, John Heilbron, and Robin Rider (eds.). *The Quantifying Spirit in the Eighteenth Century*. Berkeley and Los Angeles: University of California Press, 1990.

Frank, Robert. "Science, Medicine and the Universities of Early Modern England: Background and Sources," *History of Science* 11 (1973), 194–216, 239–69.

Frei, E. H. "Medical Applications of Magnetism," *Bulletin of the Atomic Scientists* 28(8) (1972), 34–40.

Frei, Hans W. *The Eclipse of Biblical Narrative*. New Haven: Yale University Press, 1974.

Freke, John. *A Treatise on the Nature and Property of Fire*. London, 1752.

Friedman, Robert M. *Appropriating the Weather: Vilhelm Bjerknes and the Construction of a Modern Meteorology*. Ithaca: Cornell University Press, 1989.

Froom, Leroy E. *The Prophetic Faith of Our Fathers*, 4 vols. Washington: Review & Herald, 1950–4.

Frost, Alan. "Science for Political Purposes: European Explorations of the Pacific Ocean, 1764–1806." In *Nature in Its Greatest Extent*, ed. Roy MacLeod and Philip Rehbock, pp. 27–44. Honolulu: University of Hawaii Press, 1988.

Frost, J. William. *The Records and Recollections of James Jenkins*. New York: Edwin Mellen Press, 1984.

Frost, Thomas. *The Lives of the Conjurors*. London: Tinslay Brothers, 1876.

Fujimura, Joan H. "Crafting Science: Standardised Packages, Boundary Objects, and 'Translation,'" In *Science as Practice and Culture*, ed. Andrew Pickering, pp. 168–211. Chicago: University of Chicago Press, 1992.

Fuller, Samuel. *A Mathematical Miscellany: In Four Parts*. Dublin, 1730.

Funkhouser, H. G. "Historical Development of the Graphical Representation of Statistical Data," *Osiris* 3 (1938), 269–404.

Fuss, Nicholas. *Observations et expériences sur les aimans artificiels, principalement sur la meilleure manière de les faire*. St. Petersburg: L'Académie impériale des sciences, 1778.

Fyfe, Gordon J. "Art Exhibitions and Power during the Nineteenth Century." In *Power, Action and Belief: A New Sociology of Knowledge?*, ed. John Law, pp. 20–45. London: Routledge & Kegan Paul, 1986.

Gale, John. *Gale's Cabinet of Knowledge; or, Miscellaneous Recreations*. London, 1797.

Galison, Peter. "History, Philosophy and the Central Metaphor," *Science in Context* 2 (1988), 197–212.

———. "History of Physics: The Material Culture of the Laboratory." Paper presented at MIT Program in Science, Technology, and Society workshop "New Approaches to the History and Social Study of Science and Technology," St. Petersburg, June 1994.

Galison, Peter, and Alexi Assmus, "Artificial Clouds, Real Particles." In *The Uses of Experiment: Studies in the Natural Sciences*, ed. David Gooding, Trevor Pinch, and Simon Schaffer, pp. 225–74. Cambridge: Cambridge University Press, 1989.

Gardiner, Richard. *Memoirs of the Life and Writings (Prose and Verse) of R-CH--D G-RD-N-R, Esq. Alias Dick Merry-Fellow, of Serious and Facetious Memory*. London, 1782.

Garrett, Clarke. "Swedenborg and the Mystical Enlightenment in Late Eighteenth-Century England," *Journal of the History of Ideas* 45 (1984), 67–81.

Gascoigne, John. "Mathematics and Meritocracy: The Emergence of the Cambridge Mathematical Tripos," *Social Studies of Science* 14 (1984), 547–84.

———. *Cambridge in the Age of the Enlightenment: Science, Religion and Politics from the Restoration to the French Revolution*. Cambridge: Cambridge University Press, 1989.

———. *Joseph Banks and the English Enlightenment: Useful Knowledge and Polite Culture*. Cambridge: Cambridge University Press, 1994.

Gauld, Alan. *A History of Hypnotism*. Cambridge: Cambridge University Press, 1992.

Genuth, Sara S. "Devils' Hells and Astronomers' Heavens: Religion, Method, and Popular Culture in Speculations about Life on Comets." In *The Invention of Physical Science*, ed. Mary Jo Nye, John Richards, and Roger Stuewer, pp. 3–26. Dordrecht: Kluwer Academic, 1992.

Gerard, Alexander. *An Essay on Taste*. Edinburgh, 1764.

Gibbon, Edward. *The Autobiography and Correspondence of Edward Gibbon, the Historian*. London: Alex Murray & Son, 1869.

Gilbert, William. *On the Magnet*, trans. P. Fleury Mottelay. New York: Basic Books, 1958. Originally published as *De Magnete*, 1660.

Gillespie, Richard. "Ballooning in France and Britain, 1783–1786," *Isis* 75 (1984), 249–68.

Gilman, Sander L., Helen King, Roy Porter, George S. Rousseau, and Elaine Showalter. *Hysteria beyond Freud*. Berkeley and Los Angeles: University of California Press, 1993.

Gilpin, George. "Observations on the Variation, and on the Dip of the Magnetic Needle, Made at the Apartments of the Royal Society between the Years 1786 and 1805, Inclusive," *Philosophical Transactions* 96 (1806), 385–419.

Goad, John. *Astro-meteorologica; or, Aphorisms and Discourses of the Bodies Celestial, Their Natures and Influences*. London, 1686.

Godwin, Joscelyn. *Athanasius Kircher: A Renaissance Man and the Quest for Lost Knowledge*. London: Thames & Hudson, 1979.

Godwin, William. *Report of Dr Benjamin Franklin, and Other Commissioners, Charged by the King of France, with the Examination of the Animal Magnetism as Now Practised at Paris*. London, 1785.

———. *Caleb Williams*. 1794. Ed. David McCracken. Oxford: Oxford University Press, 1970; repr. 1982.

Goldgar, Bertrand A. "Fielding, the Flood Makers, and Natural Philosophy: *Covent-Garden Journal* No 70," *Modern Philology* 80 (1982), 136–44.

Goldsmith, Maurice M. *Private Vices, Public Benefits: Bernard Mandeville's Social and Political Thought*. Cambridge: Cambridge University Press, 1985.

Golinski, Jan. "Utility and Audience in Eighteenth-Century Chemistry: Case Studies of William Cullen and Joseph Priestley," *British Journal for the History of Science* 21 (1988), 1–32.

———. "The Theory of Practice and the Practice of Theory: Sociological Approaches in the History of Science," *Isis* 81 (1990), 492–505.

———. "Language, Discourse and Science." In *Companion to the History of Modern Science*, ed. R. C. Olby, G. N. Cantor, J.R.R. Christie, and M.J.S. Hodge, pp. 110–23. London: Routledge, 1990.

———. *Science as Public Culture: Chemistry and Enlightenment in Britain, 1760–1820*. Cambridge: Cambridge University Press, 1992.

Gooding, David. "'Magnetic Curves' and the Magnetic Field: Experimentation and Representation in the History of a Theory." In *The Uses of Experiment: Studies in the Social Sciences*, ed. David Gooding, Trevor Pinch, and Simon Schaffer, pp. 182–223. Cambridge: Cambridge University Press, 1989.

———. *Experiment and the Making of Meaning*. Dordrecht: Kluwer Academic, 1990.

Gouk, Penelope. *The Ivory Sundials of Nuremberg*. Cambridge: Whipple Museum of the History of Science, 1988.

Gower, Richard H. *A Treatise on the Theory and Practice of Seamanship, etc*. London, 1793.

Graham, George. "An Account of the Observations Made of the Horizontal Needle at *London*, in the Latter Part of the Year 1722, and Beginning of 1723," *Philosophical Transactions* 33 (1724), 96–108.

———. "Observations of the Dipping Needle, Made at London, in the Beginning of the Year 1723," *Philosophical Transactions* 33 (1724), 332–9.

———. "Some Observations, Made during the last Three Years, of the Quantity of the Variation of the Magnetic Horizontal Needle to the Westward," *Philosophical Transactions* 45 (1748), 279–80.

Graham, James. *A Sketch; or, Short Description of Dr Graham's Medical Apparatus, &c. Erected about the Beginning of the Year 1780, in His House, on the Royal Terrace, Adelphi, London*. London, 1780.

Grant, G., and J. Klinkert. *The Ship's Compass*. London: Routledge & Kegan Paul, 1970.

Green, Charles, and James Cook. "Observations Made by Appointment of the Royal Society, at King George's Island in the South Sea," *Philosophical Transactions* 61 (1771), 397–421.

Green, Robert. *A Demonstration of the Truth and Divinity of the Christian Religion, as It Is Propos'd to Us in the Scriptures of the New Testament*. Cambridge, 1711.

———. *The Principles of Natural Philosophy, in Which Is Shewn the Insufficiency of the Present Systems, to Give Any Just Account of That Science: and the Necessity There Is of Some New Principles, in Order to Furnish Us with a True and Real Knowledge of Nature*. Cambridge, 1712.

[Greene, Robert] *The Principles of the Philosophy of the Expansive and Contractive Forces*. Cambridge: Cambridge University Press, 1727.

Greenberg, Mark. "Eighteenth-Century Poetry Represents Moments of Scientific Discovery." In *Literature and Science: Theory and Practice*, ed. Stuart Peterfreund, pp. 115–37. Boston: Northeastern University Press, 1990.

Greene, Donald. "Smart, Berkeley, the Scientists and the Poets," *Journal of the History of Ideas* 14 (1953), 327–52.

Grew, Nehemiah. *Musæum Regalis Societatis*. London, 1681.

———. *The Anatomy of Plants*. London, 1682.

———. *Cosmologia Sacra; or, A Discourse of the Universe as It Is the Creature and Kingdom of God*. London, 1701.

Griffin, Dustin. *Regaining Paradise: Milton and the Eighteenth Century*. Cambridge: Cambridge University Press, 1986.

Guerrini, Antonia. "The Tory Newtonians: Gregory, Pitcairne and Their Circle," *Journal of British Studies* 25 (1986), 288–311.

Guest, Harriet. *A Form of Sound Words*. Oxford: Clarendon Press, 1989.

Gunther, Albert E. "The Royal Society and the Foundation of the British Museum," *Notes and Records of the Royal Society of London* 33 (1978), 207–16.

———. *The Founders of Science at the British Museum, 1753–1900: A Contribution to the Centenary of the Opening of the British Museum (Natural History) on 18th April 1981*. Suffolk: Halesworth Press, 1980.

———. *An Introduction to the Life of the Rev. Thomas Birch DD, FRS 1705–1766*. Suffolk: Halesworth Press, 1984.

———. "Matthew Maty MD, FRS (1718–76) and Science at the Foundation of the British Museum, 1753–80," *Bulletin of the British Museum (Natural History)* 15 (1987), 1–58.

Gunther, Robert T. *Early Science in Oxford*, 19 vols. Oxford, 1920–45.

Habermas, Jürgen. *The Structural Transformation of the Public Sphere: An Inquiry into a Category of Bourgeois Society*. 1962. Trans. Thomas Burger. 1989. Cambridge: Polity Press, 1992.

Hackmann, Willem D. "The Relationship between Concept and Instrument Design in Eighteenth-Century Experimental Science," *Annals of Science* 36 (1979), 205–24.

———. "Scientific Instruments: Models of Brass and Aids to Discovery." In *The Uses of Experiment: Studies in the Social Sciences*, ed. David Gooding, Trevor Pinch, and Simon Schaffer, pp. 31–65. Cambridge: Cambridge University Press, 1989.

———. “Jan van der Staet (Stradanus) and the Origins of the Mariner’s Compass.” In *Learning, Language and Invention: Essays Presented to Francis Maddison*, ed. Willem D. Hackmann and Anthony J. Turner, pp. 148–79. Aldershot: Variorum, 1994.

Hale, Matthew. *Magnetismus Magnus; or, Metaphysical and Divine Contemplations on the Magnet, or Loadstone*. London, 1695.

Hall, A. Rupert, and Marie Boas. *Unpublished Scientific Papers of Isaac Newton*. Cambridge: Cambridge University Press, 1962.

Halley, Edmond. “A Theory of the Variation of the Magnetical Compass,” *Philosophical Transactions* 13 (1683), 208–21.

———. “A Discourse concerning Gravity, and Its Properties, Wherein the Descent of Heavy Bodies, and the Motion of Projects Is Briefly, but Fully Handled: Together with the Solution of a Problem of Great Use in Gunnery,” *Philosophical Transactions* 16 (1686), 3–21.

———. “An Account of the Cause of the Change of the Variation of the Magnetical Needle; with an Hypothesis of the Structure of the Internal Parts of the Earth,” *Philosophical Transactions* 16 (1692), 563–78.

———. “An Estimate of the Degrees of Mortality of Mankind Drawn from Curious Tables of the Births and Funerals at the City of Breslaw; with an Attempt to Ascertain the Price of Annuities upon Lives,” *Philosophical Transactions* 17 (1693), 596–610.

———. “Some Remarks on the Variations of the Magnetical Compass Published in the Memoirs of the Royal Academy of Sciences, with Regard to the General Chart of Those Variations Made by E. Halley; as Also concerning the True Longitude of the Magellan Streights,” *Philosophical Transactions* 29 (1714), 165–8.

———. “An Account of the Late Surprizing Appearance of the Lights Seen in the Air, on the Sixth of March Last; with an Attempt to Explain the Principal Phænomena Thereof,” *Philosophical Transactions* 29 (1716), 406–28.

———. “The Variation of the Magnetical Compass . . . with Some Remarks Thereon,” *Philosophical Transactions* 31 (1721), 173–6.

———. “Observations of Latitude and Variation, Taken on Board the Hartford, in Her Passage from Java Head to St Hellena,” *Philosophical Transactions* 37 (1732), 331–6.

Hamilton, James. *Calculations and Tables Relating to the Attractive Virtue of Loadstones*. 1729.

Hankins, Thomas L., and Robert J. Silverman. *Instruments and the Imagination*. Princeton: Princeton University Press, 1995.

Hans, N. A. *New Trends in Education in the Eighteenth Century*. London: Routledge & Kegan Paul, 1951.

Hardin, C. L. “The Scientific Work of the Reverend John Michell,” *Annals of Science* 22 (1966), 27–47.

Hargrave, Catherine P. *A History of Playing Cards and a Bibliography of Cards and Gaming*. London: Allen & Unwin, 1930.

Harley, J. B. “Maps, Knowledge and Power.” In *The Iconography of Landscape: Essays on the Symbolic Representation, Design and Use of Past Environments*, ed. Denis Cosgrove and Stephen Daniels, pp. 277–312. Cambridge: Cambridge University Press, 1988.

———. “Deconstructing the Map,” *Cartographia* 26(2) (1989), 1–20.

Harrington, Robert. *A New System on Fire and Planetary Life; . . . Also, an Elucidation of the Phænomena of Electricity and Magnetism*. London, 1796.

Harris, James. "Philological Enquiries in Three Parts." In *The Works of James Harris, Esq*, ed. Earl of Malmesbury, vol. 2, pp. 271–585. London, 1801.

Harris, John. *Astronomical Dialogues between a Gentleman and a Lady: Wherein the Doctrine of the Sphere, Uses of the Globes, and the Elements of Astronomy and Geography Are Explained, in a Pleasant, Easy and Familiar Way*. London, 1719.

———. *Lexicon Technicum; or, An Universal English Dictionary of Arts and Sciences: Explaining Not Only the Terms of Art, but the Arts Themselves*, 2 vols. London, 1736.

Harris, Joseph. "Astronomical Observations Made at Vera Cruz," *Philosophical Transactions* 35 (1728), 388–9.

———. *A Treatise of Navigation*. London, 1730.

———. "An Account of Some Magnetical Observations Made in the Months of May, June and July, 1732, in the Atlantick or Western Ocean," *Philosophical Transactions* 38 (1733), 75–9.

Harris, William S. *Rudimentary Magnetism: Being a Concise Exposition of the General Principles of Magnetical Science*, 3 pts. in 1. London: John Weale, 1850–2.

Harrison, Edward. *Idea Longitudinis: Being, a Brief Definition of the Best Known Axioms for Finding the Longitude*. London, 1696.

Harrison, J.F.C. *The Second Coming: Popular Millenarianism, 1780–1850*. London: Routledge & Kegan Paul, 1979.

Hartley, Beryl. "The Artist as Naturalist and Experimenter: Landscape Painting and Science in Britain." Ph.D. thesis, University of London, 1995.

Hartley, David. *Observations on Man, His Frame, His Duty, and His Expectations*, 2 vols. 1749. Gainsville, Fla.: Scholars' Facsimiles & Reprints, 1966.

Harvey, William. *The Works of William Harvey, MD*, trans. R. Willis. London, 1847 (for the Sydenham Society).

Haselden, Thomas. *The Seaman's Daily Assistant, Being a Short, Easy and Plain Method of Keeping a Journal at Sea*. London, 1761.

Hauksbee, Francis. "An Account of Experiments concerning the Proportion of the Power of the Load-stone at Different Distances," *Philosophical Transactions* 27 (1712), 506–11.

Hawes, Joan L. "Newton and the 'Electrical Attraction Unexcited,'" *Annals of Science* 24 (1968), 121–30.

Hawkesworth, John. *An Account of the Voyages Undertaken by the Order of His Present Majesty for Making Discoveries in the Southern Hemisphere*. London, 1773.

Heilbron, John L. *Electricity in the Seventeenth and Eighteenth Centuries*. Berkeley and Los Angeles: University of California Press, 1979.

———. *Physics at the Royal Society during Newton's Presidency*. Los Angeles: William Andrews Clark Memorial Library, University of California, 1983.

———. "A Mathematicians' Mutiny, with Morals." In *World Changes: Thomas Kuhn and the Nature of Science*, ed. P. Horwich, pp. 81–129. Cambridge: MIT Press, 1993.

Heimann, Peter M. "Newtonian Natural Philosophy and the Scientific Revolution," *History of Science* 11 (1973), 1–7.

———. "Ether and Imponderables." In *Conceptions of Ether: Studies in the History of Ether Theories, 1740–1900*, ed. Geoffrey Cantor and Michael Hodge, pp. 61–84. Cambridge: Cambridge University Press, 1981.

Heimann, Peter M., and James E. McGuire. "Newtonian Forces and Lockean Powers:

Concepts of Matter in Eighteenth-Century Thought," *Historical Studies in the Physical Sciences* 3 (1971), 233–306.

Heisler, Ron. "Behind 'The Magus': Francis Barrett, Magical Balloonist," *Pentacle* 1(4) (1985), 53–7.

Helsham, Richard. *A Course of Lectures in Natural Philosophy*. London, 1739.

Hemlow, J. *Fanny Burney*. Oxford: Clarendon Press, 1986.

Henry, John. "Occult Qualities and the Experimental Philosophy: Active Principles in Pre-Newtonian Matter Theory," *History of Science* 24 (1986), 335–81.

Hervey, Elizabeth. *The Mourtray Family*, 4 vols. London, 1800.

Hervey, James. *Theron and Aspasio; or, A Series of Dialogues and Letters, upon the Most Important and Interesting Subjects*, 2 vols. London, 1813.

Herwig, Michael. *The Art of Curing Sympathetically, or Magnetically, Proved to Be Most True Both by Its Theory and Practice* . . . London, 1700.

Hewson, J. B. *A History of the Practice of Navigation*. Glasgow: Brown, Son & Ferguson, 1983.

Higgins, Bryan. *A Philosophical Essay concerning Light*. London, 1776.

Higgins, Kathleen. "The Classification of Sundials," *Annals of Science* 9 (1953), 342–58.

Hillary, William. *The Nature, Properties and Laws of Motion of Fire Discovered and Demonstrated by Observations and Experiments*. London, 1760.

Hill, John. *Fossils Arranged According to Their Obvious Characters; with Their History and Description; under the Articles of Form, Hardness, Weight, Surface, Colour, and Qualities; the Place of Their Production, Their Uses, and Distinctive English, and Classical Latin Names*. London, 1771.

Hindle, Brooke. *The Pursuit of Science in Revolutionary America 1735–1789*. Chapel Hill: University of North Carolina Press, 1956.

Hindmarsh, Robert. *Rise and Progress of the New Jerusalem Church, in England, America, and Other Parts*. London: Hodson & Son, 1861.

Hine, Alfred. *Magnetic Compasses and Magnetometers*. London: Adam Hilger, 1968.

Hine, William. "Athanasius Kircher and Magnetism." In *Athanasius Kircher und seine Beziehungen zum gelehrten Europa seiner Zeit*, ed. J. Fletcher, pp. 79–97. Wiesbaden: Otto Harrassowitz, 1988.

Hirst, Désirée. *Hidden Riches: Traditional Symbolism from the Renaissance to Blake*. London: Eyre & Spottiswoode, 1964.

Hitchins, H. L., and W. E. May. *From Lodestone to Gyro-Compass*. London: Hutchinson, 1952.

Hobbs, William. *The Earth Generated and Anatomized*, ed. Roy Porter. Ithaca: Cornell University Press, 1981.

Hobson, Thomas. *Christianity the Light of the Moral World: A Poem*. London, 1745.

Hole, Robert. *Pulpits, Politics and Public Order in England 1760–1832*. Cambridge: Cambridge University Press, 1989.

———. "English Sermons and Tracts as Media of Debate on the French Revolution." In *The French Revolution and British Popular Politics*, ed. Mark Philp, pp. 18–37. Cambridge: Cambridge University Press, 1991.

Holland, Samuel. "Astronomical Observations Made by . . . His Majesty's Surveyor General of Lands for the Northern District of North America, for Ascertaining the Longitude of Several Places in the Said District," *Philosophical Transactions* 64 (1774), 182–3.

Holmes, Geoffrey. *Augustan England: Professions, State and Society, 1680–1730*. London: Allen & Unwin, 1982.

Holmes, Richard. *Coleridge: Early Visions*. London: Hodder & Stoughton, 1989.

Home, Roderick W. "Æpinus and the British Electricians: The Dissemination of a Scientific Theory," *Isis* 63 (1972), 190–204.

———. "'Newtonianism' and the Theory of the Magnet," *History of Science* 15 (1977), 252–66.

———. "Out of a Newtonian Straitjacket: Alternative Approaches to Eighteenth-Century Physical Science." In *Studies in the Eighteenth Century: Papers Presented at the Fourth David Nichol Smith Memorial Seminar, Canberra, 1976*, ed. R. F. Brissenden and J. C. Eade, pp. 235–49. Canberra: Australian University Press, 1979.

———. "Introduction." In *Æpinus's Essay on the Theory of Electricity and Magnetism*, ed. Roderick W. Home and P. J. Connor, pp. 1–224. Princeton: Princeton University Press, 1979.

Home, Roderick W., and P. J. Connor (eds.). *Æpinus's Essay on the Theory of Electricity and Magnetism*. Princeton: Princeton University Press, 1979.

Hooper, William. *Rational Recreations*, 4 vols. London, 1774.

Hooson, William. *The Miners Dictionary*. Wrexham, 1747.

Horne, George. *The Works of the Right Rev. George Horne, DD*, 6 vols. London, 1818.

Horne, Henry. "Observations on Sand Iron," *Philosophical Transactions* 53 (1763), 48–55.

———. *Essays concerning Iron and Steel*. London, 1773.

Hoskin, Michael A. "'Mining All Within': Clarke's Notes to Rohault's *Traité de Physique*," *Thomist* 24 (1961), 353–63.

Howard, Edward. *Remarks on the New Philosophy of Descartes*. London, 1700.

Howell, W. S. *Eighteenth-Century British Logic and Rhetoric*. Princeton: Princeton University Press, 1971.

Howse, Derek. *Greenwich Time and the Discovery of the Longitude*. Oxford: Oxford University Press, 1980.

Howson, Geoffrey. *A History of Mathematics Education in England*. Cambridge: Cambridge University Press, 1982.

Hoxton, Walter. "An Account of an Unusual Agitation in the Magnetical Needle, Observed to Last for Some Time, in a Voyage from Maryland," *Philosophical Transactions* 37 (1731), 53–4.

———. "The Variation of the Magnetic Needle, as Observed in Three Voyages from London to Maryland," *Philosophical Transactions* 41 (1739), 171–5.

Hoyles, John. *The Edges of Augustanism*. The Hague: Martinus Nijhoff, 1972.

Hughes, Arthur. "Sciences in English Encyclopaedias, 1704–1875—I," *Annals of Science* 7 (1951), 340–70.

Hull, David. "In Defense of Presentism," *History and Theory* 18 (1979), 1–15.

Humboldt, Alexander von. *Cosmos: A Sketch of a Physical Description of the Universe*, trans. E. Otté and W. Dallas, 5 vols. London: Henry Bohn, 1858.

Hunter, John. *The Works of John Hunter, FRS with Notes*, ed. J. Palmer, 5 vols. London: Longman et al., 1835–7.

Hunter, William. "Astronomical Observations Made in the Upper Provinces of Hindustan," *Asiatick Researches* 5 (1796), 413–21.

Hutchins, Thomas. "Experiments on the Dipping Needle, Made by Desire of the Royal Society," *Philosophical Transactions* 65 (1775), 129–38.

———. "An Account of the Success of Some Attempts to Freeze Quicksilver, at Albany Fort, in Hudson's Bay, in the Year 1775; with Observations on the Dipping-Needle," *Philosophical Transactions* 66 (1776), 174–81.

Hutchinson, John. *The Philosophical and Theological Works of the Late Truly Learned John Hutchinson*, 12 vols. London, 1748–9.

Hutchinson, William. *A Treatise on Practical Seamanship; with Hints and Remarks Relating Thereto.* 1777.

Hutchison, Keith. "What Happened to Occult Qualities in the Scientific Revolution?" *Isis* 73 (1982), 233–53.

Hutton, Charles. *A Mathematical and Philosophical Dictionary*, 2 vols. London, 1795–6.

———. *Recreations in Mathematics and Natural Philosophy*, 4 vols. London, 1803.

Hyde, Thomas. *Historia Religionis Veterum Persarum, Eorumque Magorum*. Oxford, 1700.

Imison, John. *The School of Arts; or, An Introduction to Useful Knowledge, Being a Compilation of Real Experiments and Improvements, in Several Pleasing Branches of Science*, 2 vols. in 1. London, 1790?.

Inchbald, Elizabeth. *Animal Magnetism: A Farce, in Three Acts, as Performed at the Theatre Royal, Covent-Garden*. Dublin, 1789.

Ingenhousz, John. "Experiments on Platina," *Philosophical Transactions* 66 (1776), 257–67.

———. "On Some New Methods of Suspending Magnetical Needles," *Philosophical Transactions* 69 (1779), 537–46.

Ingleton, Geoffrey C. *Matthew Flinders*. Guildford: Genesis Publications, 1986.

Inkster, Ian. "The Public Lecture as an Instrument of Science Education for Adults—The Case of Great Britain c.1750–1850," *Paedogogica Historica* 20 (1980), 80–170.

———. *An Introduction to the Study of Philosophy: Exhibiting a General View of All the Arts and Sciences, for the Use of Pupils*. London, 1744.

Jack, Malcolm. *Corruption and Progress: The Eighteenth-Century Debate*. New York: AMS Press, 1989.

Jacob, Margaret C. *The Newtonians and the English Revolution 1689–1720*. Hassocks: Harvester Press, 1976.

Jacobus, Mary. *Romanticism Writing and Sexual Difference: Essays on "The Prelude."* Oxford: Clarendon Press, 1989.

James, Frank. "Davy in the Dockyard: Humphry Davy, the Royal Society and the Electro-chemical Protection of the Copper Sheeting of His Majesty's Ships in the Mid 1820s," *Physis* 29 (1992), 205–25.

James, Robert. *A Medicinal Dictionary; Together with a History of Drugs*, 3 vols. London, 1743–5.

Jardine, Nicholas. "Writing Off the Scientific Revolution," *Journal of the History of Astronomy* 22 (1991), 311–18.

Jewson, N. D. "Medical Knowledge and the Patronage System in 18th Century England," *Sociology* 8 (1974), 369–85.

———. "The Disappearance of the Sick-Man from Medical Cosmology, 1770–1870," *Sociology* 10 (1976), 225–44.

Johns, Adrian. "Wisdom in the Concourse: Natural Philosophy and the History of the Book in Early Modern England." Ph.D. thesis, University of Cambridge, 1992.

Johnson, Samuel. *A Dictionary of the English Language: In Which the Words Are Deduced from Their Originals, and Illustrated in Their Different Significations by Examples from the Best Writers*. London, 1755.

Johnson, Samuel. *The Rambler*, 3 vols., ed. W. Bate and A. Strauss. New Haven: Yale University Press, 1969.

Jones, W., and S. Jones. "A Catalogue of Optical, Mathematical, and Philosophical Instruments." In *An Essay on Vision*, by G. Adams. London, 1792.

———. "A Catalogue of Optical, Mathematical and Philosophical Instruments, Made and Sold by W. & S. Jones." In *A Treatise on Magnetism, in Theory and Practice, with Original Experiments*, by T. Cavallo. London, 1795.

———. "A Catalogue of Optical, Mathematical and Philosophical Instruments, Made and Sold by W. & S. Jones." In *A Treatise on Magnetism, in Theory and Practice, with Original Experiments*, by T. Cavallo. London, 1800.

Jones, W. Powell. *The Rhetoric of Science: A Study of Scientific Ideas and Imagery in Eighteenth-Century English Poetry*. London: Routledge & Kegan Paul, 1966.

Jones, William. *An Essay on the First Principles of Natural Philosophy: Wherein the Use of Natural Means, or Second Causes, in the Oeconomy of the Material World, Is Demonstrated from Reason, Experiments of Various Kinds, and the Testimony of Antiquity.* Oxford, 1762.

———. *Physiological Disquisitions; or, Discourses on the Natural Philosophy of the Elements*. London, 1781.

———. *A Course of Lectures on the Figurative Language of the Holy Scripture, and the Interpretation of It from the Scripture Itself.* London, 1787.

———. *The Theological, Philosophical and Miscellaneous Works of the Rev. William Jones, MA FRS*, 6 vols. London, 1801.

Jordanova, Ludmilla. "Earth Science and Environmental Medicine." In *Images of the Earth: Essays in the History of the Environmental Sciences*, ed. Ludmilla Jordanova and Roy Porter, pp. 119–46. St. Giles: British Society for the History of Science, 1979.

Kafker, F. A. *Notable Encyclopaedias of the Seventeenth and Eighteenth Centuries; Nine Predecessors of the Encyclopédie*. Oxford: Voltaire Foundation, 1981.

Keith, George. *The Magick of Quakerism; or, The Chief Mysteries of Quakerism Laid Open*. London, 1707.

Keller, Evelyn Fox. "Secrets of God, Nature, and Life," *History of the Human Sciences* 3 (1990), 229–42.

Kelly, Suzanne. *The "De Mundo" of William Gilbert*. Amsterdam: Menno Hertzberger, 1965.

Ken, Thomas. *The Works of the Right Reverend, Learned and Pious Thomas Ken, DD*, 4 vols. London, 1721.

Kenrick, William. *A Lecture on the Perpetual Motion*. London, 1771.

King-Hele, Desmond. *The Essential Writings of Erasmus Darwin*. London: MacGibson & Kee, 1968.

Kippis, Andrew. *Biographia Britannica*, 5 vols. London, 1778–93.

Kirby, William. *On the Power Wisdom and Goodness of God as Manifested in the Creation of Animals and in Their History Habits and Instincts*, 2 vols. London: William Pickering, 1835.

Kircher, Athanasius. *Magneticum Naturæ Regnum sive Disceptatio Physiologica*. Rome, 1667.

Kirwan, Richard. "Thoughts on Magnetism," *Nicholson's Journal of Philosophy* 4 (1801), 90–4, 133–5.

———. *Elements of Mineralogy*. London, 1784.

———. *Elements of Mineralogy*, 2 vols. London, 1794–6.

Knight, David. "Science and Professionalism in England, 1770–1830," *Proceedings of the XIVth International Congress of the History of Science* 1 (1974), 53–67.

Knight, Gowin. "An Account of Some Magnetical Experiments Shewed before the Royal Society," *Philosophical Transactions* 43 (1744), 161–6.

———. "A Letter to the President, Concerning the Poles of Magnets Being Variously Placed," *Philosophical Transactions* 43 (1745), 361–3.

———. "A Collection of the Magnetical Experiments Communicated to the Royal Society . . . in the Years 1746 and 1747," *Philosophical Transactions* 44 (1747), 656–72.

———. *An Attempt to Demonstrate, That All the Phænomena in Nature May Be Explained by Two Simple Active Principles, Attraction and Repulsion: Wherein the Attractions of Cohesion, Gravity, and Magnetism, Are Shewn to Be One and the Same; and the Phænomena of the Latter Are More Particularly Explained.* London, 1748.

———. "An Account of the Mariners Compass, That was Struck with Lightning . . . Some Further Particulars Relating to That Accident," *Philosophical Transactions* 46 (1749), 113–17.

———. "A Description of a Mariner's Compass Contrived by Gowin Knight, MB FRS," *Philosophical Transactions* 46 (1750), 505–12.

———. "An Account of the Shock of an Earthquake, Felt Feb 8 1749–50," *Philosophical Transactions* 46 (1750), 603–4.

———. "Account of a Singular Recovery from a Fever," *Medical Observations and Inquiries* 1 (1757), 35–41.

———. *A Collection of Some Papers Formerly Published in the "Philosophical Transactions," Relating to the Use of Dr Knight's Magnetical Bars with Some Notes and Additions.* London, 1758.

———. "Some Remarks on the Preceding Letter," *Philosophical Transactions* 51 (1759), 294–9.

———. "A New Method of Constructing Compasses in General Use So as to Prevent Them Being Affected by the Motion of the Ship." 1766 Patent. PRO: C54/6191.

Konvitz, Josef W. *Cartography in France 1660–1848.* Chicago: Chicago University Press, 1987.

Korshin, Paul. "The Intellectual Context of Swift's Flying Island," *Philological Quarterly* 50 (1971), 630–46.

———. *Typologies in England 1650–1820.* Princeton: Princeton University Press, 1982.

Koyré, Alexandre, and I. Bernard Cohen. *Isaac Newton's Philosophiae Naturalis Principia Mathematica*, 2 vols. Cambridge: Cambridge University Press, 1972.

Kryzhanovsky, L. N. "An Application of Bibliometrics to the History of Electricity," *Scientometrics* 14 (1988), 487–92.

Kuhn, Albert J. "Glory or Gravity: Hutchinson vs. Newton," *Journal of the History of Ideas* 22 (1961), 303–22.

———. "Nature Spiritualised: Aspects of Anti-Newtonianism," *English Literary History* 41 (1974), 400–12.

———. "Dr. Johnson, Zachariah Williams, and the Eighteenth-Century Search for the Longitude," *Modern Philology* 82 (1984), 40–52.

Kuhn, Thomas S. *The Essential Tension: Selected Studies in Scientific Tradition and Change.* Chicago: University of Chicago Press, 1977.

Kunz, George F. *The Curious Lore of Precious Stones.* Philadelphia: Lippincott, 1913.

Kunz, George F. *Rings for the Finger*. Philadelphia: Lippincott, 1917.

Labaree, L. W., and W. B. Willcox (eds.). *The Papers of Benjamin Franklin*, 23 vols. New Haven: Yale University Press, 1960–83.

Land, Stephen K. *From Signs to Propositions: The Concept of Form in Eighteenth-Century Semantic Theory*. London: Longman Group, 1974.

———. *The Philosophy of Language in Britain: Major Theories from Hobbes to Thomas Reid*. New York: AMS Press, 1986.

Landow, George P. *Images of Crisis: Literary Iconology, 1750 to the Present*. Boston: Routledge & Kegan Paul, 1982.

Langford, Paul. *A Polite and Commercial People: England 1727–1783*. Oxford: Clarendon Press, 1989.

Langrish, Browne. *A New Essay on Muscular Motion*. London, 1733.

Latour, Bruno. *Science in Action: How to Follow Scientists and Engineers through Society*. Milton Keynes: Open University Press, 1987.

———. "Drawing Things Together." In *Representation in Scientific Practice*, ed. Michael Lynch and Steve Woolgar, pp. 19–68. Cambridge: MIT Press, 1988.

———. "The Force and Reason of Experiment." In *Experimental Inquiries: Historical, Philosophical and Social Studies of Experimentation in Science*, ed. Homer Le Grand, pp. 49–80. Dordrecht: Kluwer Academic, 1990.

Laudan, Larry. "The Medium and Its Message: A Study of Some Philosophical Controversies about Ether." In *Conceptions of Ether: Studies in the History of Ether Theories, 1740–1900*, ed. Geoffrey Cantor and Michael Hodge, pp. 157–86. Cambridge: Cambridge University Press, 1981.

Laudan, Rachel. "Redefinitions of a Discipline: Histories of Geology and Geological History." In *Functions and Uses of Disciplinary Histories*, ed. Loren Graham, Wolf Lepenies, and Peter Weingart, pp.79–104. Dordrecht: Reidel, 1983.

———. *From Mineralogy to Geology: The Foundations of a Science, 1650–1830*. Chicago: University of Chicago Press, 1987.

Law, John. "On the Methods of Long-Distance Control: Vessels, Navigation and the Portuguese Route to India." In *Power, Action and Belief: A New Sociology of Knowledge?*, ed. John Law, pp. 234–63. London: Routledge & Kegan Paul, 1986.

———. "Technology and Heterogeneous Engineeering: The Case of Portuguese Expansion." In *The Social Construction of Technological Systems: New Directions in the Social Study of Technology*, ed. Wiebe Bijker, Trevor Pinch, and Thomas Hughes. Cambridge: MIT Press, 1987.

Law, John, and John Whittaker. "On the Art of Representation: Notes on the Politics of Visualisation." In *Picturing Power: Visual Depiction and Social Relations*, ed. Gordon Fyfe and John Law, pp. 160–83. London: Routledge, 1988.

Law, William. *The Works of the Reverend William Law, MA*, 9 vols. Setley, 1892–3.

Lawrence, Christopher. "The Nervous System and Society in the Scottish Enlightenment." In *Natural Order: Historical Studies of Scientific Culture*, ed. Barry Barnes and Steven Shapin, pp. 19–40. Beverly Hills, Calif.: Sage, 1979.

———. "Alexander Monro *primus* and the Edinburgh Manner of Anatomy," *Bulletin of the History of Medicine* 62 (1988), 193–214.

———. "Disciplining Disease: Scurvy, the Navy and Imperial Expansion 1750–1825." In *Visions of Empire: Voyages, Botany and Representations of Nature*, ed. David Miller and P. Reill. Cambridge: Cambridge University Press, in press.

Layton, David. "Diction and Dictionaries in the Diffusion of Scientific Knowledge: An

Aspect of the History of the Popularization of Science in Great Britain," *British Journal for the History of Science* 2 (1965), 221–34.

Le Grand, Homer. "Is a Picture Worth a Thousand Experiments?" In *Experimental Inquiries: Historical, Philosophical and Social Studies of Experimentation in Science*, ed. Homer Le Grand, pp. 241–70. Dordrecht: Kluwer Academic, 1990.

Leask, Nigel. "Shelley's 'Magnetic Ladies': Romantic Mesmerism and the Politics of the Body." In *Beyond Romanticism: New Approaches to Texts and Contexts 1780–1832*, ed. Stephen Copley and John Whales, pp. 53–78. London: Routledge, 1991.

Leatherdale, W. H. *The Role of Analogy, Model and Metaphor in Science*. Amsterdam: North-Holland, 1974.

Lefanu, William. *Betsy Sheridan's Journal*. Oxford: Oxford University Press, 1986.

Lefebvre, E., and J. G. de Bruijn. *Martinus van Marum: Life and Work*, 6 vols. Leyden: Noordhoff International, 1973–6.

Leonardus, Camillus. *The Mirror of Stones: In Which the Nature, Generation, Properties, Virtues and Various Species of More Than 200 Different Jewels, Precious and Rare Stones, Are Distinctly Described*. London, 1750.

Lesch, John E. "Systematics and the Geometrical Spirit." In *The Quantifying Spirit in the Eighteenth Century*, ed. Tore Frängsmyr, John Heilbron, and Robin Rider, pp. 73–111. Berkeley and Los Angeles: University of California Press, 1990.

———. "A Letter to Mr Benj. Robins, FRS Shewing That the Electricity of Glass Disturbs the Mariners' Compass," *Philosophical Transactions* 44 (1746), 242–5.

———. "A Letter to Mr John Ellicot, FRS on Weighing the Strength of Electrical Effluvia," *Philosophical Transactions* 44 (1746), 96–9.

Lettsom, John C. *The Works of John Fothergill, MD with Some Account of His Life*. London, 1784.

Levine, George. "One Culture: Science and Literature." In *Science and Literature*, ed. George Levine, pp. 3–32. Madison: University of Wisconsin Press, 1987.

Levine, Joseph R. *Dr. Woodward's Shield*. Berkeley and Los Angeles: University of California Press, 1977.

———. *Humanism and History: Origins of Modern English Historiography*. Ithaca: Cornell University Press, 1987.

Lewis, W. S. (ed.). *The Yale Edition of Horace Walpole's Correspondence*, 48 vols. London: Oxford University Press, 1937–83.

Lindberg, David C., and Robert S. Westman. *Reappraisals of the Scientific Revolution*. Cambridge: Cambridge University Press, 1990.

Lindqvist, Svante. "The Spectacle of Science: An Experiment in 1744 concerning the Aurora Borealis," *Configurations* 1 (1993), 57–94.

Linnaeus, Charles. *A General System of Nature, through the Three Grand Kingdoms of Animals, Vegetables, and Minerals, Systematically Divided into Their Several Classes, Orders, Genera, Species, and Varieties, with Their Habitations, Manners, Economy, Structure, and Peculiarities*, 7 vols. London, 1806.

Lippincott, Louise. *Selling Art in Georgian London: The Rise of Arthur Pond*. London: Yale University Press, 1983.

Lister, Martin. "A Journey to Paris in the Year 1698." In *A General Collection of the Best and Most Interesting Voyages and Travels in All Parts of the World, Many of Which Are Now First Translated into English*, 17 vols., ed. John Pinkerton, vol. 4, pp. 1–76. London: Longman et al., 1808–14.

Lloyd, Stephen. "The Accomplished Maria Cosway: Anglo-Italian Artist, Musician, Salon Hostess and Educationalist," *Journal of Anglo-Italian Studies* 2 (1992), 108–39.

Lloyd, T. O. *The British Empire 1558–1983*. Oxford: Oxford University Press, 1984.

Locke, John. "Mr Locke's History of Navigation, from Its Original to the Year 1704, with an Explanatory Catalogue of Voyages, Prefixed by That Learned Writer to Churchill's Collection, in Eight Vols. Folio." In *The Progress of Maritime Discovery, from the Earliest Period to the Close of the Eighteenth Century, Forming an Extensive System of Hydrography*, by J. S. Clarke, book 2, pp. 75–202. London, 1803.

———. *An Essay concerning Human Understanding*, ed. P. Nidditch. Oxford: Clarendon Press, 1979.

Lofft, Capel. *Eudosia; or, A Poem on the Universe*. London, 1781.

Looney, J. Jefferson. "Cultural Life in the Provinces: Leeds and York, 1720–1820." In *The First Modern Society: Essays in Honour of Lawrence Stone*, ed. A. L. Beier, David Cannadine, and James Rosenheim, pp. 483–510. Cambridge: Cambridge University Press, 1989.

Lorimer, John. *A Concise Essay on Magnetism; with an Account of the Declination and Inclination of the Magnetic Needle; and an Attempt to Ascertain the Cause of the Variation Thereof*. London, 1795.

Lovett, Richard. *Philosophical Essays, in Three Parts*. Worcester, 1766.

———. *The Electrical Philosopher*. Worcester, 1774.

Lucretius. *On the Nature of Things*, 6 bks. in 1, trans. John M. Good. London: Henry G. Bohn, 1851.

Lukis, W. C. *The Family Memoirs of the Rev. William Stukeley, MD and the Antiquaries and Corrrespondence of William Stukeley, Roger and Samuel Gale, etc.*, 3 vols. Durham: Surtees Society, 1882–7.

Lynch, Michael, and Steve Woolgar (eds.). *Representation in Scientific Practice*. Cambridge: MIT Press, 1990.

Macalpine, Ida, and Richard Hunter. *George III and the Mad-Business*. London: Allen Lane, 1969.

Macdonald, John. "Observations of the Diurnal Variation of the Magnetic Needle at Fort Marlborough, in the Island of Sumatra," *Philosophical Transactions* 86 (1796), 340–9.

———. "Observations of the Diurnal Variation of the Magnetic Needle, in the Island of St Helena; with a Combination of the Observations at Fort Marlborough, in the Island of Sumatra," *Philosophical Transactions* 88 (1798), 397–403.

MacDonald, Michael. "Religion, Social Change, and Psychological Healing in England, 1600–1800," *Studies in Church History* 19 (1982), 101–25.

Mack, James D. *Matthew Flinders*. Melbourne: Nelson, 1966.

Mackay, David L. "Direction and Purpose in British Imperial Policy, 1783–1801," *Historical Journal* 17 (1974), 487–501.

———. *In the Wake of Cook: Exploration, Science and Empire, 1780–1801*. London: Croom Helm, 1985.

MacKenzie, Donald, and Judy Wajcman. *The Social Shaping of Technology*. Milton Keynes: Open University Press, 1985.

Macleod, Christine. *Inventing the Industrial Revolution: The English Patent System, 1660–1800*. Cambridge: Cambridge University Press, 1988.

MacPike, Eugene F. *Correspondence and Papers of Edmond Halley*. Oxford: Clarendon Press, 1932.

Madden, Samuel. *Memoirs of the Twentieth Century*. London, 1733.

Magellan, Jean. *Collection de différens traités sur des instrumens d'astronomie, physique, etc.* Paris and London, 1780.

Magnet, William. *The Newtonian System of Philosophy*. London, 1794.

Maitland, William. *An Essay towards the Improvement of Navigation, Chiefly with Respect to the Instruments Used at Sea*. London, 1750.

Mandeville, Bernard. *The Fable of the Bees: or, Private Vices, Public Benefits*, ed. F. B. Kaye, 2 vols. Oxford: Clarendon Press, 1923.

Manuel, Franz. *The Eighteenth Century Confronts the Gods*. Cambridge: Harvard University Press, 1959.

———. *Isaac Newton: Historian*. Cambridge: Cambridge University Press, 1963.

Marcel, Arnold. "An Abstract of a Letter . . . to the Illustrious Royal Society of London," *Philosophical Transactions* 37 (1732), 294–8.

Marcovich, Anne. "Concerning the Continuity between the Image of Society and the Image of the Human Body—An Examination of the Work of the English Physician, J. C. Lettsom (1746–1815)." In *The Problem of Medical Knowledge*, ed. Peter Wright and Andrew Treacher, pp. 69–86. Edinburgh: Edinburgh University Press, 1982.

Marshall, Peter G. "Oriental Studies." In *The History of the University of Oxford*. Vol. 5, *The Eighteenth Century*, ed. L. Sutherland and L. Mitchell, pp. 551–63. Oxford: Clarendon Press, 1986.

Marshall, Peter G., and Glyndwr Williams. *The Great Map of Mankind: British Perceptions of the World in the Age of Enlightenment*. London: Dent, 1982.

Martin, Benjamin. *Philosophia Britannica: or, A New and Comprehensive System of the Newtonian Philosophy, Astronomy and Geography*, 2 vols. Reading, 1747.

———. *An Account of the New Construction and Rectification of Hadley's Quadrant*. 179?.

Mather, F. C. *High Church Prophet: Bishop Samuel Horsley (1733–1806) and the Caroline Tradition in the Later Georgian Church*. Oxford: Clarendon Press, 1992.

Mathias, Peter. *The Transformation of England: Essays in the Economic and Social Transformation of England in the Eighteenth Century*. London: Methuen, 1979.

Maurice, Thomas. *The History of Hindoostan; Its Arts, and Its Sciences, as Connected with the History of the Other Great Empires of the World*, 2 vols. London, 1795.

Maxwell, John. "An Account of the Cape of Good Hope," *Philosophical Transactions* 25 (1707), 2423–34.

May, W. E. "Historical Notes on the Deviation of the Compass," *Terrestrial Magnetism and Atmospheric Electricity* 52 (1947), 217–31.

———. "True Reading Magnetic Compasses," *Terrestrial Magnetism and Atmospheric Electricity* 53 (1948), 135–51.

———. "The Compass Makers of Deptford," *Nautical Magazine* 163 (1950), 386–90.

———. "The History of the Magnetic Compass," *Mariner's Mirror* 38 (1952), 210–22.

———. "Navigational Accuracy in the Eighteenth Century," *Journal of the Institute of Navigation* 6 (1953), 71–3.

———. "Naval Compasses in 1707," *Journal of the Institute of Navigation* 6 (1953), 405–9.

May, W. E. "The Binnacle," *Mariner's Mirror* 40 (1954), 21–32.

———. "Longitude by Variation," *Mariner's Mirror* 45 (1959), 339–41.

———. "The Last Voyage of Sir Cloudisley Shovel," *Journal of the Institute of Navigation* 13 (1960), 324–32.

———. *A History of Marine Navigation*. Henley-on-Thames: Foulis, 1973.

———. "The Log Books Used by Ships of the East India Company," *Journal of Navigation* 27 (1974), 116–18.

———. "The Mystery of Captain Lane's Compass," *Journal of Navigation* 29 (1976), 411–12.

———. "Garlic and the Magnetic Compass," *Mariner's Mirror* 65 (1979), 231–4.

McCalman, Iain. *Radical Underworld: Prophets, Revolutionaries and Pornographers in London, 1795–1840*. Cambridge: Cambridge University Press, 1988.

McClellan, James E. *Science Reorganized: Scientific Societies in the Eighteenth Century*. New York: Columbia University Press, 1985.

McConnell, Anita. *Geomagnetic Instruments before 1900*. London: Harriet Wynter, 1980.

———. "Nineteenth-Century Geomagnetic Instruments and Their Makers." In *Nineteenth-Century Scientific Instruments and Their Makers*, ed. Peter R. Declercq, pp. 29–52. Amsterdam: Rodopi, 1985.

———. "From Craft Workshop to Big Business—The London Scientific Instrument Trade's Response to Increasing Demand, 1750–1820," *London Journal* 19 (1994), 36–53.

McCormmach, Russell. "Henry Cavendish on the Proper Method of Rectifying Abuses." In *Beyond History of Science: Essays in Honor of Robert E. Schofield*, ed. Elizabeth Garber, pp. 35–51. London and Toronto: Associated University Presses, 1990.

McCulloch, Kenneth. *An Account of the New Improved Sea Compasses . . . with Reports of Their Practical Utility.* London, 1789.

McGuire, James E. "Force, Active Principles and Newton's Invisible Realm," *Ambix* 15 (1968), 154–208.

McKendrick, Neil. "The Rôle of Science in the Industrial Revolution: A Study of Josiah Wedgwood as a Scientist and Industrial Chemist." In *Perspectives in the History of Science: Essays in Honour of Joseph Needham*, ed. Mikuláš Teich and Robert Young, pp. 274–319. London: Heinemann, 1973.

McKendrick, Neil, John Brewer, and J. H. Plumb. *The Birth of a Consumer Society: The Commercialization of Eighteenth-Century England*. London: Europa Publications, 1982.

McNeil, Maureen. "The Scientific Muse: The Poetry of Erasmus Darwin." In *Languages of Nature: Critical Essays on Science and Literature*, ed. Ludmilla Jordanova, pp. 159–203. London: Free Association Books, 1986.

Melvill, Thomas. "A Letter . . . with a Discourse concerning the Cause of the Different Refrangibility of the Rays of Light," *Philosophical Transactions* 48 (1753), 261–70.

———. *Memoir of the Late Thomas Halloway; by One of His Executors; and Most Respectfully Dedicated to the Subscribers to the Engravings from the Cartoons of Raphael*. London, 1827.

Metzger, Hélène. *Newton, Stahl, Boerhaave et la doctrine chimique*. Paris: Félix Alcan, 1930.

Meyer, Herbert W. *A History of Electricity and Magnetism*. Cambridge: MIT Press, 1971.

Michell, John. *A Treatise of Artificial Magnets*. Cambridge, 1750.

Middleton, Christopher. "A New and Exact Table, Collected from Several Observations, Taken in Four Voyages to Hudson's Bay in North America from London: Shewing the Variation of the *Magnetical Needle*, or Sea Compass, in the Path-Way to the Said Bay, According to the Several Latitudes and Longitudes, from the Year 1721, to 1725," *Philosophical Transactions* 34 (1726), 73–6.

———. "A New and Exact Table Collected from Several Observations Taken from the Year 1721 to 1729, in Nine Voyages to Hudson's Bay in North-America, Shewing the Variation of the Compass According to the Latitudes and Longitudes Undermentioned, Accounting the Longitude from the Meridian of London," *Philosophical Transactions* 37 (1731), 71–5.

———. "Observations on the Weather, in a Voyage to Hudson's-Bay in North-America, in the Year 1730," *Philosophical Transactions* 37 (1731), 76–8.

———. "Observations of the Variations of the Needle and Weather, Made in a Voyage to Hudson's-Bay, in the Year 1731," *Philosophical Transactions* 38 (1733), 127–33.

———. "Observations Made of the Latitude, Variation of the Magnetic Needle, and Weather, in a Voyage from London to Hudson's-Bay, Anno 1735," *Philosophical Transactions* 39 (1736), 270–80.

———. "An Observation of the Magnetic Needle Being So Affected by Great Cold, That It Would Not Traverse," *Philosophical Transactions* 40 (1738), 310–11.

———. "The Use of a New Azimuth Compass for Finding the Variation of the Compass or Magnetic Needle at Sea, with Greater Ease and Exactness than by Any Ever Yet Contriv'd for That Purpose," *Philosophical Transactions* 40 (1738), 395–8.

———. "The Effects of Cold; Together with Observations of the Longitude, Latitude, and Declination of the Magnetic Needle, at Prince of Wales's Fort, upon Churchill-River in Hudson's Bay," *Philosophical Transactions* 42 (1742), 157–71.

Middleton, Richard. "Naval Administration in the Age of Pitt and Anson, 1755–1763." In *The British Navy and the Use of Naval Power in the Eighteenth Century*, ed. Jeremy Black and Philip Woodfine, pp. 109–27. Leicester: Leicester University Press, 1988.

Millard, D.P.T. "The Chronology of Roger North's Main Works," *Review of English Studies* 24 (1973), 283–94.

Millburn, John R. "Benjamin Martin and the Royal Society," *Notes and Records of the Royal Society of London* 28 (1973), 15–23.

———. "The London Evening Courses of Benjamin Martin and James Ferguson, Eighteenth-Century Lecturers on Experimental Philosophy," *Annals of Science* 40 (1983), 437–55.

———. *Retailer of the Sciences: Benjamin Martin's Scientific Instrument Catalogues, 1756–1782*. London: Vade-Mecum Press, 1986.

———. *Wheelwright of the Heavens: The Life and Work of James Ferguson, FRS*. London: Vade-Mecum Press, 1988.

———. "The Office of Ordnance and the Instrument-making Trades in the Mid-Eighteenth Century," *Annals of Science* 45 (1988), 221–93.

Miller, David P. "Between Hostile Camps: Sir Humphry Davy's Presidency of the Royal Society of London, 1820–1827," *British Journal for the History of Science* 16 (1983), 1–47.

———. "The Revival of the Physical Sciences in Britain, 1815–1840," *Osiris* 2 (1986), 107–34.

Miller, David P. "'Into the Valley of Darkness': Reflections on the Royal Society in the Eighteenth Century," *History of Science* 27 (1989), 155–66.

Miller, E. *That Noble Cabinet: A History of the British Museum*. London: André Deutsch, 1973.

Minto, Countess of. *Life and Letters of Sir Gilbert Elliot*, 3 vols. London: Longmans, Green, 1874.

Molyneux, William. "A Demonstration of an Error Committed by Common Surveyors in Comparing of Surveys Taken at Long Intervals of Time Arising from the Variation of the Magnetick Needle," *Philosophical Transactions* 19 (1697), 625–31.

Money, John. *Experience and Identity: Birmingham and the West Midlands, 1760–1800*. Manchester: Manchester University Press, 1977.

———. "Teaching in the Market-Place, or 'Caesar Adsum Jam Forte: Pompey Aderat': The Retailing of Knowledge in Provincial England during the Eighteenth Century." In *Consumption and the World of Goods*, ed. John Brewer and Roy Porter, pp. 335–77. London: Routledge, 1993.

Moore, John H. *The Practical Navigator and Seaman's Daily Assistant*. London, 1772.

Morrell, Jack. "Professors Robison and Playfair, and the *Theophobia Gallica*: Natural Philosophy, Religion and Politics in Edinburgh, 1789–1815," *Notes and Records of the Royal Society* 26 (1971), 43–63.

Morton, Alan Q. "Lectures on Natural Philosophy in London, 1750–1765: S.C.T. Demainbray (1720–1782) and the 'Inattention' of his Countrymen," *British Journal for the History of Science* 23 (1990), 411–34.

Morton, Alan Q., and Jane A. Wess. *Public and Private Science: The King George III Collection*. Oxford: Oxford University Press, 1993.

Moseley, Charles. *A Century of Emblems: An Introductory Anthology*. Aldershot: Scolar Press, 1989.

Mottelay, Paul F. *Bibliographical History of Electricity and Magnetism, Chronologically Arranged*. London: Charles Griffin, 1922.

Moulen, Allen. "Some Experiments on a Black Shining Sand Brought from Virginia, Supposed to Contain Iron," *Philosophical Transactions* 17 (1693), 624–6.

Mountaine, William. "An Account of Some Extraordinary Effects of Lightning," *Philosophical Transactions* 51 (1759), 286–94.

———. "A Letter Containing Some Observations on the Variation of the Magnetic Needle, Made on Board the Montagu Man of War," *Philosophical Transactions* 56 (1766), 216–23.

Mountaine, William, and James Dodson. "An Attempt to Point Out, in a Concise Manner, the Advantages Which Will Accrue from a Periodic Review of the Variation of the Magnetic Needle, throughout the Known World," *Philosophical Transactions* 48 (1755), 875–80.

———. "A Letter concerning the Variation of the Magnetic Needle; With a Sett of Tables Annexed, Which Exhibit the Result of Upwards of Fifty Thousand Observations in Six Periodic Reviews from the Year 1700 to the Year 1756, Both Inclusive; and Are Adapted to Every Five Degrees of Latitude and Longitude in the More Frequented Oceans," *Philosophical Transactions* 50 (1757), 329–49.

———. *An Account of the Methods Used to Describe Lines, on Dr Halley's Chart of the Terraqueous Globe; Shewing the Variation of the Magnetic Needle about the Year 1756, in All the Known Seas; Their Application and Use in Correcting the Longitude at Sea, with Some Occasional Observations Relating Thereto*. London, 1758.

Mukerji, Shandra. "Voir le pouvoir," *Culture Technique* 14 (1985), 208–23.

Mullan, John. "Gendered Knowledge, Gendered Minds: Women and Newtonianism, 1690–1760." In *A Question of Identity: Women, Science and Literature*, ed. Marina Benjamin, pp. 41–56. New Brunswick, N.J.: Rutgers University Press, 1993.

Multhauf, Robert. "A History of Magnesia Alba," *Annals of Science* 33 (1976), 197–200.

Multhauf, Robert, and Gregory Good. *A Brief History of Geomagnetism and a Catalog of the Collections of the National Museum of American History*. Washington, D. C.: Smithsonian Institution Press, 1987.

Murray, Nancy. "The Influence of the French Revolution on the Church of England and Its Rivals, 1789–1802." Ph.D. thesis, University of Oxford, 1975.

Musgrave, William. *Obituary Prior to 1800*, 6 vols. London: Harleian Society, 1899–1901.

Musschenbroek, Pieter van. "De Viribus Magneticus," *Philosophical Transactions* 33 (1725), 370–7.

———. *Physicae Experimentales . . . de Magnete*. Leiden, 1729.

———. "Ephemerides Meteorologicæ, Barometricæ, Thermometricæ, Epidemicæ, Magneticæ, Ultrajectinae," *Philosophical Transactions* 37 (1732), 357–84, 408–26.

———. "An Abstract of a Letter . . . concerning Experiments Made on the Indian Magnetic Sand," *Philosophical Transactions* 38 (1734), 297–302.

———. *The Elements of Natural Philosophy*, trans. J. Colson. London, 1744.

———. *Cours de physique expérimentale*, 3 vols. Paris, 1769.

Musson, A. E. *Science, Technology and Economic Growth*. London: Methuen, 1972.

Myers, Greg. "Science for Women and Children: The Dialogue of Popular Science in the Nineteenth Century." In *Nature Transfigured: Science and Literature, 1700–1900*, ed. John Christie and Sally Shuttleworth, pp. 171–200. Manchester and New York: Manchester University Press, 1989.

Nairne, Edward. "Experiments on Two Dipping-Needles, Which Dipping-Needles Were Made Agreeable to a Plan of the Reverend Mr Mitchell . . . and Executed for the Board of Longitude," *Philosophical Transactions* 62 (1772), 476–80.

Nairne, Edward, and Thomas Blunt. *The Description and Use of Nairne's Patent Electrical Machine; with the Addition of Some Philosophical Experiments and Medical Observations*. London, 1783.

Nangle, Benjamin C. *The Monthly Review First Series 1749–89*. Oxford: Clarendon Press, 1934.

Neve, Michael, and Roy Porter. "Alexander Catcott: Glory and Geology," *British Journal for the History of Science* 10 (1977), 37–60.

Newton, Isaac. *Opticks; or, A Treatise of the Reflexions, Refractions, Inflexions and Colours of Light*, ed. I. B. Cohen and D.H.D. Roller from 4th ed. of 1730. New York: Dover Publications, 1952.

———. *Philosophiæ Naturalis Principia Mathematica*, 3 vols., ed. Thomas Le Seur and François Jacquier. Geneva, 1742.

Nichols, John. *Literary Anecdotes of the Eighteenth Century*, 9 vols. London, 1812.

———. *Illustrations of the Literary History of the Eighteenth Century*, 8 vols. London, 1817–58.

Nicholson, William. *An Introduction to Natural Philosophy*, 2 vols. London, 1782.

Nicolson, Marjorie Hope, and Nora M. Mohler. "The Scientific Background of Swift's *Voyage to Laputa*," *Annals of Science* 2 (1937), 299–334.

Nicolson, Marjorie Hope, and Nora M. Mohler. "Swift's 'Flying Island' in the *Voyage to Laputa*," *Annals of Science* 2 (1937), 405–30.

Nieuwentyt, Bernard. *The Religious Philosopher; or, The Right Use of Contemplating the Works of the Creator . . . Designed for the Conviction of Atheists and Infidels*, 3 vols. London, 1718–19.

Noad, Henry M. *Manual of Electricity.* London: Lockwood, 1859.

Norman, Robert. *The Newe Attractive, Shewing the Nature, Propertie, and Manifold Vertues of the Loadstone*. London, 1720.

Norris, John. *Discourses upon Several Divine Subjects*, 3 vols. London, 1711.

———. *The Poems of John Norris of Bemerton: For the First Time Collected and Edited after the Original Texts*, ed. Alexander Grosart. Miscellanies of the Fuller Worthies' Library, 1871.

Nugent, P. R. "A New Theory, Pointing Out the Situation of the Magnetic Poles, and a Method of Discovering the Longitude," *Philosophical Magazine* 5 (1799), 378–92.

O'Brien, Patrick, and Roland Quinault (eds.). *The Industrial Revolution and British Society*. Cambridge: Cambridge University Press, 1993.

Olson, Richard. G. *Scottish Philosophy and British Physics 1750–1880*. Princeton: Princeton University Press, 1975.

———. "Tory–High Church Opposition to Science and Scientism in the Eighteenth Century." In *The Uses of Science in the Age of Newton*, ed. John Burke, pp. 171–204. Berkeley and Los Angeles: University of California Press, 1983.

Ovenell, R. F. *The Ashmolean Museum 1683–1894*. Oxford: Clarendon Press, 1986.

Ozanam, Jacques. *Recreations Mathematical and Physical*. London, 1708.

Palter, Robert. "Early Measurements of Magnetic Force," *Isis* 63 (1972), 544–58.

Park, James A. *Memoirs of William Stevens, Esq*. London, 1812 (for the Philanthropic Society).

Parkhurst, John. *An Hebrew and English Lexicon, without Points: In Which the Hebrew and Chaldee Words of the Old Testament Are Explained in Their Leading and Derived Senses, the Derivative Words Are Ranged under Their Respective Primitives, and the Meanings Assigned to Each Authorised by References to Passages of Scripture*. London, 1762.

Pasquin, Anthony. *Memoirs of the Royal Academicians; Being an Attempt to Improve the National Taste*. London, 1796.

Passmore, John. *Man's Responsibility for Nature*. London: Duckworth: 1980.

———. *The Patent: A Poem . . . by the Author of the Graces*. London, 1776.

Patterson, D. "Hebrew Studies." In *The History of the University of Oxford*. Vol. 5, *The Eighteenth Century*, ed. L. Sutherland and L. Mitchell, pp. 535–50. Oxford: Clarendon Press, 1986.

Paulson, Ronald. *Hogarth's Graphic Works*, 2 vols. New Haven: Yale University Press, 1970.

———. *Hogarth: His Life, Art and Times*, 2 vols. New Haven and London: Yale University Press, 1971.

———. *Popular and Polite Art in the Age of Hogarth and Fielding*. Notre Dame: University of Notre Dame Press, 1979.

———. *Representations of Revolution*. New Haven: Yale University Press, 1983.

Peat, Thomas. *A Short Account of a Course of Mechanical and Experimental Philosophy and Astronomy*. Nottingham, 1744.

Pedersen, Olaf. "The 'Philomaths' of 18th Century England," *Centaurus* 8 (1963), 238–62.

Pemberton, Henry. *A View of Sir Isaac Newton's Philosophy*. London, 1728.

Penrose, Francis. *An Essay on Magnetism; or, An Endeavour to Explain the Various Properties and Effects of the Loadstone: Together with the Causes of the Same*. Oxford, 1753.

———. *Animadversions on a Late Sermon Presented before a Bishop and a Congregation of Clergy within the Diocese of Oxford: Together with Some Remarks on the Charge That Followed It*. London, 1756.

Pera, Marcello. *The Ambiguous Frog: The Galvani-Volta Controversy on Animal Electricity*. Princeton: Princeton University Press, 1992.

Pettus, John. *Fleta Minor. The Laws of Art and Nature, in Knowing, Judging, Assaying, Fining, Refining and Inlarging the Bodies of Confin'd Metals*, 2 pts. in 1. London, 1683.

Phillips, Hugh. *Mid-Georgian London: A Topographical and Social Survey of Central and Western London about 1750*. London: Collins, 1964.

Phillips, William. *An Elementary Introduction to the Knowledge of Mineralogy: Including Some Account of Mineral Elements and Constituents; Explanations of Terms in Common Use; Brief Accounts of Minerals, and of the Places and Circumstances in Which They Are Found*. London, 1816.

Philp, Mark (ed.). *The French Revolution and British Popular Politics*. Cambridge: Cambridge University Press, 1991.

Pickergill, Richard. "A Track of His Majesty's Armed Brig Lion . . . Also the Variation of the Compass and Dip of the Needle, as Observed during the Said Voyage in 1776," *Philosophical Transactions* 68 (1778), 1057–63.

Pickering, Andrew (ed.). *Science as Practice and Culture*. Chicago and London: University of Chicago Press, 1992.

Piggott, Stuart. *William Stukeley: An Eighteenth-Century Antiquary*. London: Thames & Hudson, 1985.

———. *Ancient Britons and the Antiquarian Imagination*. London: Thames & Hudson, 1989.

Pinch, Trevor, and Wiebe Bijker. "The Social Construction of Facts and Artefacts: Or How the Sociology of Science and the Sociology of Technology Might Benefit Each Other," *Social Studies of Science* 14 (1984), 399–441.

Pinetti, Giuseppe. *The Conjurer's Repository*. London, 1793.

Plato. *Ion*. In *Symposium and Other Dialogues*, ed. John Warrington, pp. 63–77. London: Dent, 1964.

Playfair, William. *An Inquiry into the Permanent Causes of the Decline and Fall of Powerful and Wealthy Nations, Illustrated by Four Engraved Charts*. London, 1805.

———. *The Political Magnet; or, An Essay in Defence of the Late Revolution*. 1750.

Pomet, Pierre. *A Compleat History of Druggs*. London, 1712.

Pomian, Krzysztof. *Collectors and Curiosities: Paris and Venice, 1500–1800*. Cambridge: Polity Press, 1990.

Pool, Bernard. "Navy Contracts in the Last Years of the Navy Board, 1780–1832," *Mariner's Mirror* 50 (1964), 161–76.

Pope, Alexander. *The Correspondence of Alexander Pope*, ed. G. Sherburn, 5 vols. Oxford: Clarendon Press, 1956.

Porter, Roy. "The Industrial Revolution and the Rise of the Science of Geology." In *Changing Perspectives in the History of Science: Essays in Honour of Joseph Needham*, ed. Mikuláš Teich and Robert Young, pp. 320–43. London: Heinemann, 1973.

Porter, Roy. *The Making of Geology: Earth Science in Britain, 1660–1815*. Cambridge: Cambridge University Press, 1977.

———. "The Terraqueous Globe." In *The Ferment of Knowledge: Studies in the Historiography of Eighteenth-Century Science*, ed. George S. Rousseau and Roy Porter, pp. 285–324. Cambridge: Cambridge University Press, 1980.

———. "Science, Provincial Culture and Public Opinion in Enlightenment England," *British Journal for Eighteenth-Century Studies* 3 (1980), 20–46.

———. "The Enlightenment in England." In *The Enlightenment in National Context*, ed. Roy Porter and Mikulá{{scaron}} Teich, pp. 1–18. Cambridge: Cambridge University Press, 1981.

———. "William Hunter: A Surgeon and a Gentleman." In *William Hunter and the Eighteenth-Century Medical World*, ed. William Bynum and Roy Porter, pp. 7–34. Cambridge: Cambridge University Press, 1985.

———. "The Language of Quackery in England, 1660–1800." In *The Social History of Language*, ed. Peter Burke and Roy Porter, pp. 73–103. Cambridge: Cambridge University Press, 1987.

———. *Edward Gibbon: Making History*. London: Weidenfeld & Nicolson, 1988.

———. "Before the Fringe: 'Quackery' and the Eighteenth-Century Medical Market." In *Studies in the History of Alternative Medicine*, ed. Roger Cooter, pp. 1–27. London: Macmillan, 1988.

———. *Health for Sale: Quackery in England 1660–1850*. Manchester: Manchester University Press, 1989.

———. "The History of Science and the History of Society." In *Companion to the History of Modern Science*, ed. Robert Olby, G. N. Cantor, J.R.R. Christie, and M.J.S. Hodge, pp. 32–46. London: Routledge, 1990.

———. "Barely Touching: A Social Perspective on Mind and Body." In *The Languages of Psyche: Mind and Body in Enlightenment Thought*, ed. George S. Rousseau, pp. 45–80. Berkeley and Los Angeles: University of California Press, 1990.

———. *Doctor of Society: Thomas Beddoes and the Sick Trade in Late-Enlightenment England*. London: Routledge, 1992.

Porter, Roy, Jim Bennett, Simon Schaffer, and Olivia Brown. *Science and Profit in 18th-Century London*. Cambridge: Whipple Museum of the History of Science, 1985.

Pratt, Mary L. *Imperial Eyes: Travel Writing and Transculturation*. London: Routledge, 1992.

Pressly, William L. *The Life and Art of James Barry*. New Haven and London: Yale University Press, 1981.

Preston, Thomas R. "Biblical Criticism, Literature, and the Eighteenth-Century Reader." In *Books and their Readers in Eighteenth-Century England*, ed. Isabel Rivers, pp. 97–126. Leicester: St. Martin's Press, 1982.

Prickett, Stephen. *Words and "The Word": Language, Poetics and Biblical Interpretation*. Cambridge: Cambridge University Press, 1986.

Priestley, Joseph. *Experiments and Observations on Different Kinds of Air, and Other Branches of Natural Philosophy, Connected with the Subject*, 3d ed., 3 vols. Birmingham, 1790.

———. *Heads of Lectures on a Course of Experimental Philosophy, Particularly including Chemistry, Delivered at the New College in Hackney*. London, 1794.

———. *Lectures on History, and General Policy; to Which Is Prefixed, an Essay on a Course of Liberal Education for Civil and Active Life*. London, 1826.

———. *The History and Present State of Electricity*, 3d ed., 3 vols. 1775. New York: Johnson Reprint, 1966.

Pringle, John. "A Discourse on the Different Kinds of Air," *Philosophical Transactions* 63 (1774), Appendix.

Prior, Matthew. "Alma; or, The Progress of the Mind." In *Poems on Several Occasions*, ed. A. Waller, pp. 209–54. Cambridge: Cambridge University Press, 1905.

Pumfrey, Stephen. "William Gilbert's Magnetic Philosophy, 1580–1684: The Creation and Dissolution of a Discipline." PhD thesis, University of London, 1987.

———. "Mechanising Magnetism in Restoration England—the Decline of Magnetic Philosophy," *Annals of Science* 44 (1987), 1–22.

———. "'O Tempora, O Magnes!': A Sociological Analysis of the Discovery of Secular Magnetic Variation in 1634," *British Journal for the History of Science* 22 (1989), 181–214.

———. "Who Did the Work? Experimental Philosophers and Public Demonstrators in Augustan England," *British Journal for the History of Science* 28 (1995), 131–56.

Purdy, John. *Memoir, Descriptive and Explanatory, to Accompany the New Chart of the Atlantic Ocean . . .* London, 1829.

Quarles, Francis. *Francis Quarle's Emblems and Hieroglyphicks of the Life of Man, Modernised.* London, 1764.

———. *Quarles' Emblems, Divine and Moral: Together with Hieroglyphics of the Life of Man.* London, ca. 1790.

Quarrell, William, and Margaret Mare. *London in 1710 from the Travels of Zacharias Conrad von Uffenbach.* London: Faber & Faber, 1934.

Quarrie, P. "The Christ Church Collection Books." In *The History of the University of Oxford*, Vol. 5, The Eighteenth Century, ed. L. Sutherland and L. Mitchell, pp. 493–512. Oxford: Clarendon Press, 1986.

Quill, Humphrey. *John Harrison.* London: John Baker, 1966.

Quinn, Arthur. "Repulsive Force in England, 1706–1744," *Historical Studies in the Physical Sciences* 13 (1982), 109–28.

Radelet-de-Grave, Patricia. "Magnetismus." In *Die gesammelten Werke der Mathematiker und Physiker der Familie Bernouilli: Herausgegeben von der Naturforschenden Gesellschaft in Basel*, vol. 7, pp. 1–162. Basel, Boston, and Berlin: Birkhäuser Verlag, 1994.

Randier, Jean. *Marine Navigation Instruments*, trans. J. Powell. London: Murray, 1980.

Ranft, B. M. "Labour Relations in the Royal Dockyards in 1739," *Mariner's Mirror* 47 (1961), 281–91.

Raven, James. *Judging New Wealth: Popular Publishing and Responses to Commerce in England, 1750–1800.* Oxford: Clarendon Press, 1992.

Ravenhill, William. "Churchman's Contours?" *Map Collector* 34 (1986), 22–5.

Ray, John. *The Wisdom of God Manifested in the Works of the Creation.* London, 1691.

Réaumur, René. "Expériences qui montrent avec quelle facilité le fer & l'acier s'aimantent, même sans toucher l'aimant," *Mémoires de mathématique et de physique de l'Académie royale des sciences*, 1723, pp. 81–105.

Rediker, Marcus. *Between the Devil and the Deep Blue Sea: Merchant Seamen, Pirates, and the Anglo-American Maritime World, 1700–1750.* Cambridge: Cambridge University Press, 1987.

Rees, Abraham. *Cyclopædia; or, An Universal Dictionary of Arts and Sciences*, 5 vols. London, 1778.

Regnault, Noël. *Philosophical Conversations; or, A New System of Physics, by Way of Dialogue*, trans. T. Dale, 3 vols. London, 1734.

Reid, Christopher. "Burke, the Regency Crisis, and the 'Antagonistic World of Madness,'" *Eighteenth Century Life* 16 (1992), 59–75.

Reid, William H. *The Rise and Dissolution of the Infidel Societies of This Metropolis . . . from the Publication of Paine's Age of Reason until the Present Period*. London, 1800.

Revel, Jacques. "Knowledge of the Territory," *Science in Context* 4 (1991), 133–61.

Reynolds, John. *A View of Death, or, The Soul's Departure from the World*. London, 1725.

Richards, Joan. "Rigor and Clarity: Foundations of Mathematics in France and England, 1800–1840," *Science in Context* 4 (1991), 297–319.

Richardson, Ruth. *Death, Dissection and the Destitute*. London: Penguin, 1988.

Richardson, Samuel. *The History of Sir Charles Grandison; in a Series of Letters*, 7 vols. London, 1781.

Richeson, A. W. *English Land Measuring to 1800: Instruments and Practices*. Cambridge: MIT Press, 1966.

Rippy, Frances M. *Matthew Prior*. Boston: Twayne, 1986.

Ritchie, G. S. *The Admiralty Chart: British Naval Hydrography in the Nineteenth Century*. London: Hollis & Carter, 1967.

Rivoire, P. *Traités sur les aimans artificiels*. Paris, 1752.

Robb-Smith, A.H.T. "Medical Education at Oxford and Cambridge Prior to 1850." In *The Evolution of Medical Education in Britain*, ed. F.N.L. Poynter, pp. 19–52. London: Pitman Medical, 1966.

Roberts, John M. *The Mythology of the Secret Societies*. London: Secker & Warburg, 1972.

Roberts, Lissa. "Setting the Table: The Disciplinary Development of Eighteenth-Century Chemistry as Read through the Changing Structure of Its Tables." In *The Literary Structure of Scientific Argument*, ed. Peter Dear, pp. 99–132. Philadelphia: University of Pennsylvania Press, 1991.

Robertson, George. *The Discovery of Tahiti: A Journal of the Second Voyage of HMS Dolphin round the World, under the Command of Captain Wallis RN, in the Years 1766, 1767, and 1768*, ed. Hugh Carrington. London: Hakluyt Society, 1948.

Robertson, James. "Observations on the Permanency of the Variation of the Compass at Jamaica," *Philosophical Transactions* 96 (1806), 348–56.

Robertson, John. *The Elements of Navigation*, 2 vols. London, 1780.

Robinson, Arthur H. "The Genealogy of the Isopleth," *Cartographic Journal* 8 (1971), 49–53.

———. *Early Thematic Mapping in the History of Cartography*. Chicago: University of Chicago Press, 1982.

Robinson, Bryan. *A Dissertation on the Æther of Sir Isaac Newton*. Dublin, 1743.

Robinson, Thomas. *An Essay towards a Natural History of Westmorland and Cumberland*. London, 1709.

———. *A Vindication of the Philosophical and Theological Exposition of the Mosaick System of the Creation*. London, 1709.

Robison, John. *Proofs of a Conspiracy against All the Religions and Governments of Europe, Carried On in Secret Meetings of Free Masons, Illuminati, and Reading Societies*. London, 1797.

———. "Magnetism." In *Supplement to the Third Edition of the Encyclopædia Bri-*

tannica; or, A Dictionary of Arts, Sciences and Miscellaneous Literature, ed. G. Gleig, vol. 2, pp. 112–56. Edinburgh, 1801.

———. *A System of Mechanical Philosophy*, 4 vols. Edinburgh, 1822.

Rodger, Nicholas A. M. *The Wooden World: An Anatomy of the Georgian Navy*. London: Collins, 1986.

Rogers, Pat. "Gulliver and the Engineers," *Modern Language Review* 70 (1975), 260–70.

Rohault, Jacques. *Rohault's System of Natural Philosophy, Illustrated with Dr Samuel Clarke's Notes Taken Mostly Out of Sir Isaac Newton's Philosophy*, 2 vols. London, 1723.

Ronan, Colin E. *Edmond Halley*. London: Macdonald, 1970.

Rooke, Laurence. "Directions for Sea-Men, Bound for Far Voyages," *Philosophical Transactions* 1 (1666), 140–3.

Rose, Alexander. "Extract of Two Letters to Dr Murdoch, FRS," *Philosophical Transactions* 60 (1770), 444–50.

Rosenberg, Charles E. "Medical Text and Social Context: Explaining William Buchan's *Domestic Medicine*," *Bulletin of the History of Medicine* 57 (1983), 22–42.

Rossi, Paolo. *The Dark Abyss of Time: The History of the Earth and the History of Nations from Hooke to Vico*. Chicago: University of Chicago Press, 1984.

Rouse, Joseph. "Philosophy of Science and the Persistent Narratives of Modernity," *Studies in the History and Philosophy of Science* 22 (1991), 141–62.

Rousseau, George S. "Science and the Discovery of the Imagination in Enlightened England," *Eighteenth-Century Studies* 3 (1969–70), 108–35.

———. "Science Books and Their Readers in the Eighteenth Century." In *Books and Their Readers in Eighteenth-Century England*, ed. Isabel Rivers, pp. 197–255. Leicester: St. Martin's Press, 1982.

———. *The Letters and Private Papers of Sir John Hill*. New York: AMS Press, 1982.

———. "Mysticism and Millenarianism: 'Immortal Dr Cheyne,'" In *Millenarianism and Messianism in English Literature and Thought 1650–1800*, ed. Richard Popkin, pp. 81–126. Leiden: E. J. Brill, 1988.

Rousseau, George S., and Roy Porter (eds.). *The Ferment of Knowledge: Studies in the Historiography of Eighteenth-Century Science*. Cambridge: Cambridge University Press, 1980.

Rousseau, Jean-Jacques. *Emile*, trans. B. Foxley. London: Dent, 1974.

Rowbottom, Margaret, and Susskind, Charles. *Electricity and Medicine: History of Their Interaction*. San Francisco: San Francisco Press, 1984.

Rudwick, Martin. "The Emergence of a Visual Language for Geological Science 1760–1840," *History of Science* 14 (1976), 149–95.

———. *The Meaning of Fossils: Episodes in the History of Palaeontology*. Chicago: University of Chicago Press, 1985.

Ruggles, R. I. "Governor Samuel Wegg: Intelligent Layman of the Royal Society 1753–1802," *Notes and Records of the Royal Society of London* 32 (1978), 181–99.

Rupp, Gordon. *Religion in England 1688–1791*. Oxford: Clarendon Press, 1986.

Russell, Colin A. *Science and Social Change 1700–1900*. N.p.: Macmillan, 1983.

Rutt, John T. *The Theological and Miscellaneous Works, &c. of Joseph Priestley, LLD, FRS, &c*, 2 vols. London, 1817–31.

St. Clair, William. *The Godwins and the Shelleys: The Biography of a Family*. London: Faber & Faber, 1989.

Salmon, William. *The Compleat English Physician; or, The Druggist's Shop Opened*. London, 1693.

Sanderson, William. "Observations upon the Variation of the Needle Made in the Baltick," *Philosophical Transactions* 31 (1720), 120.

Sargent, John. *The Mine: A Dramatic Poem*. London, 1785.

Savery, John. "Account of the Savery Family." British Library Additional Manuscripts 44058.

Savery, Servington. "Magnetical Observations and Experiments," *Philosophical Transactions* 36 (1730), 295–340.

———. *The Sceptic*. Retford, 1800.

Schaffer, Simon. "Halley's Atheism and the End of the World," *Notes and Records of the Royal Society* 32 (1977), 17–40.

———. "Natural Philosophy and Public Spectacle in the Eighteenth Century," *History of Science* 21 (1983), 1–43.

———. "Occultism and Reason." In *Philosophy, Its History and Historiography*, ed. A. J. Holland, pp. 117–43. Dordrecht: Reidel, 1985.

———. "Scientific Discoveries and the End of Natural Philosophy," *Social Studies of Science* 16 (1986), 387–420.

———. "Godly Men and Mechanical Philosophers: Souls and Spirits in Restoration Philosophy," *Science in Context* 1 (1987), 55–85.

———. "Newton's Comets and the Transformation of Astrology." In *Astrology, Science and Society: Historical Essays*, ed. Patrick Curry, pp. 219–43. Woodbridge, Suffolk, and Wolfeboro, N.H.: Boydell Press, 1987.

———. "Defoe's Natural Philosophy and the Worlds of Credit." In *Nature Transfigured: Science and Literature, 1700–1900*, ed. John Christie and Sally Shuttleworth, pp. 13–44. Manchester: Manchester University Press, 1989.

———. "Newtonianism." In *Companion to the History of Modern Science*, ed. Robert Olby, G. N. Cantor, J.R.R. Christie, and M.J.S. Hodge. pp. 610–26. London: Routledge, 1990.

———. "Measuring Virtue: Eudiometry, Enlightenment and Pneumatic Medicine." In *The Medical Enlightenment of the Eighteenth Century*, ed. Andrew Cunningham and Roger French, pp. 281–318. Cambridge: Cambridge University Press, 1990.

———. "A Social History of Plausibility: Country, City and Calculation in Augustan Britain." In *Rethinking Social History: English Society 1570—1920 and Its Interpretation*, ed. Adrian Wilson, pp. 128–57. Manchester and New York: Manchester University Press, 1990.

———. "The Consuming Flame: Electrical Showmen and Tory Mystics in the World of Goods." In *Consumption and the World of Goods in the Eighteenth Century*, ed. John Brewer and Roy Porter, pp. 489–526. London: Routledge, 1992.

———. "Machine Philosophy: Demonstration Devices in Georgian Mechanics," *Osiris* 9 (1994), 157–82.

———. "Self Evidence." In *Questions of Evidence: Proof, Practice, and Persuasion across the Disciplines*, ed. James Chandler, Arnold Davidson, and Harry Harootnian, pp. 56–91. Chicago: University of Chicago Press, 1994.

———. "The Show That Never Ends: Displays of Perpetual Motion in the Early Eighteenth Century," *British Journal for the History of Science* 28 (1995), 157–89.

Schenk, Jerome. "The History of Electrotherapy," *American Journal of Psychiatry* 116 (1959), 463–4.

Schiebinger, Londa. *The Mind Has No Sex? Women in the Origins of Modern Science*. Cambridge: Harvard University Press, 1989.
Schofield, Robert E. "John Wesley and Science in Eighteenth-Century England," *Isis* 44 (1953), 331–40.
———. *The Lunar Society of Birmingham*. Oxford: Clarendon Press, 1963.
———. *Mechanism and Materialism: British Natural Philosophy in an Age of Reason*. Princeton: Princeton University Press, 1970.
Schuchard, M. Keith. "Freemasons, Secret Societies, and the Continuity of the Occult Traditions in English Literature." Ph.D. thesis, University of Texas at Austin, 1975.
———. "Swedenborg, Jacobitism and Freemasonry." In *Swedenborg and His Influence*, ed. E. J. Brock, pp. 359–79. London: Swedenborg Society, 1988.
Schuster, John A. "Methodologies as Mythic Structures: A Preface to the Future Historiography of Method," *Metascience* 1(2) (1984), 15–36.
Schuster, John A., and Graeme Watchirs. "Natural Philosophy, Experiment and Discourse: Beyond the Kuhn/Bachelard Problematic." In *Experimental Inquiries: Historical, Philosophical and Social Studies of Experimentation in Science*, ed. Homer Le Grand, pp. 1–47. Dordrecht: Kluwer Academic, 1990.
Schwartz, Hillel. *Knaves, Fools, Madmen, and That Subtile Effluvium: A Study of the Opposition to the French Prophets in England, 1706–1710*. University of Florida Monographs, Social Sciences no. 62. Gainesville: University Presses of Florida, 1978.
Secord, James A. "Newton in the Nursery: Tom Telescope and the Philosophy of Tops and Balls, 1761–1838," *History of Science* 23 (1985), 127–51.
Sekora, John. *Luxury: The Concept in Western Thought, Eden to Smollett*. Baltimore and London: Johns Hopkins University Press, 1977.
Sennett, Richard. *The Fall of Public Man*. Cambridge: Cambridge University Press, 1977.
———. *Flesh and Stone: The Body and the City in Western Civilization*. London and Boston: Faber & Faber, 1994.
Serres, Michel. *Hermes: Literature, Science, Philosophy*. Baltimore and London: Johns Hopkins University Press, 1982.
Seymour, W. A. *A History of the Ordnance Survey*. Folkestone: Dawson & Sons, 1980.
Shapin, Steven. "Of Gods and Kings: Natural Philosophy and Politics in the Leibniz-Clarke Disputes," *Isis* 72 (1981), 187–215.
———. "The House of Experiment in Seventeenth-Century England," *Isis* 79 (1988), 373–404.
———. "Robert Boyle and Mathematics: Reality, Representation, and Experimental Practice," *Science in Context* 2 (1988), 23–58.
Shapin, Steven, and Simon Schaffer. *Leviathan and the Air-Pump: Hobbes, Boyle and the Experimental Life*. Princeton: Princeton University Press, 1985.
Shelley, Mary. *Frankenstein or The Modern Prometheus*. 1818. Ed. Maurice Hindle. London: Penguin, 1992.
Shelley, Percy B. *The Works of Percy Bysshe Shelley in Verse and Prose*, 8 vols., ed. H. B. Forman. London: Reeves & Turner, 1880.
Shiach, Morag. *Discourse on Popular Culture: Class, Gender and History in Cultural Analysis, 1730 to the Present*. Cambridge: Polity Press, 1989.
Shorr, Philip. *Science and Superstition in the Eighteenth Century*. New York: Columbia University Press, 1932.

Shuckburgh, George. "On the Variation of the Temperature of Boiling Water," *Philosophical Transactions* 69 (1779), 362–75.

Sibly, Ebenezer. *A Key to Physic, and the Occult Sciences*. London, 1794.

Singer, D. W. "Sir John Pringle and His Circle," *Annals of Science* 6 (1949), 127–80.

Sisco, Anneliese G. *Réaumur's Memoirs on Steel and Iron*. Chicago: University of Chicago Press, 1956.

Skempton, A. W. *John Smeaton, FRS*. London: Thomas Telford, 1981.

Smart, Christopher. *The Poetical Works of Christopher Smart*, 4 vols., ed. Karina Williamson. Oxford: Clarendon Press, 1980–7.

Smeaton, John. "An Account of Some Improvements of the Mariners Compass, in Order to Render the Card and Needle, Proposed by Dr Knight, of General Use," *Philosophical Transactions* 46 (1750), 513–17.

Smeaton, John. *Reports of the Late John Smeaton, Frs*, 4 vols. London: Longman, 1812–14.

Smiles, Samuel. *Lives of the Engineers: Rennie and Smeaton*. London: John Murray, 1891.

Smith, Adam. *Essays on Philosophical Subjects*, ed. W.P.D. Wightman and J. C. Bryce. Oxford: Clarendon Press, 1980.

Smith, Crosbie, and M. Norton Wise. *Energy and Empire: A Biographical Study of Lord Kelvin*. Cambridge: Cambridge University Press, 1989.

Smith, David. *Antique Maps of the British Isles*. London: Batsford, 1982.

Smith, Olivia. *The Politics of Language 1791–1819*. Oxford: Clarendon Press, 1984.

Smith, Roger. *Inhibition: History and Meaning in the Sciences of Mind and Brain*. Berkeley and Los Angeles: University of California Press, 1992.

Smollett, Tobias. *The Expedition of Humphry Clinker*, ed. Lewis M. Knapp, rev. Paul-Gabriel Boucé. Oxford: Oxford University Press, 1984.

Solkin, David H. *Painting for Money: The Visual Arts and the Public Sphere in Eighteenth-Century England.* New Haven and London: Yale University Press, 1993.

Solomon, Harry M. *Sir Richard Blackmore*. Boston: Twayne, 1980.

Southey, Robert. *Letters from England*, ed. J. Simmons. London: Cresset Press, 1951.

Spadafora, David. *The Idea of Progress in Eighteenth-Century Britain.* New Haven and London: Yale University Press, 1990.

Spears, M. K. "Matthew Prior's Attitude toward Natural Science," *Publications of the Modern Language Association* 63 (1948), 485–507.

Spinney, David. *Rodney*. London: Allen & Unwin, 1969.

Stafford, Barbara M. *Body Criticism: Imaging the Unseen in Enlightenment Art and Medicine*. Cambridge: MIT Press, 1991.

———. *Artful Science: Enlightenment Entertainment and the Eclipse of Visual Education*. Cambridge and London: MIT Press, 1994.

Star, Susan L., and Jane R. Griesemer. "Institutional Ecology, 'Translations' and Boundary Objects: Amateurs and Professionals in Berkeley's Museum of Vertebrate Zoology, 1907–39,' *Social Studies of Science* 19 (1989), 385–420.

Stearns, Raymond P. "The Course of Capt. Edmond Halley in the Year 1700," *Annals of Science* 1 (1936), 294–301.

———. *Science in the British Colonies of America*. Urbana: University of Illinois Press, 1970.

Steven, Margaret. *Trade, Tactics and Territory: Britain in the Pacific, 1783–1823*. Melbourne: Melbourne University Press, 1983.

Stevenson, B. *The Home Book of Quotations*. New York: Dodd, Mead, 1967.

Stewart, Larry. "Samuel Clarke, Newtonianism and the Factions of Post-Revolutionary England," *Journal of the History of Ideas* 42 (1981), 53–72.

———. "Texts and Contextualists: The Hunting of Newtonianism," *Historical Studies in the Physical and Biological Sciences* 19 (1988), 193–7.

———. *The Rise of Public Science: Rhetoric, Technology, and Natural Philosophy in Newtonian Britain, 1660–1750*. Cambridge: Cambridge University Press, 1992.

Stirling, James, Peter Brown, William Watts, and William Dearn. *A Course of Mechanical and Experimental Philosophy*. 1727.

Stone, Edmund. *The Construction and Principle Uses of Mathematical Instruments. Translated from the French of M. Bion*. London, 1758.

Stone, Lawrence. "The Size and Composition of the Oxford Student Body 1580–1910." In *The University in Society*, vol. 1, ed. Lawrence Stone, pp. 3–111. Princeton: Princeton University Press, 1975.

Stukeley, William. *Stonehenge a Temple Restor'd to the British Druids*. London, 1740.

Sturgeon, William. *Scientific Researches, Experimental and Theoretical, in Electricity, Magnetism. . .* Bury: Thomas Crompton, 1850.

Sudduth, William M. "Eighteenth-Century Identifications of Phlogiston with Electricity," *Ambix* 25 (1978), 131–47.

Sulivan, Richard J. *A View of Nature, in Letters to a Traveller among the Alps*, 6 vols. London, 1794.

Sutherland, L. S., and L. G. Mitchell (eds.). *The History of the University of Oxford*. Vol. 5, *The Eighteenth Century*. Oxford: Clarendon Press, 1986.

Sutton, Geoffrey. "Electric Medicine and Mesmerism," *Isis* 72 (1981), 375–92.

Swedenborg, Emanuel. *The Principia; or, The First Principles of Natural Things*, 2 vols., trans. J. Rendell and I. Tansley. London: Swedenborg Society, 1912.

Swift, Jonathan. *Gulliver's Travels*. 1738. London: Dent, 1966.

Sydenham, Peter H. *Measuring Instruments: Tools of Knowledge and Control*. London: Science Museum, 1979.

Talon, Henri. *Selections from the Journals and Papers of John Byrom*. London: Rockliff, 1950.

Tatar, Maria. *Spellbound: Studies on Mesmerism and Literature*. Princeton: Princeton University Press, 1978.

Tavor, Eve. *Scepticism, Society and the Eighteenth-Century Novel*. London: Macmillan, 1987.

Taylor, Brook. "An Account of an Experiment Made by Dr Brook Taylor Assisted by Mr Hawkesbee, in Order to Discover the Law of the Magnetical Attraction," *Philosophical Transactions* 29 (1715), 294–5.

———. "Extract of a Letter from Dr *Brook Taylor FRS* to Sir Hans Sloan, Dated 25. June, 1714. Giving an Account of Some Experiments Relating to Magnetism," *Philosophical Transactions* 31 (1721), 204–8.

Taylor, E.G.R. *The Mathematical Practitioners of Tudor and Stuart England*. Cambridge: Cambridge University Press, 1954.

———. *The Haven-finding Art: Navigation from Odysseus to Cook*. London: Hollis & Carter, 1956.

———. "A Reward for the Longitude," *Mariner's Mirror* 45 (1959), 59–66.

———. *The Mathematical Practitioners of Hanoverian England*. Cambridge: Cambridge University Press, 1966.

Telescope, Tom. *The Newtonian System of Philosophy Adapted to the Capacities of Young Gentlemen and Ladies, and Familiarised and Made Entertaining by Objects with Which They Are Intimately Acquainted*. London, 1779.

Telford, John (ed.). *The Letters of the Rev. John Wesley, AM*, 8 vols. London: Epworth Press, 1960.

Temple, William. *The Works of Sir William Temple, Bart*. London, 1814.

Templeman, Peter. *Curious Remarks and Observations in Physics, Anatomy, Chirurgery, Chemistry, Botany and Medicine. Extracted from the History and Memoirs of the Royal Academy of Sciences at Paris*. London, 1753.

Thacker, Jeremy. *The Longitudes Examin'd*. London, 1714.

Thackray, Arnold. *Atoms and Powers: An Essay on Newtonian Matter-Theory and the Development of Chemistry*. Cambridge: Harvard University Press, 1970.

Thomas, Keith. *Religion and the Decline of Magic*. Harmondsworth: Penguin, 1971.

Thompson, E. P. "The Moral Economy of the English Crowd in the Eighteenth Century," *Past and Present* 50 (1971), 76–136.

———. "Eighteenth-Century English Society: Class Struggle without Class?" *Social History* 3 (1978), 133–65.

———. *Customs in Common*. London: Penguin, 1991.

Thorndike, Lynn. "L'Encyclopédie and the History of Science," *Isis* 6 (1924), 361–86.

Thrower, Norman J. "Edmond Halley and Thematic Geo-cartography." In *The Compleat Plattmaker: Essays on Chart, Map, and Globe Making in England in the Seventeenth and Eighteenth Centuries*, ed. Norman Thrower, pp. 195–228. Berkeley and Los Angeles: University of California Press, 1978.

———. *The Three Voyages of Edmond Halley in the Paramore 1698–1701*, 2 vols. London: Hakluyt Society, 1981.

Tiles, Mary. *Bachelard: Science and Objectivity*. Cambridge: Cambridge University Press, 1984.

Tilling, Laura. "The Interpretation of Observational Errors in the Eighteenth and Early Nineteenth Centuries." Ph.D. thesis, University of London, 1973.

———. "Early Experimental Graphs," *British Journal for the History of Science* 8 (1975), 193–213.

Todd, Dennis. "Laputa, the Whore of Babylon, and the Idols of Science," *Studies in Philology* 75 (1978), 93–120.

Todd, Janet. *Sensibility: An Introduction*. London: Methuen, 1986.

Torlais, Jean. *Un physicien au siècle des lumières: l'abbé Nollet*. Paris: SIPUCO, 1954.

Townson, Robert. *Philosophy of Mineralogy*. London, 1798.

Toynbee, Paget, and L. Whibley (eds.). *Correspondence of Thomas Gray*, 3 vols. Oxford: Clarendon Press, 1935.

Trapp, Joseph. *Lectures on Poetry Read in the Schools of Natural Philosophy at Oxford*. London, 1742.

Trenchard, John. *The Natural History of Superstition*. 1709.

Tucker, Abraham. *The Light of Nature Pursued*, 2 vols. London, 1768.

Turnbull, H. W. (ed.). *The Correspondence of Sir Isaac Newton*, 3 vols. Cambridge: Cambridge University Press, 1959–61.

Turner, Daniel. "A Defence of the XIIth Chapter of the First Part of a Treatise *De Morbis Cutaneis*." In *A Discourse concerning Gleets*, by Daniel Turner, pp. 67–162. London, 1729.

Turner, Gerard L'E. "The Physical Sciences." In *The History of the University of Ox-*

ford. Vol. 5, *The Eighteenth Century*, ed. L. Sutherland and L. Mitchell, pp. 659–82. Oxford: Clarendon Press, 1986.

Turner, Gerard L'E., and Trevor H. Levere. *Martinus van Marum*. Vol. 4, *Van Marum's Scientific Instruments in Teyler's Museum*. Leiden: Noordhoff, 1973.

Tyacke, Sarah. "Map-Sellers and the London Map Trade c1650–1710." In *My Head Is a Map: Essays and Memoirs in Honour of R. V. Tooley*, ed. Helen Wallis and Sarah Tyacke, pp. 62–80. London: Francis Edwards & Carta Press, 1973.

———. *The Vanity of Philosophick Systems*. London, 1761.

Varley, ?. "On the Irregularity in the Rate of Going of Time-Pieces Occasioned by the Influence of Magnetism," *Philosophical Magazine* 1 (1798), 16–21.

Verschuur, Gerrit L. *Hidden Attraction: The Mystery and History of Magnetism*. Oxford: Oxford University Press, 1993.

Vince, Samuel. *A Plan of a Course of Lectures on the Principles of Natural Philosophy*. Cambridge, 1793.

Vincent, William. *The Commerce and Navigation of the Ancients in the Indian Ocean*, 2 vols. London, 1807.

Voltaire (François Marie Arouet). *Letters on England*, trans. Leonard Tancock. London: Penguin, 1980.

Waddell, J. "A Letter . . . concerning the Effects of Lightning in Destroying the Polarity of a Mariners Compass," *Philosophical Transactions* 46 (1749), 111–12.

Waddington, Robert. *An Epitome of Theoretical and Practical Navigation*. London, 1777.

Wagner, Peter. "The Discourse on Sex—Or Sex as a Discourse: Eighteenth-Century Medical and Paramedical Erotica." In *Sexual Underworlds of the Enlightenment*, ed. George S. Rousseau and Roy Porter, pp. 46–68. Manchester: Manchester University Press, 1987.

———. *Eros Revived: Erotica of the Enlightenment in England and America*. London: Paladin, 1990.

Wakeley, Andrew. *The Mariner's Compass Rectified*. London, 1761.

Wales, William, and William Bayly. *The Original Astronomical Observations, Made in the Course of a Voyage towards the South Pole*. London, 1777.

Walker, Adam. *A System of Familiar Philosophy: In Twelve Lectures; Being the Course Usually Read by Mr A. Walker*, 2 vols. London, 1802.

Walker, Arthur. *William Law: His Life and Thought*. London: SPCK, 1973.

Walker, D. P. *The Decline of Hell*. London: Routledge & Kegan Paul, 1964.

Walker, Ralph. *The Memorial of Ralph Walker to the Honorable the Board of Longitude*. 1794.

———. *A Treatise on Magnetism, with a Description and Explanation of a Meridional and Azimuth Compass, for Ascertaining the Quantity of Variation, without Any Calculation Whatever, at Any Time of the Day*. London, 1794.

Walker, William. *The Life and Times of the Rev. John Skinner MA*. London: Skeffington & Son, 1883.

Wallis, Helen. "Maps as a Medium of Scientific Communication." In *Studia z dziejów geografii i kartografii*, ed. Józef Babicz, pp. 251–62. Breslau: Polska Academia Nauk, 1973.

———. "Geographie Is Better than Divinitie: Maps, Globes, and Geography in the Days of Samuel Pepys." In *The Compleat Plattmaker: Essays on Chart, Map and Globe Making in the Seventeenth and Eighteenth Centuries*, ed. Norman Thrower, pp. 1–43. Berkeley and Los Angeles: University of California Press, 1978.

Wallis, Helen, and H. Robinson. *Cartographical Innovations: An International Handbook of Mapping Terms to 1900*. Map Collector Publications in association with the International Cartographic Association, 1987.

Wallis, John. "A Second Letter of Dr Wallis . . .: and, Some Magnetick Affairs," *Philosophical Transactions* 22 (1701), 1030–8.

———. "A Letter of Dr Wallis to Captain Edmund Halley; Concerning the Captain's Map of *Magnetick Variations*; and Some Other Things Relating to the Magnet," *Philosophical Transactions* 23 (1702), 1106–12.

Wallis, Peter J. "British Philomaths—Mid-Eighteenth Century and Earlier," *Centaurus* 17 (1972), 301–14.

Walsh, James W. *Cures*. New York: Appleton, 1923.

Walters, Alice N. "Tools of Enlightenment: The Material Culture of Science in Eighteenth-Century England." Ph.D. thesis, University of California at Berkeley, 1992.

Walton, Christopher. *Notes and Materials for an Adequate Biography of the Celebrated Divine and Theosopher, William Law*. London, 1854.

Ward, Ned. *The London-Spy Compleat*. London: Casanova Society, 1924.

Ward, W. R. *Georgian Oxford*. Oxford: Clarendon Press, 1958.

Wargentin, Pehr. "A Letter from the Secretary of the Royal Academy of Sciences in Sweden concerning the Variation of the Magnetic Needle," *Philosophical Transactions* 47 (1750), 126–31.

Warner, Deborah J. "What Is a Scientific Instrument, Where Did It Become One, and Why?" *British Journal for the History of Science* 23 (1990), 83–93.

———. "Terrestrial Magnetism: For the Glory of God and the Benefit of Mankind," *Osiris* 9 (1994), 67–84.

Warter, John (ed.). *Southey's Common-Place Book*, 4 vols. London: Longman et al., 1849–51.

Wasserman, Earl R. "The Sympathetic Imagination in Eighteenth-Century Theories of Acting," *Journal of English and German Philology* 46 (1947), 265–72.

———. "Nature Moralised: The Divine Analogy in the Eighteenth Century," *Journal of English Literary History* 20 (1953), 39–76.

Waters, David W. "Nautical Astronomy and the Problem of Longitude." in *The Uses of Science in the Age of Newton*, ed. John Burke, pp. 143–69. Berkeley and Los Angeles: University of California Press, 1983.

Watt, Robert. *Bibliotheca Britannica; or, A General Index to British and Foreign Literature*, 4 vols. Edinburgh, 1824.

Weatherill, Lorna. *Consumer Behaviour and Material Culture in Britain 1660–1760*. London: Routledge, 1988.

Webster, A. B. *Joshua Watson: The Story of a Layman 1771–1855*. London: SPCK, 1954.

Webster, Charles. "The Medical Faculty and the Physic Garden." In *The History of the University of Oxford*. Vol. 5, *The Eighteenth Century*, eds. L. Sutherland and L. Mitchell, pp. 683–724. Oxford: Clarendon Press, 1986.

Weisinger, Herbert. "English Treatment of the Relationship between the Rise of Science and the Renaissance, 1740–1840," *Annals of Science* 7 (1951), 248–74.

Weld, Charles R. *A History of the Royal Society, with Memoirs of the Presidents*, 2 vols. London: John W. Parker, 1848.

Werenfels, Samuel. *A Dissertation upon Superstition in Natural Things*. London, 1748.

Westfall, Richard S. *Force in Newton's Physics: The Science of Dynamics in the Seventeenth Century*. New York: Science History Publications, 1971.

Weyant, Robert G. "Protoscience, Pseudoscience, Metaphors and Animal Magnetism." In *Science, Pseudo-science and Society*, ed. Marsha Hanen, Margaret Osler, and Robert G. Weyant, pp. 77–114. Waterloo: Wilfrid Laurier University Press, 1980.

Wheatland, David P. *The Apparatus of Science at Harvard 1765–1800*. Cambridge: Harvard University Press, 1968.

Whewell, William. *A History of the Inductive Sciences, from the Earliest to the Present Time*, 3 vols. London, 1857.

Whiston, William. *Astronomical Principles of Religion, Natural and Reveal'd*. London, 1717.

———. *The Longitude and Latitude Found by the Inclinatory or Dipping Needle; Wherein the Laws of Magnetism Are Also Discover'd*. London, 1721.

———. *A New Theory of the Earth, from Its Original, to the Consummation of All Things*. London, 1722.

———. *The Calculation of Solar Eclipses without Parallaxes*. London, 1724.

———. *The Eternity of Hell Torments Considered: or, A Collection of Texts of Scripture, and Testimonies of the Three First Centuries, Relating to Them*. London, 1740.

———. *Memoirs of the Life and Writings of Mr William Whiston*, 2 vols. London, 1753.

White, John. *A Rich Cabinet, with Variety of Inventions, Unlock'd and Open'd, for the Recreation of Ingenious Spirits*. London, 1689.

Widmalm, Sven. "Accuracy, Rhetoric, and Technology: The Paris-Greenwich Triangulation, 1784–88." In *The Quantifying Spirit in the Eighteenth Century*, ed. Tore Frängsmyr, John Heilbron, and Robin Rider, pp. 179–206. Berkeley and Los Angeles: University of California Press, 1990.

Wilde, Christopher B. "Hutchinsonianism, Natural Philosophy and Religious Controversy in Eighteenth-Century Britain," *History of Science* 18 (1980), 1–24.

———. "Matter and Spirit as Natural Symbols in Eighteenth-Century British Natural Philosophy," *British Journal for the History of Science* 15 (1982), 99–131.

Williams, Carolyn. "The Changing Face of Change: Fe/male In/constancy," *British Journal for Eighteenth-Century Studies* 12 (1989), 13–28.

Williams, Glyndwyr. *Documents Relating to Anson's Voyage round the World 1740–44*. London, 1967 (for the Navy Records Society).

Williams, Zachariah. *The Mariners Compass Compleated; or, Tthe Expert Seaman's Best Guide*, 2 pts. in 1. London, 1745.

———. *An Account of an Attempt to Ascertain the Longitude at Sea, by an Exact Theory of the Variation of the Magnetical Needle*. London, 1755.

Williamson, Karina. "Smart's *Principia*: Science and Anti-science in *Jubilate Agno*," *Review of English Studies* 30 (1979), 409–22.

Wilson, Adrian. "Foundations of an Integrated Historiography." In *Rethinking Social History: English Society 1570–1920 and Its Interpretation*, ed. Adrian Wilson, pp. 293–335. Manchester and New York: Manchester University Press, 1993.

———. "The Politics of Medical Improvement in Early Hanoverian London." In *The Medical Enlightenment of the Eighteenth Century*, ed. Andrew Cunningham and Roger French, pp. 4–39. Cambridge: Cambridge University Press, 1990.

Wilson, Andrew. *The Principles of Natural Philosophy: With Some Remarks upon the*

Fundamental Principles of the Newtonian Philosophy; in an Introductory Letter to Sir Hildebrand Jacob, Bart. London, 1754.

Wilson, Benjamin. *A Treatise on Electricity.* London, 1750.

———. "Account of Dr Knight's Method of Making Artificial Loadstones," *Philosophical Transactions* 69 (1779), 51–3.

———. "Autobiography." 1783. Unpublished typescript. National Portrait Gallery.

Wilson, Lindsay. *Women and Medicine in the French Enlightenment: The Debate over "Maladies des Femmes."* Baltimore: Johns Hopkins University Press, 1993.

Wilson, Philip. "'Out of Sight, Out of Mind?': The Daniel Turner–James Blondel Dispute over the Power of the Maternal Imagination," *Annals of Science* 49 (1992), 63–85.

Winkler, John H. *Elements of Natural Philosophy Delineated*, 2 vols. London, 1757.

Winter, Alison. "'The Island of Mesmeria': The Politics of Mesmerism in Early Victorian Britain." Ph.D. thesis, University of Cambridge, 1992.

Winter, Heinrich. "The Pseudo-Labrador and the Oblique Meridian," *Imago Mundi* 2 (1937), 61–73.

Wollstonecraft, Mary. *Vindication of the Rights of Woman: With Strictures on Political and Moral Subjects.* London, 1792.

Wonnacott, W. "Martin Clare and the Defence of Masonry," *Ars Quatuor Coronatorum* 28 (1915), 80–110.

Wood, Alexander, and Frank Oldham. *Thomas Young: Natural Philosopher 1783–1829.* Cambridge: Cambridge University Press, 1954.

Woodcroft, Bennett. *Subject-Matter Index of Patents of Invention*, 2 vols. London: Eyre & Spottiswoode, 1854.

Woodward, John. *Fossils of All Kinds, Digested into a Method, Suitable to Their Mutual Relation and Affinity; with the Names by Which They Were Known to the Antients, and Those by Which They Are at This Day Known: and Notes Conducing to the Setting Forth the Natural History, and the Main Uses, of Some of the Most Considerable of Them.* London, 1728.

Woolf, Harry. *The Transits of Venus.* Princeton: Princeton University Press, 1959.

Wormhoudt, Arthur. "Newton's Natural Philosophy in the Behmenistic Works of William Law," *Journal of the History of Ideas* 10 (1949), 411–29.

Worster, Benjamin. *A Compendious and Methodical Account of the Principles of Natural Philosophy: As They Are Explain'd and Illustrated in the Course of Experiments, Perform'd at the Academy in Little Tower-Street.* London, 1722.

Worster, Benjamin, and Thomas Watts. *A Course of Experimental Philosophy.* 1725?.

Wotton, William. *Reflections upon Ancient and Modern Learning.* London, 1705.

Wrightson, Keith. "The Enclosure of English Social History." In *Rethinking Social History: English Society 1570–1920 and Its Interpretation*, ed. Adrian Wilson, pp. 59–67. Manchester and New York: Manchester University Press, 1993.

Wynter, Harriet. *A Catalogue of Scientific Instruments Related to the Sea.* London: Author, 1977.

Yeo, Richard. "Genius, Method and Morality: Images of Newton in Britain, 1760–1860," *Science in Context* 2 (1988), 257–84.

———. "Reading Encyclopedias," *Isis* 82 (1991), 24–49.

Yolton, John W. *Thinking Matter: Materialism in Eighteenth-Century Britain.* Oxford: Basil Blackwell, 1983.

———. "Schoolmen, Logic and Philosophy." In *The History of the University of Ox-*

ford. Vol. 5: *The Eighteenth Century*, eds. L. Sutherland and L. Mitchell, pp. 565–92. Oxford: Clarendon Press, 1986.

Young, Matthew. *An Analysis of the Principles of Natural Philosophy*. Dublin, 1800.

Young, Thomas. *A Course of Lectures on Natural Philosophy and the Mechanical Arts*, 2 vols. London, 1807.

Zilsel, Edgar. "The Origins of William Gilbert's Scientific Method," *Journal of the History of Ideas* 2 (1941), 1–32.

Zirkle, Conway. "The Theory of Concentric Spheres: Edmond Halley, Cotton Mather, and John Cleves Symmes," *Isis* 37 (1947), 155–9.

Zwicker, Steven N. *Politics and Language in Dryden's Poetry: The Arts of Disguise*. Princeton: Princeton University Press, 1984.

INDEX

Abercorn, Earl of, 40, 47, 50, 125
adamant, 150, 151, 161–3, 255n.14. *See also* diamond
Adams, George, 21, 54, 249n.29; on French, 130–1, 170; Hutchinsonian ideas of, 28, 170; as instrument seller, 42, 55–7, 83, 136, 217; as Knight's agent, 43, 81, 83, 87, 135, 210
Admiralty, 75, 105, 109–10, 161; attitude towards invention, 72, 78, 82–5, 86–90; criticisms of, 75; Knight's negotiations with, 52, 79–88. *See also* maritime practitioners, Royal Navy
Aepinus, Franz, 16, 18, 129, 130, 184
America, 42, 93, 96–7, 109, 131, 215
Anderson, Adam, 166
animal magnetism, 6, 9, 19, 61, 175, 195–207
Anson, George: at Admiralty, 73, 82, 84, 87, 228n.40; as circumnavigator, 66, 73, 96
Arden, James, 55
Arderon, William, 52–3, 139–40, 168, 254n.87
artificial magnets: of Knight, 15, 16, 18, 42–3, 52–3, 55, 84, 161; methods of making, 40–1, 52–3, 55, 118–19, 128, 143–4; prices of, 55, 125; significance of, 16, 42, 50–3, 84
attraction. *See* God, gravitational attraction, inverse-square law, love, sexual attraction, sympathetic attraction
aurora borealis, 99–100, 101, 135, 143, 182; Halley on, 154, 178, 256nn. 27, 28

Babbage, Charles, 4
Bacon, Francis, 151, 163
Bacon, Roger, 166
Baconian ideology, 22, 51–2, 79, 99, 163, 164, 184, 209, 210
Baker, Henry, 19, 42, 44, 52, 139, 140, 168
Banks, Joseph, 53, 89, 90, 103–5, 124; empire of, 37, 96, 144
Baratier, John, 215
Barlow, Peter, 4, 89, 116, 246n.76
Barrow, John, 212, 213
Barry, James, frontispiece, 3, 5, 69, 166, 209, 213
Bate, Julius, 27, 176–7, 180
Beddoes, Thomas, 199, 204
Behmenism, 28–30, 173, 183, 211; literary influence of, 158, 187–90
Bell, John, 196–7, 206
Bennet, Abraham, 137
Bennett, John, 51, 118
Berkeley, George, 48; on mathematics, 27, 163; religious beliefs of, 27, 172; *Siris*, 27, 183; son of (George), 27
Bernoulli, Daniel, 17
Bewley, William, 64
Billingsley, Case, 169
binnacle, 67, 78, 86, 145
Birch, Thomas, 44, 45, 60
Blackmore, Richard, 158–9
Blake, William, 28, 212
Bligh, William, 88, 105
Blunt, Thomas, 134
Boehme, Jakob. See Behemenism
Boerhaave, Herman, 182, 183
Boscovich, Roger, 181
Boyle, Robert, 23, 124, 141, 142, 175–6
brass, 80, 139–40, 252n.62
Breslaw, Phillip, 61–2
Brewer, John, 7, 33
British Museum, 37, 43, 44–6
Brooke, Frances, 192
Brooke, Henry, 189
Brothers, Richard, 204
Brown, Thomas, 159–61
Browning, John, 179
Bryan, Margaret, 147, 170, 260n.14
Buchan, William, 197
Burke, Edmund, 28, 200, 202, 204, 209; on mathematics, 30, 122, 131
Burnet, Thomas, 153, 155
Burney, Frances, 191, 194, 206, 212
Bute, Earl of, 28
Butler, Bishop, 157–8, 172
Butterfield, Herbert, 4
Byron, George Gordon, 192–3

cabinet collections, 6, 47–50, 139
Calandrini, Jean-Louis, 126, 249n.25
Cambridge, 27, 50, 129, 180, 184, 185, 211
Campbell, George, 122
Canterbury, Archbishop of, 24

Canton, John, 22, 25, 87, 145; artificial magnets of, 118–19, 125, 143–4, 161; and Knight, 46, 53, 79; and terrestrial magnetism, 132, 135, 142
Cartesian. *See* Descartes
Catcott, Alexander, 27, 155
Cavallo, Tiberius, 15, 20, 22, 61, 140; on magnetic therapies, 194, 200; on terrestrial magnetism, 132, 141; *Treatise on Magnetism*, 54, 121, 128, 141
Cavendish, Henry, 25, 137
Chamberlen, ?Paul, 37
Chambers, Ephraim, 11, 142
Chapple, William, 145
charts, navigational, 91–117; of Halley, 93–4, 108–13; influence of on natural philosophy, 17, 95, 105–6, 108–9, 112–17
chemistry, 140, 180, 182; discipline of, 106, 204, 213; lecturers in, 61, 98
Cheyne, George, 36, 166
childbirth, 6, 150–1
Christ's Hospital, 35
Churchman, John, 109, 156, 215
circulation, 177, 178, 182, 199, 209
Clark, Jonathan, 33
class, 9, 34–6, 64–5, 122–3, 171–2, 193–5, 201–2, 209–12
classification, 106, 138–40; Linnean system of, 45, 251n.57. *See also under* encyclopedias
Claudian, 151, 152, 161
Cleland, John, 190
Cockburn, Patrick, 155–6
Coleridge, Samuel Taylor, 96, 146, 152, 153, 157; on animal magnetism, 202–4, 206
Colley, Linda, 7, 34
Collinson, Peter, 42, 44
commercialization, 7–8, 31–9, 46–60, 69, 91–2, 98–9, 141, 208–9, 225n.6
compass dials, 50, 118
compasses
—attitudes towards, 67–70, 72, 75–9, 98–9, 152
—azimuth: design of, 67–8, 77, 79–81, 135–8, 215–17; prices of, 81–3, 85; use of, 69–70, 77–8, 82–6
—dip needle, 67, 132–5, 215–17
—as metaphor, 4, 69–70
—in mines, 19, 31, 50
—origins of, 9, 152, 164–60
—steering: design of, 67, 76, 79–81; prices of, 81–3, 85; use of, 69–70, 77–8, 82–6
—surveying, 97–8, 142
—variation, 135–8
Comus, 61
conjuring, 47, 56, 60–5
consumerism. *See* commercialization
Cook, James, 3, 46, 86, 88, 102, 103, 104, 105, 135; on compasses, 84, 85
Cooke, William, 168, 184
Copley Medal, 23, 39, 41, 53, 96, 99
Cornwall, Captain, 107
Cosway, Maria, 200, 206
Cosway, Richard, 197, 200
Cottle, Joseph, 69
Coulomb, Charles, 16, 53, 140, 221n.17
Cox, James, 91, 92
Croker, Temple Henry, 58–60, 87, 234n.106

da Costa, Emmanuel, 46
Dahl, Michael, 154
Dalton, John, 182
Darwin, Erasmus, 61, 199; *Botanic Garden*, 53, 61, 66, 192, 193, 213
Davy, Humphry, 74, 89, 204
Day, Thomas, 62
de Loutherbourg, Philip, 197, 202–3, 205
de Mainauduc, John, 197–200, 202, 204, 205, 206
declination. *See* variation
Defoe, Daniel, 60, 98, 149, 165, 166
Delaval, Francis, 61–2
Demainbray, Stephen, 56
Derham, William: magnetic experiments of, 133–4; as natural theologian, 100, 157, 160–1, 167
Desaguliers, John, 39, 56, 118, 122, 128, 169; magnetic experiments of, 21, 126, 177; as Newtonian, 15, 21, 121, 180
Descartes, René, 18; theories of, 16, 17, 72, 174, 178–9
diamond, 150, 151, 180. *See also* adamant
dictionaries. *See* encyclopedias
Digby, Kenelm, 150
dip, 67, 99, 103, 114, 132–5, 215–17, 245n.56
Dodson, James, 102, 103, 108, 112, 113
Douglas, Charles, 88
Dryden, John, 189–90
Dubois, Dorothea, 185–6

East India Company, 17, 95, 105
Eberhart, Christopher, 215
Edgeworth, Maria, 64, 213
Edgeworth, Richard Lovell, 61, 62, 64
effluvia, 14, 24, 147, 149, 170, 177–9, 197, 198, 261n.222. *See also* ethers, fluids
electricity, 5, 7, 181, 182; comparisons of

with magnetism, 11, 13, 18–20, 47, 121; Leiden jar, 18, 20, 115. *See also* electromagnetic field
electromagnetic field, 147, 184; as metaphor, 6, 9–10, 174
encyclopedias, 118, 149, 194; Chambers' *Cyclopædia*, 11, 142; classification of knowledge in, 11–13, 20, 29, 60, 141–4; *Encyclopædia Britannica*, 15, 20, 28, 62–4, 130, 143, 200, 224n.58; l'*Encyclopédie*, 11. *See also* Falconer; Harris, John; Johnson; maps of knowledge
Enlightenment: in England, 8, 11, 15, 40; rhetoric of, 15, 29, 60–1, 98, 101, 122–3, 194, 201, 206
ethers, 43–4, 147, 150, 177, 179–85, 197. *See also* effluvia, fire, fluids
Euler, Leonhard, 17, 142
experimentation, 8, 16–20, 21–2, 79–81, 82–4, 118–45, 181–2. *See also* Royal Society, experiments at.
exploration, 91–105, 109–12, 116–17, 212–15

Falconer, William:; *Dictionary of the Marine*, 85, 89; *The Shipwreck*, 66, 69, 144
Faraday, Michael, 4, 12, 115–16, 130, 184, 229n.48, 248n.17
Ferguson, James, 37, 46, 56
Fergusson, James, 95–6, 144
Fielding, Henry, 21
Fiennes, Celia, 47
fire, 176, 182, 183, 188, 215. *See also* ethers, fluids
fiscal-military state, 7, 30
Flamsteed, John, 127
Fletcher, Andrew, 36, 166–7
Flinders, Matthew, 14, 89, 105
fluids, 16, 24, 147, 173, 176, 181–5, 206. *See also* ethers
Folkes, Martin, 40, 44, 74–5, 82
Forbes, Duncan, 168
Fothergill, John, 42, 45
France, 96, 97; comparisons of with England, 13, 15–18, 24, 61–2, 72, 102, 128–31, 182, 195, 200, 221n.21, 249n.25; criticisms of, 15–16, 121, 128–31, 145, 178, 185, 202, 204. *See also* Paris
Franklin, Benjamin, 18, 58, 74, 115, 184, 195; and Knight, 42, 44, 45
Freemasonry, 58, 118, 130, 149, 224n.61. *See also* Robison
Freke, John, 28, 183, 188, 189
French, John, 215
Galileo Galilei, 124
Galvani, Luigi, 204
garlic, 77, 145
Gay, John, 191
gender: and instrument imagery, 93, 185–7, 189–93; and natural philosophy, 64, 93, 194–5, 209, 212–14; and social structure, 4, 7, 34, 123, 161, 185–7, 189–95, 209. *See also* women
genius, 17, 206–7, 208
geology, 5, 22, 112–13, 139
George II, 16
George III, 27, 37, 41, 199, 200, 202
Gerard, Alexander, 208, 213
Gibbon, Edward, 40, 92, 151
Gilbert, William: cosmology of, 14, 127, 149, 171, 190; as hero, 210; influence of, 18, 50, 149, 150, 184; instruments of, 21, 22, 131, 133–4; *De Magnete*, 14–15, 21, 60
Gioia, Flavio, 165, 166
God: as divine magnet, 4, 30, 185–93, 213; as divine planner, 25, 35, 69, 77, 100, 139, 146–7, 152–9, 167–70, 172, 183–4
Godwin, William, 196, 201, 206, 213
Golinski, Jan, 6, 35, 37
gout, 6, 51, 150, 203. *See also* medicine
Graham, George, 22, 135
Graham, James, 56–8, 60, 194, 200, 203
Graham, W., 215
gravitational attraction: criticisms of, 176–7; and magnetic attraction, 12, 125–8, 149–50, 174, 179–80, 190. *See also* Newton, Newtonianism
Gray, Thomas, 46
Greville, Frances, 192
Grew, Nehemiah, 12, 149, 176
Gulliver's Travels. *See* Swift

Hackney Phalanx, 28
Hadley, John, 66
Hale, Matthew, 146
Hales, Stephen, 66
Halley, Edmond, 4, 22, 125–7, 210; Atlantic voyages of, 73, 101–3, 109–12, 133; charts of, 93–4, 106, 108–17; influence of, 8, 14, 109, 115–17, 141, 143, 155–6; magnetic theories of, 14, 99, 124, 132, 152–5, 163, 168, 177–8; as Newtonian polemicist, 118–19, 127; religious views of, 100, 127, 152–5
Hamilton, James. *See* Abercorn, Earl of
Hardwicke, Earl of, 44
Harrington, Robert, 182

Harris, John: *Astronomical Dialogues*, 191, 194; *Lexicon Technicum*, 142, 150
Harris, Joseph, 96
Harrison, Edward, 74, 75, 77, 95
Harrison, John, 66, 74
Hartley, David, 157–8
Harvey, William, 150, 177
Hauksbee, Francis, 18, 126–7
Hauxley, Edward, 216
Helsham, Richard, 126
heroes, 24, 68, 89, 120, 209–12
Herschel, John, 143
Hervey, Elizabeth, 191
Hervey, James, 146
Higgins, Bryan, 129
High Church, 24–9, 130–1, 156, 172–3, 211, 214
Hill, John, 45, 251n.57
Hillary, William, 182, 183
historiography: of eighteenth century, 4–6, 13–15, 24, 27, 29–30, 33–7, 120–1, 124, 208, 211; of invention, 68, 70–4, 132–3. *See also* Whiggishness
Hobson, Thomas, 158
Hodges, Walter, 27–8
Hogarth, William, 21, 25
Holland, 16, 40, 42, 95, 109, 160, 169, 182
Holloway, John, 197
Hooke, Robert, 21–2, 79, 120, 125–7, 131, 174
Hooper, William, 62–4
Horne, George, 28, 184, 187
Horne, Henry, 22
horseshoe magnets, 6, 56
Humboldt, Alexander von, 109, 115
Hume, David, 170
Hunter, John, 198–9
Hunter, William, 39, 52, 197
Hutcheson, Frances, 190
Hutchins, Thomas, 135
Hutchinson, John: as anti-Newtonian, 25–6, 151–2, 188; cosmology of, 183–4; on language, 152, 183–4, 187; on magnetic phenomena, 151–2, 155, 168, 246n.61. *See also* Hutchinsonianism
Hutchinson, William, 81–2, 144, 173
Hutchinsonianism: anti-Newtonianism of, 24–9, 176–7, 211; followers of, 27–9, 170, 176–7, 183–4; and language, 173, 183–4, 187, 211. *See also* Hutchinson, John
Hutton, James, 181
Hyde, Thomas, 167, 168

imagination, 151, 199, 208, 213
imperialism, 7, 8, 69, 91–117, 165–6
Inchbald, Elizabeth, 201
inclination. *See* dip
instrument makers: and marketing, 17, 22, 35, 50, 86–8, 118, 133; and polite culture, 46–65; status of, 8, 17, 22–3, 123, 133, 210. *See also* inventors; London, instrument trade; Royal Society, instrument makers
invention, 35, 66, 71–9; and patents, 74, 85, 87; and progress, 3, 146, 164–70, 209
inventors, 70–1, 135, 215–17; changing status of, 73, 86–8. *See also* instrument makers, invention.
inverse-square law, 125–8
Ireland, 44, 128, 220n.11
iron: filings, 18, 47, 50, 131, 177, 179, 184, 187; ore, 22, 50, 138–40, 181, 182

Jackson, Joseph, 216
Jefferson, Thomas, 206
Johnson, Samuel, 3, 4, 17, 66, 217; comments on natural philosophy, 19, 27, 51–2, 119–20, 122, 161, 163; *Dictionary*, 150, 165, 171–2, 177, 186, 189, 193
Jones, W. and S., 83
Jones, William (of Nayland), 177, 210, 211; as Hutchinsonian, 24, 27–8, 172–3
Jones, William (orientalist), 166, 262n.38
Jurin, James, 128

Katterfelto, Gustavus, 56–8, 60, 61, 194, 203
Ken, Thomas, 190
Kepler, Johannes, 127
Keppell, Augustus, 84
keys, 55, 120, 125, 208; symbolism of, 118–19, 131, 145
Kidby, John, 46
King, Erasmus, 54
Kirby, William, 28
Kircher, Athanasius, 63–4, 141, 147–8, 149, 189
Knight, Gowin: artificial magnets of, 15, 16, 18, 40–3, 48, 52, 84; book by, 43–4, 151, 156, 172, 180–2; at British Museum, 44–6; comments on, 44, 45, 52, 60, 68–9, 161, 181, 210; compasses of, 35, 42, 68, 69, 70, 79–88, 135–7, 142, 143, 216; life of, 36, 37–46, 61, 122; magnetic theories of, 119, 139, 141, 169, 177, 180–2; patents of, 85, 228n.39; portraits of, 37–8, 226n.19; and Royal Navy, 66–90; at Royal Society, 8,

41–3, 72, 79, 84; self-promotion of, 24, 41–6, 52, 54–5, 79–88
Kuhn, Thomas, 128, 210

laboratories, 6, 105
Lambert, Johann, 184
Langford, Paul, 33
Langrish, Browne, 179, 180
language, 7, 9, 171–207
Latour, Bruno, 72, 113
Law, William, 28, 30, 158, 173, 187–8, 189
laws of nature, 98–100, 190, 193, 212; and magnetic science, 99–100, 105–9, 125–8, 141–2, 209–10; and women, 209, 212
Leadbetter, Charles, 106
lecturers, 6, 8–19, 54–65, 98, 118, 146
Leiden jar. *See* electricity
Lichtenberg, Georg, 30
light, 5, 7, 172–3, 179, 181, 182, 256n.28
Lister, Martin, 49
loadstone: as divine gift, 146–52, 157–63; in navigation, 31–2, 50–1, 77; as ore, 138–40, 141; specimens of, 40, 48–50; types of, 6, 47–50; value of, 47–9, 125. *See also* terrellae
Locke, John, 129, 165, 172, 173, 175; ideas of, 29, 163, 173, 175, 177, 185, 186
Lofft, Capel, 92
London: comparisons of with other cities, 13, 16, 19, 27, 129, 135; instrument trade of, 6, 17, 33, 40, 87–8, 98, 133–8, 150, 216; as metropolis, 23, 30, 91, 108, 149, 159
longitude, 66–7, 99, 119–20, 152, 222n.29. *See also* Longitude Board
Longitude Board, 25, 88, 99, 103, 136, 144; establishment of, 66–7, 215–17; Knight and, 68, 82, 84. *See also* longitude
Lorimer, John, 113, 115
love, 116, 150. *See also* sexual attraction
Lovelace, William, 40
Lovett, Richard, 99, 152, 169, 216
Lucretius, 158, 163, 164
Lunar Society, 61, 62
luxury, 35–6, 92, 99, 166–7, 170, 176
Lyell, Charles, 4

Macartney, Lord, 166
Madden, Samuel, 164
Magellan, Jean, 43, 61, 87
magic. *See* conjuring, miracles
magnet. *See* artificial magnets, horseshoe magnets, loadstone, sand, terrellae
Magnet, William. *See* Telescope
magnetism, science of, 219n.4; in encyclopedias, 20, 140–3; as example of disciplinary science, 4–6, 8–13, 151, 208–14; language of, 172–5, 194–5; and social authority, 22, 65, 124, 140–5, 194–5, 208–14. *See also* animal magnetism, sciences, terrestrial magnetism
Mahomet, 31, 64–5
Mandeville, Bernard, 69; ideas of, 29, 33, 35, 100
maps, 91–8, 106; of dip, 134, 245n.56; and Swift, 110, 162. *See also* charts, maps of knowledge
maps of knowledge, 11–12, 20, 142–4, 170, 175, 193. *See also* encyclopedias
maritime practitioners: practices of, 50, 67–8, 75–8, 86, 88–9, 143–5; relationships of with natural philosophers, 69–70, 73–9, 85–90, 95–6, 105, 132–8, 141–5. *See also* Admiralty, Royal Navy
Martin, Benjamin, 56, 83, 126
Maskelyne, Nevil, 74, 144, 251n.49
Mason, Abraham, 59, 119, 247n.3
mathematics: attitudes towards, 27, 30, 120, 123–31, 144–5, 163, 185; calculus, 4, 124; data manipulation, 105–12; probability, 109, 112, 122; quantification and, 30, 120, 123–4
Mather, Cotton, 153
Maty, Matthew, 161
Maxwell, John, 96, 243n.13
McCulloch, Kenneth, 83, 251n.49
McKendrick, Neil, 34, 37
measurement, 30, 78, 95–105, 120–3, 123–8, 131–40
medicine, practice of, 38–9, 42, 54, 197–8, 200; and magnetic therapies, 17, 19, 50, 51, 100, 150, 159–61, 194. *See also* animal magnetism, gout, quacks
Mesmer, Franz, 9, 175, 195–6, 197, 204, 206. *See also* animal magnetism
metaphor, 7, 9, 39, 69–70, 171–207. *See also* symbolism
meteorology, 22, 102, 106, 113, 131
Michell, John, 46, 50, 61, 74, 121–2, 126
Middleton, Christopher, 96, 107, 131
millenarianism, 29, 153, 173, 197
Milton, John, 171, 190, 194
mineralogy, 12, 109, 138–40
mining, 89–90, 113, 138–9, 182; compasses for, 19, 31, 50
Minto, Countess of, 196
miracles, 157–8, 169
Mitchell, John, 45

models, 20, 23, 105–17, 131–3, 177, 180. *See also* terrellae
Molyneux, William, 97
Montagu, Mary, 49
More, Hannah, 202
Mount and Page, 103, 112
Mountaine, William, 102, 103, 108, 112, 113
Moxon, Joseph, 109
Musschenbroek, Pieter van, 40, 42, 126, 128, 131, 141, 142
mythology, 3, 4, 151, 156, 168

Nairne, Edward, 83, 87, 134, 137
nationalism, 3, 7–8, 15–18, 130–1, 165–6, 202, 209
natural theology, 24, 35, 69, 100, 139, 146–7, 156–63
navigation: charts for, 91–5, 108–17; compasses for, 65–9, 74–8, 79–86, 132–7; dangers of, 66–7, 69–70, 75–7, 83, 84–5, 95–6; techniques of, 65–9, 75–7, 143–5. *See also* Admiralty, maritime practitioners, Royal Navy
Neoplatonism, 29, 138, 149, 189, 200, 206
Newton, Isaac, 18, 35, 48, 149, 151, 157, 168–9, 183, 210–12; on magnets, 16, 125–7, 174, 178, 179–80, 248n.22; *Opticks*, 22, 24, 43–4, 120, 179–80; *Principia*, 22, 24, 25, 119, 120, 127–8, 170; references to, 15, 58, 92, 119, 131, 149, 154, 155, 168, 181, 183, 211. *See also* Newtonianism
Newtonianism: criticisms of, 12, 24–9, 151–2, 158–9, 179–80, 183–5, 186–8, 211, 214; as rhetorical term, 11–12, 24–9, 37, 68, 118, 121, 142, 190, 211–12. *See also* Newton
Nollet, Jean, 62
Norman, Robert, 133–4
Norris, John, 185, 187, 190
North, Roger, 27, 180
Northumberland, Lord, 44
Nourse, John, 44
Nugent, P. R., 216

Oxford: Ashmolean Museum, 48, 49; Hutchinsonians at, 27, 184; University of, 40, 48, 49, 129, 153, 156, 158

Paisley, Lord. *See* Abercorn, Earl of
Paley, William, 129
Paris, 9, 13, 16, 175, 182, 195, 196, 200, 202. *See also* France
Parkhurst, John, 184
Peat, Thomas, 55
Pemberton, Henry, 60
Penrose, Francis, 183
Pepys, Samuel, 31, 73
perpetual motion, 58–60, 67, 119, 216, 234n.108
Petty, William, 47, 96–7, 109, 143
Phipps, Constantine, 88, 105
phlogiston, 182, 204
Pitt, William, 202–4
Plato: *Ion*, 189, 206; rings of, 47, 56, 148, 163, 189
Playfair, John, 165
Playfair, William, 165–6
Pliny, 125, 150
plurality of worlds, 153–5, 156
pokers, 118, 119, 125, 208
polite culture, natural philosophy in, 31–6, 46–65, 118–19, 122–3, 140–3, 209
Polo, Marco, 166
Pond, Arthur, 38, 39, 41
Pope, Alexander, 21
Porter, Roy, 54
Prescott, Ann, 202
Price, Richard, 18, 131
Priestley, Joseph, 37; colleagues of, 53, 62, 64, 115, 199; on electricity, 18, 19; on progress, 100, 165; satires on, 204
Prince of Wales, 197
Pringle, John, 45, 99
Prior, Matthew, 180
probability. *See under* mathematics
professionalization, 33–9, 64–5, 73–4, 210
progress: and magnetic inventions, 98–9, 146–7, 152, 163–70; rhetorics of, 3, 9, 35–6, 69, 100, 122–3, 140, 165–6
provincial centers, 27–8, 34, 55–6, 64–5, 146, 152, 155, 179
public sphere, 6, 8, 164, 170, 208
Purcell, Henry, 185
Pyefinch, Henry, 83

quacks, 17, 42, 47, 56–8, 194, 195–207, 211
Quakers, 44, 45, 149, 197, 202
quantification. *See* mathematics, quanitificaton and; measurement
Quarles, Francis, 176, 185–6, 191–2

Rainsford, General, 197
Ray, John, 146
Réaumur, René, 42, 182
Rees, Abraham, 142
revelation, 27–9, 151–2, 156–9, 167–70, 182–4
Reynolds, John, 149, 150, 191

Ribright, Thomas, 201, 266n.79
Richardson, Samuel, 28, 37, 189
Roberts, James, 216
Robinson, Bryan, 43–4
Robison, John, 193, 210; criticizes French, 130–1; and *Encyclopædia Britannica*, 15, 143; magnetic diagrams of, 184–5; and Masonic conspiracies, 130, 204
Rodney, George, 84, 85
Rohault, Jacques, 179
Rousseau, Jean-Jacques, 16, 62
Royal Navy, 17, 70, 71–90, 112, 135. *See also* Admiralty, maritime practitioners
Royal Society: and British Museum, 44–5; elections at, 44, 45–6; experiments at, 18, 19, 54–5, 58–60, 127, 132, 137, 142; ideology of, 22–3, 37–9, 53, 98, 105, 172; instrument makers at, 17, 22, 73, 87, 123, 124, 144, 210; Knight at, 8, 24, 39, 41, 43, 60, 79, 81–2; lectures at, 14, 40, 88, 98, 140, 152–3; and navigational improvement, 35, 74–5, 82, 93, 96, 98–9, 101–5, 135; possessions of, 12, 23, 93, 109, 113, 159; satires on, 19, 20–1, 45, 51, 157, 159–63, 172; status of, 4, 6, 8, 36–9, 65, 70, 74, 90, 104–5, 122. *See Also* Copley Medal
Russia, 101–2

Sabine, Edward, 90
sand, magnetic, 80, 180, 239n.48, 252n.65
satires: on Banks, 103–4; on Hutchinsonians, 25–6; on medical practitioners, 57–8, 159–61, 200–7; on projectors, 67, 74, 161–3; on Royal Society, 19, 20–1, 45, 51, 157, 159–63, 172
Saunders, Samuel, 87
Saussure, Henri de, 135
Savery, Servington, 40, 228n.32
Schaffer, Simon, 22
sciences: construction of, 4–6, 9, 20, 35, 175, 193–207, 208–14; in modern society, 4–6, 10, 173–4, 208, 213–14. *See also* magnetism
Scotland, 220n.11; *Encyclopædia Britannica* in, 130, 143; Hutchinsonianism in, 28, 168; philosophers from, 130, 165–6, 184, 190, 208
secrecy, 52, 53, 118, 120
sex, 58, 61, 159–61, 197, 209
sexual attraction, 150–1, 174, 190–3, 197, 209. *See also* love
Shaftesbury, Earl of, 176, 190
Shapin, Steven, 22
Shelley, Mary, 105, 156, 189, 213–14
Shelley, Percy B., 28, 62, 189, 190–1, 206, 213
Sheridan, Betsy, 197
Shovell, Cloudesley, 66, 76
Sibly, Ebenezer, 149, 197, 198, 206
Sisson, Jeremiah, 59
Smart, Christopher, 159, 187–8
Smeaton, John: azimuth compass of, 79, 81; as colleague of Knight, 39, 46, 81, 83
Smith, Adam, 165, 177, 179
Smith, Roger, 7
Smollett, Tobias, 75, 192
Snow, C. P., 173
Society for the Encouragement of Arts, frontispiece, 3, 46, 166
Solander, Daniel, 104
Southey, Robert, 204, 206
Spearman, Robert, 27
Sprat, Thomas, 22
Stafford, Barbara, 54
Stahl, Georg, 182
steel, 41, 53, 80, 138, 182; manufacture of, 17, 40, 181–2, 221n.21
Stephens, William, 126, 128, 220n.11
Stewart, Larry, 6, 35, 37
Stone, Edmund, 52, 70
Stukeley, William, 168, 255n.12
Swedenborgians, 28, 202
Swift, Jonathan, 21, 27, 67, 100, 164; *Gulliver's Travels*, 161–3, 172
symbolism: of images, 3–4, 22–3, 31–3, 209; importance of, 7, 9, 30, 174–5, 214; of keys, 118–19, 131, 145; of Kircher, 63–4, 147–8. *See also* metaphor
sympathetic attraction, 4, 6, 64, 147–51, 164, 189–93, 197–200, 206, 208, 214

Taylor, Brook, 126–7
Telescope, Tom, 16, 260n.14
Temple, William, 164
terrellae, 20, 23, 131–2, 177; of Earl of Abercorn, 40, 50
terrestrial magnetism, 14, 91–117, 128, 131–8, 142–3; theories of, 106, 110, 113–16. *See also* charts, Halley, variation
theories, magnetic, 113–16, 152–6, 175–85
Thomas, Keith, 60
Thompson, E. P., 9
Tories, 25–9, 130–1, 156, 158–9, 172–3, 180, 202, 211, 214
touch, 61, 197, 200, 203
Trapp, Joseph, 156–7
Trenchard, John, 149
Tucker, Abraham, 129

Tulloch, John, 216
Tuttell, Thomas, 31, 32, 34, 50, 64
Tytler, James, 143, 224n.58

Ussher, Archbishop, 168

variation, 67, 101, 152–3; charts of, 93–5, 108–17; measurements of, 67–8, 76–7, 88–9, 106–8, 131–8, 215–17. *See also* terrestrial magnetism
Vico, Charles de, 216
virtue, 31–3, 52, 149, 176, 177, 187, 190, 209
vitalism, 183, 198–9, 213
Voltaire, 13

Wager, Mr., 84
Wales, William, 88
Walker, Adam: influence on Shelley, 190; lectures of, 62, 64–5, 150, 194; magnetic theories of, 115–16, 182
Walker, Ralph, 89, 97, 99, 144–7, 217; compass of, 83, 135–6
Walpole, Horace, 202
Walpole, Robert, 98
Warburton, William, 42
Ward, Ned, 23, 159
Watts, Thomas, 12
weapon salves, 13, 150
Wedgwood, Josiah, 37
Wegg, Samuel, 45, 103
Werner, Abraham, 138, 139
Wesley, John, 19, 172
Westmorland, Countess of, 48, 49
Wheeler, Peter, 201
Whewell, William, 4, 12
Whiggishness, 5, 10, 120, 210. *See also* historiography
Whigs, 25, 202, 261n.22
Whiston, William, 25, 126, 141, 155, 156, 217; *Longitude and Latitude*, 43, 99, 151, 155, 215; magnetic experiments of, 74, 99, 103, 134, 155, 248n.19
White, Gilbert, 100
Whytt, Robert, 199
Wilkes, John, 204
Williams, Zachariah, 99, 119–20, 161, 217
Willis, Francis, 200
Wilson, Andrew, 177
Wilson, Benjamin: as friend of Knight, 37, 38, 39, 42, 44, 45, 53; mathematical ignorance of, 129–30; at Royal Society, 23, 53
Wollstonecraft, Mary, 202
women: status of, 64, 123, 161, 194–5, 209, 212–14. *See also* gender
Woodward, John, 139, 160–1
Wordsworth, William, 92, 212
Worster, Benjamin, 12–13, 149, 220n.4, 248n.19
Wotton, William, 164–5
Wren, Christopher, 23
Wright, Catherine, 197

Yeldell, Dr., 205
Young, Thomas, 113–14, 129, 131, 145, 184

ABOUT THE AUTHOR

Patricia Fara is a Research Fellow at Darwin College, Cambridge, and lectures in the History and Philosophy of Science Department at Cambridge University. She is the author of several popular books on science and invention, and is coeditor of *The Changing World* (Cambridge).